Advances in Computing Science

R. Albrecht (ed.)

Systems: Theory and Practice

Springer-Verlag Wien GmbH

Univ.-Prof. Dr. Rudolf Albrecht
Informatik, Universität Innsbruck
Innsbruck, Austria

Typesetting: Camera-ready by authors

Graphic design: Ecke Bonk

Printed on acid-free and chlorine-free bleached paper
SPIN: 10688397

With 143 Figures

ISSN 1433-0113

ISBN 978-3-211-83206-6   ISBN 978-3-7091-6451-8 (eBook)
DOI 10.1007/978-3-7091-6451-8

Preface

There is hardly a science that is without the notion of "system". We have systems in mathematics, formal systems in logic, systems in physics, electrical and mechanical engineering, architectural-, operating-, information-, programming systems in computer science, management- and production systems in industrial applications, economical-, ecological-, biological systems, and many more.

In many of these disciplines formal tools for system specification, construction, verification, have been developed as well as mathematical concepts for system modeling and system simulation. Thus it is quite natural to expect that systems theory as an interdisciplinary and well established science offering general concepts and methods for a wide variety of applications is a subject in its own right in academic education. However, as can be seen from the literature and from the curricula of university studies - at least in Central Europe-, it is subordinated and either seen as part of mathematics with the risk that mathematicians, who may not be familiar with applications, define it in their own way, or it is treated separately within each application field focusing on only those aspects which are thought to be needed in the particular application. This often results in uneconomical re-inventing and re-naming of concepts and methods within one field, while the same concepts and methods are already well introduced and practiced in other fields.

The fundamentals on general systems theory were developed several decades ago. We note the pioneering work of M. A. Arbib, R. E. Kalman, G. J. Klir, M. D. Mesarovic, R. Rosen, Y. Takahara, A. W. Wymore, L. A. Zadeh, B. P. Zeigler, just to list a few names. There are many others, and good monographs are available.

The goal of this book is to promote systems theory, to give some insight into present theoretical approaches and to show relationships between concepts and methods employed in various application fields. The book is neither a "state of the art" nor a literature report nor an introductory textbook, although the level of the articles presented is such that graduates should easily be able to read them.

The inevitably rather limited number of selected topics, as submitted by the authors about half a year ago, deal with abstract modeling of complex systems, formal specification, mathematical systems theory, Zeigler's discrete event formalisms, aspects of microsystems, decision support systems as parts of management information systems, object-oriented software development, software engineering of interactive systems, modeling of fault-tolerant systems, some applications of artificial neural networks, the control of manipulator robots, and with manufacturing algebra and dynamics as the outcomes of an EC research project, developed by production engineers.

May 1998

Rudolf F. Albrecht

Contents

Abstractly modelling complex systems

Charles Rattray

1 Introduction

Systems theory means different things to different people. Here, we mean the mathematical study of abstract representations of systems, keeping in mind the problems that seem most important for real systems and the aspects of the theory which seem most amenable to mathematical development.

The first step is the mathematical formulation of the basic concepts. In engineered software systems, for instance, many basic concepts have previously been formulated with insufficient generality or have not been formulated at all. We are trying here to be completely general in our explications of the basic concepts. A mathematical explication takes a vaguely defined concept from the everyday language of the system under study and gives a precise meaning which includes all or nearly all important related situations; it therefore *clarifies* the concept. Our choice of concepts takes *system*, *object*, *interconnection* or *link*, and *behaviour* as basic, and provides the explications within category theory.

The application of the category approach in systems theory is not new. Rosen (1958a, 1958b), in 1958, used it to discuss a relational theory of biological systems; Mesarovic and Takahara (1989) published their well-known book on abstract systems theory , in which category theory played an important role. More recently, Ehresmann and Vanbremeersch (1985, 1986) have used category theory in modelling evolving systems in the medical field. Historically, category theory played an important role in mathematics during the period approximately from 1945 to 1975. Four achievements of this development can be singled out: (i) concepts were clarified by phrasing them in categorical terms; (ii) uniform definitions of constructions were given in categorical terms. Specialisations of these constructions in particular categories showed how seemingly ad hoc constructions were "really" the same as well understood constructions in other contexts; (iii) abstraction led to simplification and generalisation of concepts and proofs. By analysing the fundamental aspects of a proof and providing categorical descriptions of the constructions involved in the proof, a much more general categorical proof could be obtained, applicable in quite different circumstances to yield new results; (iv) complicated computations were guided by general categorical results. A general construction often makes apparent a lengthy computation in a particular category.

Our needs in this paper are altogether much simpler than those above. Rather than concentrate on categories of abstract systems we shall also consider a category itself *as an abstract system*.

Most of the work covered can be understood with only five "basic doctrines" aimed at providing an initial intuition of the elementary categorical concepts (Goguen 1973, 1991):

1. any (species of) mathematical structure is represented by a category;
2. any mathematical construction is represented by a functor;
3. any natural correspondence of one construction to another is represented by a natural transformation;
4. any construction forming a complex structure from a pattern of related simple structures is represented by a colimit;
5. given a species of structure, then a species of structure obtained by enriching the original is represented by a comma category.

Many areas of system theory are susceptible to the application of category theory. System definition, system specification, system analysis, system design, and even system implementation may well benefit from a categorical description and analysis. Only a small range of aspects of these system activities are briefly considered, herein, in order to give a flavour of the categorical approach to problem areas familiar to system theorists and practitioners.

2 Systems as categories and categories of systems

Many representations of systems are possible. For instance, von Bertalanffy (1956) gives the view of a system as a set of units with relationships among them. The units themselves may be complex objects and the relationships between them should respect the structure of the objects. Such structure-preserving relations are normally called morphisms. With a minimum of conditions on the *morphisms* the collection of complex objects and morphisms form a *category*.

A **category** is a directed graph together with a composition operation on composable edges (or arrows). To each arrow a is assigned a pair of graph nodes $dom(a)$ and $cod(a)$, called the *domain* and *codomain* of a, such that a is the arrow from $dom(a)$ to $cod(a)$. Each pair of arrows a and b, which are composable ($cod(a) = dom(b)$), is assigned an arrow called the *composition* of a and b, written $b \circ a$. Composition is associative. Composition requires identities, one for each graph node, so that if A is a graph node then there is an arrow $id_A : A \to A$ such that if $a: A \to B$ then

$$a = a \circ id_A = id_B \circ a$$

The nodes of a category are usually called *objects* and the arrows are called *morphisms*. An *object*, in our sense, is considered to be a primitive concept (Ginali & Goguen 1978) and permits objects to represent almost anything. This freedom from the limiting effect of developing the theory for a particular class of objects means that each application can be determined by a class of objects tailored to its needs, without the necessity of awkwardly identifying its objects with some standard class of given objects. Goguen points out that "this feature for general systems theory should not be underestimated".

A *diagram* in a category **G** is a set of objects of **G** together with a set of arrows between these objects; these sets may be empty.

A *path* in a diagram D is a finite sequence of arrows $(a_1, a_2, \ldots, a_n)$ of D such that $cod(a_i) = dom(a_{i+1})$, $1 \le i \le n-1$. The *length* of the path is n. Diagram D commutes if, for any paths $(a_1, a_2, \ldots, a_n)$, $(b_1, b_2, \ldots, b_m)$ such that $m \ge 2$ or $n \ge 2$ (or both), $dom(a_1) = dom(b_1)$ and $cod(a_n) = cod(b_m)$,

$$a_n \circ a_{n-1} \circ \ldots \circ a_1 = b_m \circ b_{m-1} \circ \ldots \circ b_1.$$

The category **Set** of sets, for which the collection of objects is the collection of all sets and the collection of morphisms is the collection of all functions between sets, is perhaps the most common category. To any directed graph G we can associate a category of directed paths $\mathbf{P}_G$ in G; the objects of $\mathbf{P}_G$ are just the nodes of G, the paths in G are the morphisms, composition is concatenation of paths (which is associative), and identities are zero-length paths associated with each node. For any category, **G**, there is an obvious underlying graph, U_G.

2.1 Systems as categories

The nature of many systems representations is diagrammatic/graphical. Design representations for software systems, for instance, may be in the form of flowcharts, structure charts, data flow diagrams, SADT "blueprints", statecharts, etc. These are all essentially (directed) graphs with various properties or peculiarities; for information systems, all of the above have been used as have *entity-relation-attribute* diagrams (ERA). Obviously, none of these repesentations form a category as they stand. For each, one must "embed" the representation in a suitable category. Rosen (1958b) in developing his relational theory of biological systems discusses (**M**, **R**)-systems; from these he constructs *abstract block diagrams* which are diagrams in the category **Set**. A similar approach was taken by Rattray and Phua (1990) in considering Data-Flow-Diagrams (DFD). Ehresmann and Vanbremeersch (1992), in developing a categorical view of general systems, give an illustrative neural network example in which the category in question is the path category of the network. More interestingly, but still following the same general line, Dampney, *et al* (1992) and Johnson and Dampney (1993) construct the category associated with ERA diagrams as the *classifying category* or *theory* for the information system being analysed. Generating the category requires that the ERA diagram be normalised, ie. many-many relationships be transformed into a pair of many-one relationships by the introduction of a new entity. The resulting diagram is a directed graph with entities and attributes as nodes and arrows as relations, directed in the "many" to "one" direction, together with arrows from entities to attributes. The many-one relations have real world counterparts so that real world compositions must appear in the diagram. Particular attention must be paid to the paths with common domains and common codomains to determine if the diagram commutes as these represent system constraints. Johnson and Dampney make the point strongly that current ERA-modelling technologies ignore the question of commuting diagrams, yet identifying these often leads to modification of the ERA diagram to produce a better model. The category constructed is just the presentation (specification) for the canonical classifying category. For an information system, certain basic "exactness" properties need to be satisfied. In particular, the classifying category will be *lextensive* (Carboni et al 1992).

Thus, we can see that a system can be modelled by a suitable category.

2.2 Categories of systems

Mesarovic and Takahara (1989), in developing their Abstract System Theory, say that it could be taken as a "system theory of systems theories". Starting with a minimal mathematical structure to describe basic systems concepts, and adding to that structure in order to define needed systems properties, the theory aims at revealing "the essence of various concepts and properties", and acts as a "theory of structures". Thus, a system is taken as a proper relation over the objects which represent the constitutent parts of the system. Basic to much of the developed theory is the notion of input/output or terminal systems. These are usually represented as the relation S with respect to the input object A and the output object B, i.e.

$$S \subset A \times B$$

where A is the set of system objects representing the influence from the environment, and B is the set of system objects representing the influence from the system to the environment.

Similarity between systems, e.g.similarity in structure and similarity in behaviour, is an important concept. For instance, similarity in structure may be dealt with in the following way. Let $S \subset A \times B$ and $S' \subset A' \times B'$ be two input-output systems. A *modelling relation* $h : S \rightarrow S'$ is a pair of mappings

$$h = (h_1, h_2),\ h_1 : A \rightarrow A' \text{ and } h_2 : B \rightarrow B'$$

such that $(a, b) \in s$ and $(h_1(a), h_2(b)) \in S'$.

For S and S' as functions then this "modelling relation" can be represented in the commutative diagram

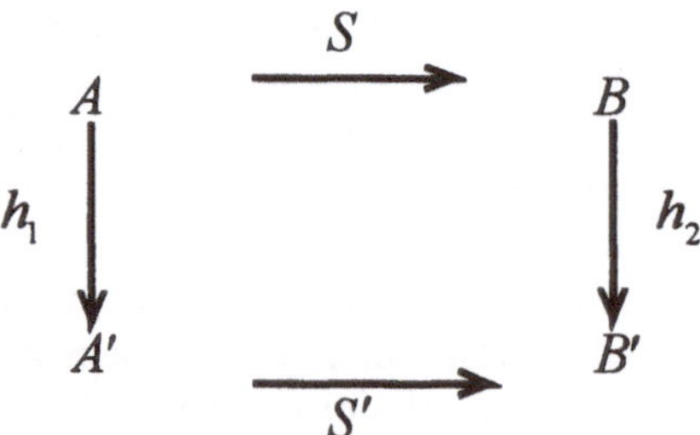

Modelling relations are composable. A category of input-output systems is now definable and this provides a framework in which "*to integrate various specialised systems theories various specialised systems theories* such as discrete time systems, continuous time systems, automata, etc." The first basic doctrine of category theory is achieved. Now, common properties of input-output systems can be studied using the other doctrines as a guide.

Abstract systems theory may lead to different types of input-output systems by introducing more structure to the system objects. For example, choosing a suitable categorical structure or a suitable algebraic structure deepens our understanding of systems and broadens the applicability of the abstract systems theory results to other fields of interest.

2.3 Category of complex systems

Systems evolve with time. For instance, a software system goes through a number of phases: conception, requirements, specification, design, implementation, and use. This latter phase is sometimes referred to in the literature as software evolution or software maintenance. It represents a period in which the system , as constructed, is modified to match the changing environment and the changing needs of the user.

The categorical representation of a system is in no sense absolute, rather it represents the state of the system at some time t. The evolving system then moves through a sequence of states to reach some goal state, determined by the requirements specified for the modified system to match its new environment. Unfortunately, not only does this process move the system to a new state but, because of the complexity of the internal structure of the state, it also changes the system structure and its categorical representation. That is, we may be constructing a new category or, at the very least, an incremental change to that of the previous state. The determination of an appropriate strategy to achieve the new goal state may be undermined by timing constraints. Reaching a goal state may have to accomplished within a given time period. The chosen strategy will affect the internal structuring and functioning of the system state. The interaction between the external time period available to achieve the goal state and the internal timescales for system component changes can be quite subtle.

To reflect state transitions, these are represented as functors.

Let $\mathbf{G}$ and $\mathbf{H}$ be categories. A functor $F: \mathbf{G} \to \mathbf{H}$ is essentially a graph homomorphism which preserves composition and identities. That is, F is a function which assigns to each object A in $\mathbf{G}$ an object $F(A)$ in $\mathbf{H}$, and to each arrow $a: A \to B$ in $\mathbf{G}$ an arrow $F(a) : F(A) \to F(B)$ in $\mathbf{H}$ such that

- if $A \xrightarrow{a} B \xrightarrow{b} C$ in $\mathbf{G}$ then $F(b \circ a) = F(b) \circ F(a)$
- $F(id_A) = id_{F(A)}$ for all objects A in $\mathbf{G}$.

Mesarovic and Takahara show that the free representation of a time system becomes a dynamical system under the condition of stationarity. Constructing a corresponding time invariant model by the Nerode realisation procedure can be formalised as a functor as a concrete representation of the stationarity functor. This is an example of the second basic doctrine of category theory, namely that constructions can be represented by functors.

The collection of all system states and transitions is a category **Cat**, the category of small categories and functors. The sequence of states and their transitions representing the evolution of a system is just a diagram in **Cat**.

To capture this more fully, we need to define a category of time. Consider a total order on the set of positive reals. This defines a category: objects are the real numbers representing instances of time, and the unique arrow (t, t') from instance t to instance t', $t \leq t'$, represents the period of time between t and t'. This is the category **Time**.

A **time-varying complex system** (TVCS) is a functor S: $\mathbf{T} \to \mathbf{Cat}$, where $\mathbf{T}$ is a subcategory of **Time**. The category $S(t) = \mathbf{S}_t$ over the instance t is the category representing the state of the system at time t.

One of the "basic doctrines" suggests that to study any structure we should determine the category in which it resides. Indeed, in understanding the development process of any complex system, we need to compare possible developments of the system, compare different development processes, or simply determine invariant properties of a single process. Modelling such processes as TVCSs identifies the need for the notion of morphisms between them. A possible way of dealing with this is to compare states of the systems at "corresponding" times. Thus, the timescales of each system must be compared.

Time-varying complex systems are represented as functors. The category of TVCSs has functors as objects, and morphisms in this category must "transform" one functor into another. In categorical terms this transformation is a *natural transformation*.

Consider two functors F, F': $\mathbf{G} \to \mathbf{H}$ for categories **G** and **H**. Then, for every morphism a: $A \to B$ in **G**, a *natural transformation* η: $F \to F'$ is a collection of morphisms $\{\eta_A : FA \to F'A\}$, indexed by the objects of **G**, such that the following diagram commutes:

$$\begin{array}{ccc} FA & \xrightarrow{\eta_A} & F'A \\ {\scriptstyle F(a)}\downarrow & & \downarrow{\scriptstyle F'(a)} \\ FB & \xrightarrow[\eta_B]{} & F'B \end{array}$$

For TVCSs, the idea is to compare the states of the systems (by way of their defining functors) at "corresponding" times. To do this, the timescales of the systems must first be compared, ie., the "corresponding" times must be identified by a *change of time* functor since any two TVCSs operate (possibly) on different timescales. The natural correspondence between the constructions of the state representations of the two TVCSs at their corresponding time points is then determined by a natural transformation. This is an example of the third "basic doctrines".

Let S and R be two TVCSs over **T** and **U**, respectively, where **T** and **U** are subcategories of **Time**. *A morphism from* S *to* R is a pair (F, ϕ) such that

- ϕ: $\mathbf{T} \to \mathbf{U}$ is a *change of time* functor
- $F : S \to R \circ \phi$ is a *natural transformation* between S and R such that

– for each object t in **T**, $F(t)$ (usually written as F_t) is an arrow in **Cat**

$$F_t : \mathbf{S}_t \rightarrow \mathbf{R}_{\phi(t)}$$

– for each period (t, t^*) in **T**, the diagram

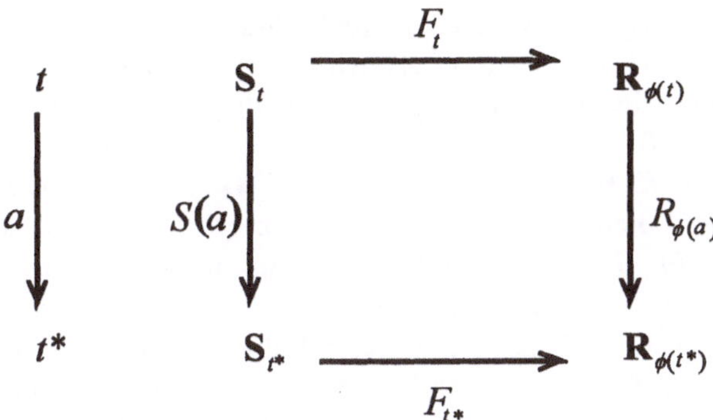

commutes.

Composition of morphisms requires composition of the time change functors, and the correspondences at relevant time instances. With this defined, we now have the *category of time-varying complex systems.*

Time-varying complex systems have brought together in one framework the notions of a category of systems and systems as a category. These systems are complex, in the sense of Mesarovic and Takahara, as they are seen as systems "whose components are systems in their own right".

3 Using basic categorical properties

Having formed the approriate category for the problem area under consideration, we are now in a position to look at using basic categorical constructions. For a category of systems, system realisation is an important aspect; for a system as a category, in the context of time-varying complex systems, investigation of the structure of the system state introduces further basic categorical notions.

3.1 System realisation

Much of the content of the treatise by Mesarovic and Takahara centres around two fundamental problems in system theory, viz., the *characterisation problem* and the *realisation problem*. The former is concerned with identifying the set of characteristic properties which define the type of system under consideration, ie., given a system, what properties its set of input-output pairs should have so that a given type of model represents the system completely. In this case, the input-output behaviour of the system and the model must be isomorphic. For the realisation problem, it suffices that the input-output behaviour of the system can be embedded in the input-output behaviour of the model.

The Nerode construction (Nerode 1958) gives a canonical realisation and it was Goguen (1972, 1973) who gave conditions which characterise minimal realisation up to isomorphism, these conditions being *universal* in the category theory sense. This universality can be described in terms of one object "best approximating" another object in some category of objects.

Given two categories **G** and **H**, and a functor K: **G** $\rightarrow$ **H**, an approximation to an object B in **H** is the pair (f, A) with A in **A** and f: $B \rightarrow KA$.

An possible alternative is to proceed using a dual notion of approximation, namely,.$(f\colon KA \rightarrow B, A)$.

For functor K: **G** $\rightarrow$ **H**, an object A of **G** is said to be *K-universal* for an object B of **H** if there exists an approximation $(f\colon B \rightarrow KA, A)$ such that, for each approximation $(g\colon B \rightarrow KA', A')$, with A' an object of **A**, there exists a unique morphism $h\colon A \rightarrow A'$ in **G** with $g = K(h) \circ f$.

If every object of **H** has a K-universal object in **G**, then **G** is said to be K-universal in **H**. If **G** is K-universal in **H** then there exists a unique functor L: **H** $\rightarrow$ **G** such that every LB is the K-universal object for B. In this circumstance, the functors K and L are called adjoint functors where K is the *right adjoint* of L, or L is the *left adjoint* of K.

Goguen showed that minimal realisation is a right adjoint to behaviour, as functors between categories of machines and behaviours. In order to investigate this idea we have to define the category of machines to be considered. Rather than follow Goguen directly here machines are defined as *coalgebras*. Consider, firstly, an input-output system as an automaton, represented as a tuple (A,B,S,α,β) where A is the input alphabet, B is the output alphabet, S is the set of states, $\alpha : S \times A \rightarrow S$ is the state transition function, and $\beta\colon S \rightarrow B$ is the output function. The state transition function may be rewritten as $\alpha : S \rightarrow S^A$ and the product of α and β formed to give

$$\delta = (\alpha, \beta) = S \rightarrow S^A \times B$$

The automaton may now be represented as a pair (S, δ) where S^A is the set of all functions from A to S.

Underlying this is the concept of coalgebra. For any "polynomial" functor T: **Sets** $\rightarrow$ **Sets**, a coalgebra (or T-coalgebra) is a pair (X,c) with a carrier set X and a structure function $c\colon X \rightarrow T(X)$. In the automaton case, $T(X) = X^A \times B$ and $c = (\alpha, \beta)$ the (state transition, output) pair of operations. By suitable choice of functor T, a variety of different automata can be obtained. The "modelling relation" is now captured by a (homo)morphism from T-coalgebras (X,c) to (Y,d) as a function $f\colon X \rightarrow Y$ between the carrier sets such that $d \circ f = T(f) \circ c$. The polynomial functors on the category **Sets** is the least class containing the identity functor, the constant functors, product and coproduct functors, and the exponent functor.

Having come this far, we can now follow Jacobs (1996a) and define a category of systems, **Sys**, of (deterministic) automata, and a corresponding category of (deterministic) behaviours, **Beh**.

Category **Sys** has automata as objects, $(S, S \rightarrow S^A \times B)$, and morphisms,

$$(S, S \rightarrow S^A \times B) \rightarrow (R, R \rightarrow R^C \times D),$$

as triples

$$(f\colon S \to R,\ g\colon B \to D,\ h\colon C \to A)$$

such that *f* preserves the initial state, $\beta(f(s)) = g(\beta(s))$ and $d(f(s),c) = f(d(s,f(c)))$. Category **Beh** has input-output functions, (A,B,p), as objects where A, B are sets and P maps finite sequences of As to B, ie. $p : A^* \to B$.. Morphisms for **Beh**, $(A,B,p) \to (C,D,q)$, are pairs $(h\colon C \to D,\ g\colon B \to D)$, with $h^*\colon C^* \to A^*$, such that $q \circ h^* = g \circ p$.

Jacobs (1996a) then proves that the behaviour functor *Behaviour* : **Sys** ⌫ **Beh** has a right adjoint realisation functor *Realisation* : **Beh** ⌫ **Sys**. This kind of relationship can be extended to other classes of systems and, indeed, Jacobs (1996b) does this for hybrid systems.

This area of interest has been revived (in part) in recent years by the computing science community developing formal specification techniques. Structured transition systems, or non-deterministic automata, are used in the specification and modelling of many computing systems, including concurrent systems modelled by Petri nets, process algebras, etc. The new set of problems occurring has engendered an interest for category theorists and the conjoining of the two groups has led to increased activity. The development of the object oriented paradigm (where objects carry the notion of "state"), through object oriented design and object oriented programming languages, demands a more theoretical underpinning. Various attempts to provide suitable semantics for these have often been based on the application of category theory. Some of these attempts have made use of the concept of coalgebras (Jacobs 1995), a concept now having an impact in formal specification. Hill (1998), this volume, gives a clear account of the uses of algebra (but not coalgebras) and logic in formal specification.

3.2 System fragments

An individual has a view of a system and only understands certain aspects of the whole system. Colleagues have their own views and they too are likely to understand some, but not necessarily all, of the system. Certainly, the "intersection" of these views of the system will not be empty; there will some shared understanding. What is required is to have some mechanism to "compose" the different views into a more comprehensive understanding of the system as a whole. A categorical way of dealing with this is the following.

A *control centre* (CC) of a TVCS *S* is represented by a sub-TVCS with its own timescale. The objects of CC are what we call *agents.*

To model the notion of an agent's view of the system state S and a system approximation over a control centre, we introduce a *slice category*, a special case of a *comma category* (MacLane 1971).

If **S** is a category and A any object of **S**, the slice category $\mathbf{S}{\downarrow}A$ is described in terms of

- an object of $\mathbf{S}{\downarrow}A$ is an arrow $b\colon B \to A$ of **S** for some object B
- an arrow of $\mathbf{S}{\downarrow}A$ from $b\colon B \to A$ to $c\colon C \to A$ is an arrow $f\colon B \to C$ such that

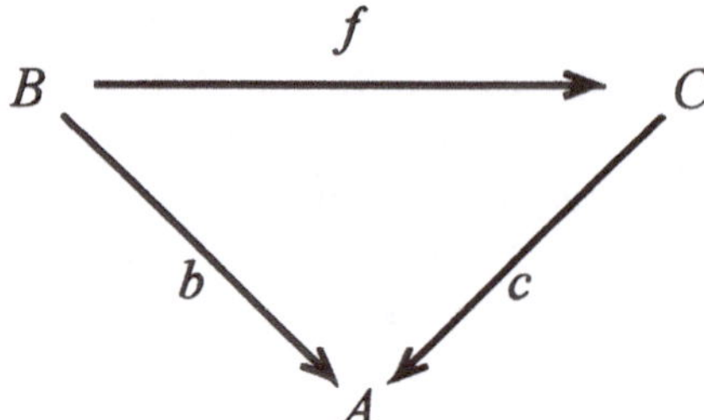

commutes

- the composition of $f: b \to c$ and $g: c \to d$ is $g \circ f$.

The *fragment* for agent A at time t is the category $\mathbf{S}_t{\downarrow}A$ of objects over A in $\mathbf{S}_t$, with $b\colon B \to A$ an *aspect* of $\mathbf{S}_t$ recognised by A. Any arrow $\alpha\colon A \to A'$ between agents in CC induces a functor $\alpha^*.\mathbf{S}_t{\downarrow}A{\to}\mathbf{S}_t{\downarrow}A'$ between fragments.

A *system approximation* for the control centre CC at t is a *colimit* of the collection of fragments over the control centre CC: its objects are equivalences of aspects b linked by a *zig-zag* of arrows connecting agents. This approximation supports a distortion functor to $\mathbf{S}_t$. For a TVCS, a corresponding system approximation forms a TVCS.

A colimit construction provides a means of forming complex objects from patterns (diagrams) of simpler objects, where a pattern can be considered as a graph homomorphism P from graph G (its "shape") to the underlying graph of some category, S. Pattern P defines a *pattern of linked (related) objects* in **S**. The colimit object binds together the component objects according to their internal organisation determined by the corresponding pattern.

The category **S** models the environment of the pattern P. Modification of the environment category may take various forms. For instance, enlarging **S** to **S′** or blurring the distinction between two **S** objects in **S′** may make retaining the limits difficult. It may be that a pattern P cannot be bound to a colimit in **S** but can be forced to have a colimit in an extended environment **S′**.

By using colimits, a hierarchical structure can be imposed upon the system states. This provides a convenient abstraction concept to make the understanding and development of complex systems more amenable to analysis and construction. We should be conscious of the fact, though, that such abstract objects as modelled by a colimit do not necessarily exist in real systems. We may define a *hierarchical system* to be a category **S** in which the objects are distributed on levels $(0,1,\ldots,p)$ such that each object of level $n+1$ $(n<p)$ is the limit in **S** of a pattern P of linked objects on level n. In such a hierarchical system, the system components are associated with levels corresponding to increasing complexity of their internal organisation. Any object at level $n+1$ is the colimit of a pattern of linked objects at level n but it may form part of a pattern of linked objects whose colimit is at level $n+2$.

Control centres may occur on any level, either singly or as multiple CCs. Higher level CCs will normally be colimits of patterns of CCs at lower levels.

Thus, we see, in understanding the structure of complex systems, the role played by colimits and comma categories. The uses of both these constructions are in keeping with the fourth and fifth "basic doctrines".

4 Conclusions

Category theory has a number of attractions for systems theory. It provides basic notation and language, in much the same way that set theory does, for discussing, investigating, and explaining systems theory concepts and constructions. It allows this from a viewpoint different to that of set theory in which an object is described in terms of its "internal"structure. In categorical terms, the only structure that an object has is by virtue of its interaction with other objects. Without such interactions little can be said about the object. This, of course, is in keeping with the von Bertalanffy view of a system as a collection of units with relationships among them. However, the categorical framework has basic concepts, some of which we have indicated earlier, which have a natural analogy to systems theory concepts. By abstracting and generalising away from the specific, classes of systems problems can be described categorically and common features of the class discovered. The categorical framework then provides a methodology to guide further investigation of the class of problems through determination of standard properties and constructions. The universality of such properties and constructions are then available to investigate other, more specific classes of problems and systems. In this sense, categorical analysis is unlikely to provide deep theorems about specific systems but it does provide deep insights to such systems and gives a meaningful path to follow in the investigation.

We have discussed a number of basic aspects of categories and have emphasised the notion of "first find your category". Whether your category is a category of systems or a system as a category, the further analysis required is guided by the search for properties that have categorical purpose and system significance. Two areas, machine behaviour and time-varying complex systems, have been used to illustrate the ideas. Of course, there is a wide range of other areas that could have been considered; for example, Pavel (193) used category theory in pattern recognition and Srinivas (1992) has shown how the analysis of a well-known algorithm, when viewed categorically, led to a generalisation which greatly broadened its applicability. Considerable application of the categorical approach has already had a strong impact on our understanding and development of systems of interest in the field of computing science.

References

von Bertalanffy, L. (1956): Les Problemes de la Vie, Gallimard, Paris.

Carboni, A., Lack, S., Walters, R.F.C. (1992): Introduction to Extensive and Distributive Categories, Technical Report, Pure Mathematics 92-9, University of Sydney.

Dampney, C.N.G., Johnson, M., Munro, G.P. (1992): An Illustrated Mathematical Foundation for ERA, in The Unified Computation Laboratory (ed. C.M.I. Rattray, R.G.Clark), Oxford University Press, Oxford.

Ehresmann, A.C., Vanbremeersch, J-P. (1985): Systemes Hierarchique s Evolutifs: une modelisation des systemes vivants, Prepublication No 1, Universite de Picardie.

Ehresmann, A.C., Vanbremeersch, J-P. (1986): Systemes Hierarchique s Evolutifs: modele d'evolution d'un systeme ouvert par interaction avec des agent, Prepublication No 2, Universite de Picardie.

Ehresmann, A.C., Vanbremeersch, J-P. (1992): Outils Mathematiques Utilises pour Modeliser les Systemes Complexes, Cahiers de Topologie et Geometrie Differentielle Categoriques, XXX111, 2.

Ginali, S., Goguen, J. (1978): A Categorical Approach to General Systems, in Applied General Systems Research: recent developments and trends (ed. GJ Klir), Plenum Press.

Goguen, J. (1972): Minimal Realization of Machines in Closed Categories, Bull. Amer. Math. Soc., 78, 5.

Goguen, J. (1973): Realization is Universal, Maths. Sys. Theory, 6, 4.

Goguen, J.A. (1991): A Categorical Manifesto, Mathematical Structures in Computer Science, 1.

Hill, G. (1998): Formal Specification, this volume.

Jacobs, B. (1995): Objects and Classes, coalgebraically, Report CS-R9536, Computer Science, CWI, Amsterdam.

Jacobs, B. (1996a): Automata and Behaviours in Categories of Processes, Report CS-R9607, Computer Science, CWI, Amsterdam.

Jacobs, B. (1996b): Hybrid Systems of Coalgebras plus Monoid Actions, Report CS-R9614, Computer Science, CWI, Amsterdam.

Johnson, M., Dampney, C.N.G. (1993): On the Value of Commutative Diagrams in Information Modelling, in Algebraic Methodology and Software Technology AMAST'93 (Editors: M Nivat, C Rattray, T Rus, G Scollo), Workshops in Computing Series, Springer-Verlag.

MacLane, S. (1971): Categories for the Working Mathematician, Springer-Verlag.

Mesarovic, M.D., Takahara, Y. (1989): Abstract Systems Theory, Lecture Notes in Control and Information Sciences, 116, Springer-Verlag.

Nerode, A. (1958): Linear Automaton Transformations, Proc. Amer. Math. Soc., 9.

Rattray, C., Phua, G. (1990): A Meta--Model for Software Processes: Complex System Modelling by Categories --- Data Flow Diagrams, Tech Rpt CSET/90.3, University of Stirling.

Rosen, R. (1958a): A Relational Theory of Biological Systems, Bulletin of Mathematical Biophysics, 20.

Rosen, R. (1958b): The representation of Biological Systems from the standpoint of the Theory of Categories, Bulletin of Mathematical Biophysics, 20.

Formal Specification

Gillian Hill

1 Introduction

We identify both the mathematical structures and the logic that underpin the main approaches to the formal specification of systems. Our aim is to present a theoretical framework for system specification that is built on precise mathematical foundations. Being faced with the syntactic details of particular specification languages can be confusing for someone who lacks mathematical experience. For this reason we prefer to present the important theoretical concepts that underlie system construction in the familiar notation of logic and mathematics. Once an understanding of these concepts is achieved, the notations that are currently in use can be mastered without difficulty. We see an analogy with the need to understand the fundamental principles of programming before becoming too involved with the fussy, and often confusing, syntactic details of concrete mechanical codes. Understanding is best achieved when an *abstract* programming language is used.

In our view the activity of specification involves building theories and reasoning about them in a formal system. Formal languages for specification are designed over a formal deductive system and offer structuring mechanisms in order to gain power of expression. We present the concept of an abstract data type in Sec. 3 as a primitive object in a system and specify the behaviour of example abstract data types in a simple semi-formal language of sets and functions.

A many-sorted first-order theory of an abstract data type is then presented as a formal specification. First-order logic is a powerful and natural language for specification; its expressiveness brings advantages to the client who has required the development of the system. Specifications in the algebraic approach also present the properties of abstract data types as axioms, but use a restricted form of first-order logic, equational logic, for reasoning about these properties.

In summary, the logic and algebraic approaches to specification take an abstract view of a system and specify the properties of its data structures by logical axioms. Although the alternative abstract model approach, based on mathematical structures as models, is described as 'abstract', it produces specifications that come closer to describing the concrete representation of the data in a system. Specifications in the Vienna Development Method (VDM) and Z, currently the most popular specification languages, require the construction of an explicit model for the state of a software system. Clearly the difference between specification approaches are deeper than the notational variations that become obvious when they are used. The need to match the theoretical frameworks that are available for specification to the problems of system construction that exist is an urgent topic for research. We concentrate on comparing the theory underlying the different specification approaches and do not attempt to prescribe the suitability of a particular approach to the type of system that is to be constructed.

2 Specification

The activity of specification is traditionally informal and has produced a textual description of system requirements, or an informal statement of what the system actually does. The formalization of this activity, by the use of formal languages for specification, introduces soundly based reasoning into the activity of system construction.

2.1 The activity of specification

The traditional waterfall model of the software life cycle divides the production of a software system into distinct sequential stages. There are two problems with such a simple model, however. The first is a lack of control as production moves down the 'waterfall'; the second is a lack of communication between the client and the software engineer. A more useful model involves the integration of control into the software life cycle. By formalizing the activity of specification, the software engineer brings precision into the development process; by actively co-operating throughout a series of small development steps the customer brings understanding into the development process. The simple model which expresses the development of one large-scale system needs to be replaced by a model that breaks the development process into a series of small and well-defined steps. Each step should involve negotiation, with a clear contract entered into between the software engineer and the customer. Although we refer specifically to a software system, the ideas are applicable, in general, to the construction of any complex system.

The activity of specification takes place at each stage in the software life cycle. To specify means 'to give details of' or 'to indicate precisely'. Formal languages as 'precisely defined unambiguous notations', are therefore suitable for expressing system specifications. By making specification a formal activity, control is incorporated into the new software life cycle and verification becomes a parallel activity with construction.

A further advantage of formalizing the activity of specification in order to provide a clear contract is the communication of information to *all* involved in the project. The requirement for precision and unambiguity may conflict with the requirement for ease of communication but in our view this is not an argument against formal specification in system development.

Interaction with the client ensures that the implementation is what the client really wants and should take place at each stage of system development.

Extra information from the client about the problem must be added to the specification and not to the code as a last minute fix. In contrast, the extra detail about the implementation is given in some chosen design language, before being further refined and expressed in some chosen programming language. The final implementation will express the algorithms and data structures in a concrete mechanical code that is directly computed by a machine.

Example 2.1.1

An informal specification is given for a program that reads a list of integers from a file and outputs the largest integer in the list as well as the position at which this largest integer occurs. Additions are then made to the original specification, which is permissive, and allows design decisions to be made later.

The informal specification is: 'The program is to read a list of integers from a file and output the largest integer in the list as well as the position at which this largest integer occurs. The program prints the largest number and its position on the user's terminal'.

This specification is permissive and does not make design decisions prematurely. The following additions could be made later to this informal specification: 'The results will be output to an external file as well as to the terminal. If the largest number occurs more than once in the input file, the program prints the position of the first occurrence. If the input file does not contain any numbers, the program reports this as an error. The program should take any file, existing in the file structure, that contains a list of integers. The input file does not contain any characters, other than integers separated by spaces. The input file is a text file'.

The design specification is for a program called 'program large'. The top-level algorithm is an implementation of the informal specification. There is no need to store the list of numbers.

```
if the file is empty
then inform the user
else largest number so far is the first number read;
while there are still numbers in the file
        read in remaining numbers and
        assign each one to the largest
        number if it is greater than the
        previous largest number;
        remember the position of each number and
        which is the largest number;
print the results
```

This algorithm transcribes to the following more concrete algorithm in a single refinement step:

```
if end of file then
        printstring ("File is empty or does not exist")
else readnumber largest_number_so_far;
```

```
        position_of_largest_so_far : = 1;
        position_in_the_list : = 1
fi ;
while not end of file
do
        readnumber current number;
        position_in_the_list : = position_in_the_list + 1 ;
        if current_number ⟩ largest_number_so_far then
              largest_number_so_far : = current_number;
              position_of_largest_so_far : = position_in_the_list
        else donothing
        fi
od
printstring ("the largest number in the list is");
printnumber (largest_number_so_far);
printstring ("this value occurs at position");
printnumber (position_of_largest_so_far)
```

Transcribing into a programming language such as Pascal involves consideration of file handling and reading input.

2.2 Formal languages for specifications

The most abstract formal languages are the languages of first-order logic. The use of quantifiers provides power of expression for these languages, and a natural deduction system provides power of reasoning. Since complex systems involve operations on many different sorts of data, however, it is necessary to extend simple first-order languages to a *many-sorted* language with a set of symbols to represent each sort of data. Interpretation for a many-sorted theory involves dividing the domain of interpretation into subsets. This is exactly how many-sorted algebras can be defined. We shall present a many-sorted similarity type in a logical specification as the analogue of a many-sorted signature in an algebraic specification.

3 Abstract Data Types

In this section we identify an *abstract data type* as a key concept for the construction of systems. We present specifications of abstract data types as the primitive specifications from which the specification for the whole system is built. The key concept of a *module* is identified as the reusable instance of the textual specification of an abstract data type. As simple examples, we specify the behaviour of both queues and tables in a semi-formal language of sets and functions.

The importance of writing specifications for parts of systems was recognized by Parnas in 1972 . He specified modules in a new 'formal' notation by giving the user enough information for the module to be used without the user having any knowledge of the method of implementation. This idea has become the important design concept of *information hiding*. By specifying data abstractions, Liskov and Zilles (1974) extended information hiding to the users of data abstractions.

Definition 3.0.1 (Abstract data type) An abstract data type defines a class of abstract data objects that is completely characterized by the operations available to manipulate those objects.

The power of data abstraction is that the user is deprived of the ability to make use of implementation details. Objects can be described by their properties rather than by their representation in terms of concrete data structures such as arrays or pointers.

3.1 Primitive objects in a system

Parnas' idea of an information-hiding module and Liskov's idea of specifying an abstract data type were major signposts for the formal specification of systems. They pointed along rather different routes, however, and we see these different routes reflected in the formal specification approaches that are currently popular. It seems today that the real importance of the concepts of a module and an abstract data type lies in the role that they can play as the reusable components in a system. Specifications of abstract data types form the primitive specifications in a system from which the specification for the whole system is built, Modules can be seen as reusable instances of these textual specifications. The work of Parnas and Liskov has led to different views of the primitive objects in a system, and it seems now that both these views are useful if they are applied at different levels of abstraction.

1. *Objects with state.* Parnas thought of a module as an abstract machine with an explicit state. By operating on a module we cause the internal state of the module to change. We use this idea when we put a coin in a drinks machine and press a button to move the machine into a new state, which hopefully results in the output of our chosen drink. The procedural style of programming that includes object-oriented programming is based on the idea of an explicit state, At the level of specification, some approaches define an explicit notion of state for a system and specify a change in state in terms of pre-conditions and post-conditions.

2. *Abstract data types with values.* Liskov's concept of data abstraction has been developed at a more abstract level within the logic and algebraic approaches to specification. Operations on abstract data types are specified by functions that return values. The relationships between the operations describe the meaning, or the semantics, of the abstract data type and there is no explicit notion of state. This view is mirrored by the declarative style of programming.

The concept of an abstract data type is clearly important for specification and we view abstract data types as the primitive objects in a system. A notion of state can be built in to primitive objects as they are refined and made more concrete. For the construction of a system specification from the primitive specifications of objects, a module can be constructed as a unique reusable instance of its textual specification.

3.2 Specification of a queue

Our first choice of an abstract data type is a common structure, at the problem level, but is not a concrete data type in a programming language. It is vital to consider structures with different disciplines of storage and retrieval when we specify systems. Unfortunately, however, the data structures of programming languages are inadequate when we work at an abstract level.

We specify the behaviour of a queue in terms of how we can use the queue for storing information. In particular we consider the relationships between the operations on the queue. First we give an intuitive definition of a queue:

1. Each queue holds items of information of the same type, or sort.

2. The queue is a general data type meaning that different queues may hold collections of items that are of different types. For example, one queue may hold a collection of names, while another may hold a collection of people, and yet another a collection of integers.

3. The following operations may be performed on a queue:

   (a) Add a new item to the tail of the queue.

   (b) Delete the item at the head of the queue.

   (c) Get a copy of the item at the head of the queue.

   (d) Find out whether the queue is or is not empty.

In order to formally describe a queue in the problem domain rather than the program domain, we need to define a language that has a structure, or *syntax*, and also a *semantics* that gives a meaning to the specification. We use a simple language with a syntax defined by sets and functions and a semantics defined by equational axioms. Later we shall use formal languages, with more expressive notations, that are currently in use as specification languages.

The queue is represented as a general data type by passing the type *item* as a parameter to the type queue. The whole structure is queue[item] and a queue is said to be *generic* with respect to the type *item*. The definition of a queue forms a pattern for all queues, whatever the type of item that is stored in them. This demonstrates the power of specifying at the abstract level.

3.2.1 The syntax of the language

We represent the sorts, or kinds, of objects involved by sets and the operations on these objects by functions. Let

$$\begin{aligned} Q &= \{\text{queue1, queue2, } \ldots\} \\ I &= \{\text{item1, item2, } \ldots\} \\ E &= \{\text{no-item}\} \\ B &= \{\text{true, false}\} \end{aligned}$$

The set E contains the only error condition. We shall avoid the use of error-values as much as possible in these early attempts at specification. Error-handling has proved to be very difficult in some formal specification approaches. We define individual constants by functions and specify the empty queue as a 0-ary function. We specify the following argument and result sets for the functions:

$$\begin{aligned} &\text{create} &&: \ \to Q \\ &\text{add} &&: \ Q \times I \to Q \\ &\text{head} &&: \ Q \to I \cup E \\ &\text{delete} &&: \ Q \to Q \\ &\text{isempty} &&: \ Q \to B \end{aligned}$$

3.2.2 The semantics of the language

We present rules that relate the functions and give meaning to the operations on the queue. The rules are arrived at by a process of intuition and represent the properties of the objects and the operations in the real world. As formal rules they are called *axioms* and we use them in an equational form. Our axioms are purely functional in that they only return values and do not have any side effects on a state. We therefore do not have a notion of a *before object* as distinct from an *after object* in our specification. This means that we do not have a notion of pre-condition or post-condition as we do in a procedural programming language like Pascal. It is important to keep our set of basic axioms to a minimum; we can extend our specification later by adding more detail as we build and implement a system.

We now define in a precise way the meaning of the operations that we may perform on the abstract data type queue. We describe the relationships between the functions that represent the operations by six axioms. Our axioms are expressed so that they apply to all possible queues by quantifying universally over the sets

$$\forall q \in Q,\ i \in I$$

1. isempty(create) = true. This axiom formalizes our understanding that a newly created queue does not contain any elements.

2. isempty(add(q, i)) = false. Once an item has been stored, the queue is no longer empty.

3. head(create) = no-item. This axiom formalizes our understanding that we cannot get a copy of the item at the head of an empty queue. Note that *head* is a function that takes an argument that itself is a constant function. *Create*, as a constructor or constant function, is a representation of an empty queue.

4. head(add(q, i)) = if isempty(q) then i else head(q).

5. delete(create) = create. It is better not to over-specify with error conditions. For this reason, deleting from an empty queue merely results in an empty queue again.

6. delete(add(q, i)) = if isempty(q) then create else add(delete(q), i). The axiomatic definitions for the *head* and *delete* as functions are recursive and the axioms for head and delete on the empty queue terminate the recursion. It is important to understand the way that recursion 'counts down' the structure until it terminates with an empty structure. This contrasts with the way that structural induction 'builds up' from the basis clause as the beginning of the induction. The axiom for *head* states that if an item has just been added to an empty queue, the *head* function on that queue will return a copy of that item. If the queue was not empty, however, before the item was added, then the *head* function is applied to the non-empty queue (before the item was added).

   The axiom for *delete* is more difficult to understand intuitively. If *delete* is performed on a queue that *was* empty before an item was added, then the resulting queue is empty. If the queue on which *delete* is performed has more than one item in it, however, then deleting from that queue is the same as adding the last item *after* deleting from the queue that contains all the previous items.

A formal specification should give just enough information to define the type, but not so much that the choice of implementation is limited. For example, the formal specification of the queue does not state that the items are added to the tail of the queue and deleted from the head of the queue. This is stated in the informal specification, but is really part of the implementation decision. The formal specification relates the functions implicitly and no explicit model is required.

Example 3.2.1 We use our specification language for a queue to create an empty queue by the expression: create. Then we add Fred, Anne, Mary and Jill to the queue in stages, forming a new queue as each addition is made.

create
add(create, Fred)
add(add(create, Fred), Anne)
add(add(add(create, Fred),Anne),Mary)
add(add(add(add (create, Fred),Anne),Mary),Jill)

Next we get a copy of the item at the head of the queue.

head(add(add(add(add(create, Fred),Anne),Mary),Jill))
= if isempty(add(add(add(create,Fred),Anne),Mary)) then Jill
else head(add(add(add(create,Fred),Anne),Mary))
= if isempty(add(add(create,Fred),Anne)) then Mary
else head (add(add(create,Fred),Anne))
= if isempty(add(create,Fred)) then Anne
else head (add(create,Fred))
= if isempty(create) then Fred

The recursion terminates because the queue that was created initially is empty. Notice how the end bracket around the queue moves inward until the recursion terminates.

4 Specifications in First-Order Logic

First-order logic is simple and natural to use for specification and is becoming increasingly important as difficulties emerge with other more popular approaches. The simplicity of logic results from its power to abstract: specifications in logic are loose and descriptive with no explicit notion of state. Although model theory provides the notion of interpretation in a structure there is usually no need to build a model when we specify complex systems.

We take the view that the specifications of complex application systems are constructed by building many-sorted theories and naming them as textual specifications. The simple specifications of the primitive objects in a system are *combined* by structuring operations to form more structured specifications. A primitive operation on a specification is the *extension* of the specification by the addition of new symbols and properties. Extensions that are conservative play an important role in the refinement, or implementation, of specifications.

As detail is added by refinement, specifications become less permissive. This is because refinement, and the eventual implementation of specifications by programs, requires the addition of further symbols for operations and constants to the language of the specification. An *extension* to a specification is the specification with the new symbols added. By extending safely, or conservatively, the need for complicated correctness arguments can be avoided.

A conservative extension is safe because every property in the specification that has been extended holds in the extension to the specification *and also* because in the extension no new property, in the language of the original specification, can be deduced that did not hold in the specification before the extension was made.

Example 4.0.1 If we extend a specification for the door of a house by adding a window we should extend conservatively. A basic door that is extended to have a window should possess all the properties of a basic door: it should, for example, still open and close. We should not add the new property that all doors can be seen through. Clearly, however, in the new language with symbols that refer to both a door and a window there will be extra properties about the new extended object. These will assert, for example, that a window lets light through.

4.1 Many-sorted theories

A specification of an abstract data type such as a queue needs to describe elements that are queues as well as elements that are items. In fact, most specifications describe operations on many different sorts of data. We need to divide the domain of interpretation into subsets, according to the sorts of names that we need to quantify over. In our syntax we have a set S of many sorts, containing symbols that name each sort of object in the theory. Many-sorted theories are appropriate for expressing specifications of the varied sorts of objects in application systems. The many-sorted syntax of a logical specification is given by a many-sorted similarity type; for algebraic theory we use a many-sorted signature.

Definition 4.1.1 (Many-sorted similarity type) A many-sorted similarity type σ is defined as $\sigma = (S, ar_r, ar_f)$, where S is a non-empty set of sorts and the functions ar_r and ar_f are called arity functions. The elements in the domains of ar_r and ar_f are the relation and function symbols respectively. We write rel_σ for the set of relation symbols and func_σ for the set of function symbols.

Example 4.1.2 We use the semi-formal specification that we gave in Sec. 3 to write a logical specification of a queue as an abstract data type.

queue is queue[item]

by renaming of T_{fifo} to $T_{\mathrm{queue}} = (\sigma_{\mathrm{queue}}, \Gamma_{\mathrm{queue}})$ where,

$$\sigma_{\mathrm{queue}} = (S, ar_r, ar_f)$$

and,

$$\begin{aligned}
S &= \{q, i\} \\
\mathrm{func}\ \sigma_{\mathrm{queue}} &= \{\lambda^q, \mathrm{add}^{q,i;q}, \mathrm{delete}^{q;q}, \mathrm{head}^{q;i}\} \\
\mathrm{rel}\ \sigma_{\mathrm{queue}} &= \{\mathrm{queue}^q, \mathrm{item}^i\}
\end{aligned}$$

The formation axiom is,

$\forall y^i\ [\mathrm{item}^i(y^i) \rightarrow \exists x^q\ \mathrm{queue}^q(x^q[y^i])]$

The introduction axioms are,

$\mathrm{queue}^q(\lambda^q)$

$\forall x^q \forall y^i\ \mathrm{queue}^q(\mathrm{add}^{q,i;q}(x^q, y^i))$

$\forall x^q \forall y^i\ [\neg \mathrm{add}^{q,i;q}(x^q, y^i) =_q \lambda^q]$

$\forall x_1^q \forall x_2^q \forall y_1^i \forall y_2^i\ [x_1^q =_q x_2^q \wedge y_1^i =_i y_2^i \rightarrow \mathrm{add}^{q,i;q}(x_1^q, y_1^i) =_q \mathrm{add}^{q,i;q}(x_2^q, y_2^i)]$

The reduction axioms express the addition of items to the back of the queue and the removal of items from the front of the queue. The first axiom states that using delete on a queue to which two items have previously been added is like adding the last item to the queue to which the first item was added and then deleted.

$$\forall x^q \forall y_1^i \forall y_2^i\ [\text{delete}^{q;q}(\text{add}^{q,i;q}(\text{add}^{q,i;q}(x^q, y_1^i), y_2^i)) =_q$$

$$\text{add}^{q,i;q}(\text{delete}^{q;q}(\text{add}^{q,i;q}(x^q, y_1^i)), y_2^q)]$$

$$\forall x^q \forall y_1^i \forall y_2^i\ [\text{head}^{q;i}(\text{add}^{q,i;q}(\text{add}^{q,i;q}(x^q, y_1^i), y_2^i)) =_i \text{head}^{q;i}(\text{add}^{q,i;q}(x^q, y_1^i))]$$

$$\forall y^i\ [\text{delete}^{q;q}(\text{add}^{q,i;q}(\lambda^q, y^i)) =_q \lambda^q]$$

$$\forall y^i\ [(\text{head}^{q;i}(\text{add}^{q,i;q}(\lambda^q, y^i)) =_i y^i) \wedge \text{item}^i(y^i)]$$

endspec

4.2 Extending many-sorted theories

We have stressed that the *refinement* of logical specifications should be carried out by a process of conservative extension and is one that should involve a client in order to ensure that the evolving system meets its requirements. In order to define the conservative extension of a theory we first define the extension of a theory.

A theory can be extended in different ways: by adding new function or predicate symbols to its language and also by adding new properties for the symbols of the language. It is important to ensure that when the language of a theory is extended, every non-logical symbol of the theory remains in the language of the extension. As well as ensuring that the language of the theory is contained in its extension, it is also necessary to ensure that every property of the theory is still a property of the extension.

Definition 4.2.1 (Extension of a theory) Let $T_1 = (\sigma_1, \Gamma_1)$ and $T_2 = (\sigma_2, \Gamma_2)$ be many-sorted theories. Let A be a sentence in L_{T_1}. Then T_2 extends T_1 if for all sentences A we have if $\Gamma_1 \vdash A$ then $\Gamma_2 \vdash A$; that is every theorem of T_1 is a theorem of T_2. We write $T_1 \subseteq T_2$ if T_2 extends T_1. Although every non-logical axiom of T_1 must be a theorem of T_2, it is not necessary for it to be an axiom of T_2.

Definition 4.2.2 (Conservative extension of a theory) Let $T_1 = (\sigma_1, \Gamma_1)$ and $T_2 = (\sigma_2, \Gamma_2)$ be many-sorted theories. Let A be a sentence in L_{T_1}. An extension of a theory T_1 by another theory T_2 is a conservative extension iff the extended theory, T_2, does not allow the deduction of any more properties than T_1 in L_{T_1}. Hence $T_1 \subseteq T_2$ is conservative if for all sentences A, in the language of T_1, we have if $\Gamma_2 \vdash A$ then $\Gamma_1 \vdash A$.

5 Specifications in the Algebraic Approach

In this section we bring together ideas and results from classical and universal algebraic theory, with the concept of abstract data types from computer science. An awareness of the need for the precise specification of complex systems led to the interest in discrete mathematics. Zilles was one of the first to use the theoretical framework of universal algebras to specify the abstract data types that Liskov had identified at the programming level. The abstract data type was specified by a syntactic presentation of an algebra by sorts and operation symbols together with a set of equational axioms.

The underlying logic for these specifications is called equational logic. Although this logic lacks existential quantification and universal quantification is implicit only, it does offer the advantage that a minimal model can be constructed for a specification. This minimal model is useful for discussing the properties given by the equational axioms and has been chosen to provide an *initial* semantics for algebraic specifications. The name initial has been given to the semantics because the initial algebra in the category of all algebras is the concrete model for an abstract data type. An alternative semantics for algebraic specifications is called a *loose* semantics.

The early use of algebraic specifications revealed their unsuitability (in a pure form) for expressing the properties of complex systems, however. Although they offered the advantages of a precise notation for specification, they showed the same disadvantages that large unstructured programs in low-level mechanical codes had at the programming level. Just as procedural abstraction and data structures had been used to structure programs, so efforts were made to structure the theories that algebraic specifications expressed. It is in the *structuring* of theories, pioneered by Burstall and Goguen (1979) in their work on Clear, that the algebraic approach to specification has made important contributions to the specification of complex systems.

5.1 A specification is a theory presentation

The syntactic presentation of an algebraic structure is used to specify an abstract data type. As in the logical approach to specification, we begin by defining an algebraic specification of an abstract data type as a named theory presentation. Our definition of a theory presentation in the algebraic approach is based on equational logic and the deduction of consequences that belong to the theory.

Definition 5.1.1 (Signature) A signature $\Sigma = (\mathcal{S}, \mathcal{F})$ is a pair consisting of a set $\mathcal{S}$ whose members s_i are called sorts, and a set of function names $\mathcal{F}$. The function names each have an arity over $\mathcal{S}$, written in the form $f : s_i \times \cdots \times s_j \rightarrow s_k$. Any function name f with arity $f :\rightarrow s_k$ is said to be a constant of sort s_k. If $\mathcal{S}$ contains more than one sort, then the signature is said to be many-sorted.

Example 5.1.2 $\Sigma_1 = (\mathcal{S}, \mathcal{F})$ where $\mathcal{S} = \{s\}$, $\mathcal{F} = \{\Box, inv, e\}$ and function arities are:

$$\begin{array}{rll} \Box: & S \times S & \to S \\ inv: & S & \to S \\ e: & & \to S \end{array}$$

This is the signature for groups.

Definition 5.1.3 (Σ-algebra) For a signature $\Sigma = (\mathcal{S}, \mathcal{F})$, a Σ-algebra $\mathcal{A}$ is a pair $(\mathcal{A}_\mathcal{S}, \mathcal{A}_\mathcal{F})$ which is a model for Σ. This means that for each sort $S \in \mathcal{S}$ there is a set $A_s \in \mathcal{A}_\mathcal{S}$ (called a carrier set), and for each $f \in \mathcal{F}$ of arity $s_i \times \cdots \times s_j \to s_k$ there is a function $A_f \in \mathcal{A}_\mathcal{F}$ of arity $A_{s_i} \times \cdots \times A_{s_j} \to A_{s_k}$.

Example 5.1.4 $(\mathbb{Z}, \{+, -, 0\})$ is a Σ_1-algebra. The carrier set for S is $\mathbb{Z}$, and the function names $\Box, inv, e$ are represented by the actual functions $+, -, 0$.

Definition 5.1.5 (Theory presentation in equational logic) A theory presentation is a pair (Σ, E) where Σ is a signature and E is a set of Σ-equational axioms.

Definition 5.1.6 (Theory in equational logic) A theory T for a signature Σ is called a Σ-theory and is a theory presentation (Σ, E) where E is closed. A theory is satisfiable if it has at least one model.

The theory, T, is therefore the same as its theory presentation when the set of equations is closed. We need the concept of a theory presentation as a *specification* for a theory. The words are used interchangeably, however. Within the algebraic approach the important idea is that a theory *specifies* a set of algebras which is the set of all its models.

Definition 5.1.7 (Algebraic specification) A specification in the algebraic approach is a theory presentation (Σ, E) where Σ is a signature and E is a set of Σ-equational axioms.

Example 5.1.8 The specification *Boolean* is a named theory presentation for the theory of truth-values. This is an initial specification that must be given an initial semantics. The importance of initiality is that it imposes a condition on the model for the specification, as we see in the next section.

Initial	*Boolean*
sorts	bool
constants	true: bool
	false: bool
opns	**not**: bool $\to$ bool
eqns	**not**(true) = false
	not(**not** (a)) =a
end	

5.2 A model for a specification

The elements in the models give values to, or name, the terms that are built from the signature. In this sense the model of a specification is more concrete than the specification, which is a named theory presentation. The algebraic model for a specification can be easily characterized as the *initial* model for that specification, where the idea of initiality comes from category theory.

The existence of an initial model for a specification has contributed to the popularity of the algebraic approach. This is because it opens up the possibility of writing a specification in two different ways:

1. Specifying the properties of an abstract data type by directly giving the set of equations to form an 'abstract' specification.

2. First constructing a model for the abstract data type from some known properties and then deriving the 'constructive' specification.

The first way is close to that used for writing logical specifications in Sec. 4 but is restricted by the need to use equational logic instead of the full power of first-order logic. The restriction deprives the specifier of being able to write sentences that use quantification, negation and implication.

If algebraic specification languages are extended to permit axioms that use quantifiers then the existence of an initial model, as a minimal model, for an algebraic specification is not guaranteed. The advantage of using the restricted equational logic for specifying abstract data types is that the second way of writing a specification can be used to produce a 'constructive' specification by concentrating first on building an initial model of an abstract data type. The initial algebra always exists and satisfies the equational axioms E of an abstract algebraic structure (Σ, E), but it does not satisfy any additional properties. This is why it is described as a minimal model. The other models of (Σ, E) are 'looser' models that satisfy additional properties. They are all related to the representative initial model because there is a homomorphism *from* the initial model *to* the other 'loose' models in the class of (Σ, E)-algebras.

Now we need to explain the significance of the algebraic theory about the models of algebraic specifications that we have presented. First we recall that an algebraic specification of an abstract data type is a presentation of the theory that describes the data type. Then we present the special properties of the initial model for the specification, (Σ, E):

1. Every element in the carrier set of the model is the value of some ground Σ term in the signature of the specification. This property holds for finitely generated algebras and is described by the slogan 'no junk', meaning that there must be no unnecessary (unnamed) elements in the model.

2. The model must not satisfy any ground equation that is not in E. This property is known as 'no confusion' and means that no two terms must be given the same value unless the equations in E say they should have the same value.

Clearly the initial model provides a semantics for a specification by giving values to the terms in the language of the specification. Intuitively this meaning is 'tight', or strict, and contrasts with the permissive style of specification we suggested in Sec. 4. The semantics provided by the rest of the models in the class of (Σ, E)-algebras provide a 'looser' and more permissive semantics. It is useful to view the semantics for algebraic specifications in terms of the closure of the set of equations E in the specification (Σ, E). An algebraic specification is either a loose or tight closure of a set of equations. The keywords *loose* and *initial* are often used to indicate which closure is used.

5.3 Building theories for algebraic specification

Although the early use of algebraic specifications brought advantages for specifying simple abstract data types, they were quite unsuitable for specifying large systems. We identified abstract data types as primitive objects in a system in Sec. 3.1. The view of an abstract data type with values is clearly expressed precisely by an algebraic specification. A major problem for specification approaches, however, is to provide a theoretical framework within which specifications can be structured from simple, or primitive, specifications. By joining simple specifications together we can work towards the specification of complex systems. The extension of specifications is one of the main ways of building more structured specifications.

We give a simple example of a specification that is built by extending the even simpler specification for Boolean values. The notation is simple but explicitly states the extensions that are made.

Example 5.3.1 We build a specification for a queue by extending the theory of Booleans.

Queue = the extension of *Boolean* by:

types	queue of α
constants	nilqueue: queue of α
	add: α × queue of $\alpha \rightarrow$ queue of α
	head: queue of $\alpha \rightarrow \alpha$
	delete: queue of $\alpha \rightarrow$ queue of α
	isempty: queue of $\alpha \rightarrow$ bool
axioms	head(add(x, nilqueue)) = x
	head(add(y, (add(x, q)))) =head(add(x, q))
	delete(add(x,nilqueue))=nilqueue
	delete(add(y, (add(x, q)))) = add(y, (delete(add(x, q))))
	isempty(add(x, q)) = false
	isempty(nilqueue) = true

The algebraic specification language Clear was the first to provide theory-building operations. The semantics of combining theories is an important topic for research. The important ideas in Clear were to build structured theories by extending theories (in different ways) and to allow parameterized theories. These ideas have been continued in the specification language OBJ which is described as a *functional logic* language that integrates a specification and a programming language. The integration of the logic and the algebraic approaches is illustrated in the language Eqlog which combines OBJ with logic programming.

The algebraic specification language ACT ONE (Ehrig and Mahr, 1985) also provides structuring mechanisms for extending, combining, renaming and parameterizing small specifications in order to build large ones. The language has been modified following its use on several software specification projects without any loss of mathematical rigour in the definition of its semantics. A lot of ideas that were incorporated in the revised ACT ONE came from the use of ACTONE as the data type part of LOTOS, a formal specification language for the design of distributed and concurrent systems (Brinksma, 1988).

6 Model-Based Specifications

An alternative view of a primitive object as a building block for the construction of a system specification was suggested in Sec. 3.1. This view identified a primitive object, at a rather less abstract level, with its own internal *state* that is changed by the defined operations on that object. The incorporation of the notion of a state into system specifications is shown in this section to be a key characteristic of model-based specifications.

6.1 Construction of a mathematical model

Specifications in VDM and Z are described as *model-based*, or constructive, because they involve the explicit construction of a model of the system in terms of mathematical structures such as sets, lists, sequences and mappings. Specifying in VDM and Z can therefore be viewed as a less abstract activity than writing the axioms of a theory presentation within the logical and algebraic approaches to specification. The more concrete nature of model-based specifications may account for their greater popularity among programmers who work in software development as an industrial process.

In the abstract axiomatic approaches to specification, the relationships between operations on abstract data types are expressed by axioms, either in first-order logic or in the subset called equational logic. The behaviour of an abstract data type is therefore defined implicitly by the properties of the operations on the data type. Although no explicit model is constructed, we saw in Sec. 5 that the existence of a minimal algebraic model, the initial algebra, does provide a semantics for tight specifications. We also pointed out in some sense that the ability to construct this initial model during the activity of specification makes specification easier for programmers. However, the role of the initial model is essentially to model properties of the system by equational axioms and to give values to abstract data types by interpretation in the carrier sets.

By contrast, the mathematical model constructed in the model-based approaches to specification provides a semantics for the abstract data type and comes closer to suggesting an implementation for an abstract data type. It is because a *concrete* mathematical model is constructed for a specification in this approach that we avoid using the name 'abstract model approach' for the specifications.

VDM and Z are also described as *state-based* specification techniques because the notion of system state is central to the specification. The operations on an abstract data type are modelled by operations on mathematical structures that *process* objects by changing their values. This procedural view of a sequence of state changes, represented by a before state object and an after state object, is close to the way we think about programming in languages such as Pascal. In the specification, the abstract data type is represented by a mathematical object that *changes* when it is operated on. This contrasts with the algebraic approach in which a *new* value results from the application of a function to the set that represents the data type.

6.2 The Vienna development method (VDM)

VDM is a methodology for the systematic development of software, which was developed during the 1970s and has been applied to a variety of applications in industry. Offering more than a specification language, VDM provides rules and procedures as a framework for the stages of software development from abstract specification to implementation in a programming language. Support tools, management and training aspects have also been developed for the methodology, which has been applied to the construction of medium-sized systems.

One reason for the acceptance of VDM in industry may be that it offers a flexible approach to the development of software within a formal system. By emphasizing a rigorous approach within a formal framework, it allows the use of intuition in the construction of correctness arguments, rather than demanding complete formality for the verification of all development steps.

The activity of specification in VDM is a top-down development from an abstract specification that expresses the functions that are required of the system at the top level. Development proceeds from the abstract specification by refining the data objects in the system to more concrete *reified types* and by modelling the operations at the abstract level by operations on the concrete types. The addition of more concrete detail is described as reification, rather than refinement, by Jones (1986). At each stage of reification *retrieve functions* are defined as mappings from the concrete type back to the abstract type, and an argument is constructed to show the *adequacy* of the development step. The *adequacy proof obligation* provides an argument for correctness and can be discharged by an informal constructive argument.

A VDM specification focuses on the functions of a system that define 'what' the system does. As development proceeds a transition is made, however, from mathematical functions to operations that are executed on a state. In effect operations are textually like procedures in programs. Although VDM is state-based, the ideas of the algebraic approach to specification are carried through into 'local' specifications of abstract data types. The semantics of VDM specifications is based on the theories for the mathematical modelling structures such as sets, sequences and mappings.

6.2.1 VDM specifications

The syntactic part of a VDM specification gives the names of the variables that define the state of the system. Each name *denotes* a set of possible values that is defined by the *type* of the mathematical structure in the model. Built-in types such as *Int* and *Nat*, denoting $\mathbb{Z}$ and N, are provided and this part of the specification looks like a Pascal program. After the declaration of the state variables, the operations on the state are defined by their names, including any parameters. The syntactic specification is completed by predicates that express assumptions about the execution of each defined operation. Each predicate specifies the semantics of an operation in terms of pre-conditions and post-conditions for the operation. A more abstract VDM specification is called an *implicit* specification; a concrete specification is called a *direct definition* and satisfies some implicit specification if an associated proof obligation can be discharged. The direct definition of a function uses the notation $\triangleq$ rather than the symbol for equality, =.

Example 6.2.1 An implicit specification for a simple function that gives as a result the larger of the two integers is

$max(i : \mathbb{Z}, j : \mathbb{Z}) \qquad r : \mathbb{Z}$

pre true

post $(r = i \vee r = j) \wedge i \leq r \wedge j \leq r$

The only assumption on the arguments to the function *max* concerns its type, so the pre-condition is explicitly stated to be true. The post-condition implicitly gives the meaning of the function, with the variable r denoting the maximum of the two integers. A direct definition for the function *max* is

$max\ (i, j) \triangleq$ if $i \leq j$ then j else i

Implicit specifications of mathematical functions are extended by a notion of state in order to express programs in VDM. Any piece of program-like text is called an *operation*. The *state* of an operation is the collection of external variables that the operation can access and change. The input variables and their types are listed in brackets after the operation name. The VDM notation is extended to describe the access that an operation is allowed to variables as either read only (rd) or read and write (wr). It also marks the before object for an operation with a hook, $\leftharpoondown$, to distinguish this object from the after object which is changed by the operation.

Example 6.2.2 The abstract data type stack is modelled by a sequence in the following VDM specification:

$Stack$ = seq of El

INITIALIZE-STACK( )
ext wr $s : Stack$
post $s_0 = [\,]$

PUSH($e : El$)
ext wr $s : Stack$
post $s = [e]\,\overleftarrow{s}$

POP ( ) $e : El$
ext wr $s : Stack$
pre $s \neq [\,]$
post $\overleftarrow{s} = [e] \frown s$

ISEMPTY ( ) $r : \mathbb{B}$
ext rd $s : Stack$
post $r \leftrightarrow s = [\,]$

This specification is for an unbounded stack. A bound on the length of a stack, to say 256, could be given by an invariant:

$BStack$ = seq of El where $inv\text{-}BStack(s) \triangleq len\, s \leq 256$

The invariant properties in a specification are defined on the state and specify the relationship that *must* hold between the values of the data objects in the system.

7 Summary

In this chapter we have presented the theoretical framework for the formal specification of systems. We have shown that the activity of specification involves building theories and reasoning about them within a formal framework. Formal languages provide different structuring facilities for describing systems and it is clearly important to match the theoretical framework to the type of application system that is to be constructed.

Abstract data types have been presented as the primitive objects in a system, and we specified them initially by a semi-formal language of sets and functions. The construction of a specification for a large system from the specification of its parts appears to be a promising technique for the future.

The most abstract specification approaches express the properties of abstract data types by axioms. First-order logic provides a powerful and natural language for writing loose and descriptive specifications of theories. In the algebraic approach the more restricted form of equational logic offers a less powerful language for expressing the axioms of an abstract data type. However, it does provide the possibility for a constructive approach to specification by guaranteeing that a model of the specification can be built. This is the initial algebra that realizes the specification as its minimal model.

The popularity of VDM and Z may be due to the fact that they both belong to the model-based approach of specification. By explicitly constructing a mathematical model based on set theory these specification methods are, in this sense, more concrete than the axiomatic approaches of first-order logic and equational logic. An explicit notion of state is declared in a specification, and operations that change the state are described by predicates.

The approaches to specification have emerged from different theoretical frameworks, and this is reflected in the different paradigms of programming. However, there is now an acceptance that advantages are to be gained by bringing together the different paradigms and matching the specification approach to the type of application system that is to be constructed. Within the algebraic approach there is an increasing use of first-order logic to gain greater power of expression: the language OBJ is a *functional logic* specification and programming language; *Eqlog* combines OBJ with logic programming; FOOPS unifies OBJ with object-oriented programming; and FOOPlog unifies the functional, logic and object-oriented paradigms and has a rigorous logical basis.

A more recent solution to the problem of dealing simultaneously with different formalisms when specifying software engineering systems is proposed by Fiadeiro and Maibaum (1995). They formalise the different modelling approaches individually in the common mathematical framework of category theory and establish relationships between them using functors.

References

[1] R. M. Burstall and J. A. Goguen. The semantics of Clear, a specification language. In *Abstract Software Specifications, LNCS 86*. Springer-Verlag, 1979.

[2] E. Brinksma (ed.). Information processing systems-open systems interconnection-LOTOS. a formal description technique based on the temporal ordering of observational behaviour. *International Standard, ISO 8807*, 1988.

[3] H. Ehrig and B. Mahr. *Fundamentals of Algebraic Specification 1: Equations and Initial Semantics*. Springer-Verlag, 1985.

[4] J. Fiadeiro and T. Maibaum. Interconnecting formalisms: Supporting modularity, reuse and incrementality. *Association for Computing Machinery*, 1995.

[5] B. H. Liskov and S. N. Zilles. Programming with abstract data types. In *Proc. ACM SIGPLAN Conf. on Very High Level Languages*, pages 50–59, April 1974.

[6] D. Parnas. A technique for software module specification with examples. *Communications of the ACM*, 15(5):330–336, May 1972.

On mathematical systems theory

Rudolf F. Albrecht

1 Introduction

Intuitively, systems are, frequently hierarchical, aggregations of physical or mental objects, composed either by nature or composed by us for a certain purpose or for logical reasons according observable and distinguishable "properties". Properties can be physical or logical qualities and quantities, location in physical space, behavior in physical time, physical or logical relationships between objects.

To model a system, we extract certain objects, facts and features, sufficient for the purpose intended, represent them by "symbols" and manipulate the symbols to describe structural and behavioral properties of the system within the precision chosen. To this end, abstract concepts and formal theories are provided by mathematics and logic.

In this article we are not concerned with the problem what and how to model in a specific application case nor with particular interpretations of abstract results, though the selection of the theoretical topics we consider is inspired by important classes of applications. The intention is to give a general and uniform mathematical framework for description, modeling and simulation of systems by mathematical structures. We use classical set theory, algebra, topology and logic and follow partly the line marked by the basic work of Mesarovic and Takahara (e.g. M.D.Mesarovic, Y. Takahara 1989).

The subsequent chapters are dealing with sets, relations, functions, variables, and hierarchies of these, giving emphasis to parameterized representations as needed in most applications, with compositions of sets and relations, valuated objects, logics, topological structures, the concepts of "time" and "processes", specification by "properties", undirected, directed and time dependent structures, modeling of physical systems, and some applications like approximation of functions, adaptive systems, hierarchical and topological control structures, computer architecture, knowledge bases. Statistical aspects and classical differential equation systems are not considered.

Many of the concepts presented in the following are extensions and generalizations of those given in literature. Moreover, parts of the general theory are treated under new aspects. The article is based on earlier publications (R. F. Albrecht 1994 a,b,c, 1995, 1996a,b,c, 1997, R. F. Albrecht and G. Németh 1997).

2 Mathematical foundations

2.1 Sets, relations, functions

Given a set S, $S \neq \varnothing$, of objects s, then the power set of S is defined by $\text{pow}\, S =_{def} \{U \mid U \subseteq S\}$, i. e. the set of all subsets of S. Given an "index set" I, $I \neq \varnothing$, and an "indexing" by a function $\text{ind}: I \to S$ with $\bigwedge i \in I\ (i \mapsto s_{[i]})$, then $(s_{[i]})_{i \in I}$ is a "family". In other notation $s_{[i]} = \text{ind}(i)$. For $I = \varnothing$ the family is empty, written $(\,)$. If $\mathcal{S}$ is a set of sets S, K an index set, $\text{ind}: K \to \mathcal{S}$ an indexing and $(S_{[k]})_{k \in K}$ a family of sets $S_{[k]}$, then the set product is defined by $\prod_{k \in K} S_{[k]} =_{def} \{(s_{[k]})_{k \in K} \mid \bigwedge k \in K\ (s_{[k]} \in S_{[k]})\}$ ($\bigwedge$ denotes the universal, $\bigvee$ the existential quantifier). If $K \neq \varnothing$ and $S_{[k]} \neq \varnothing$ then $\prod_{k \in K} S_{[k]} \neq \varnothing$. $\text{pow}\, S$ and $\prod_{k \in K} S_{[k]}$ are well known constructs in classical set theory. In particular, if $\bigwedge k,k' \in K\ (S_{[k]} = S_{[k']} = S)$, then $\prod_{k \in K} S_{[k]}$ is the exponentiation $S^{[K]}$ of S by K, and if K is a finite set of ordinals, $K = \{1\text{st}, \ldots \text{n-th}\}$, the set product specializes to the cartesian product $\times_{k \in K} S_{[k]} = S_{[1\text{st}]} \times \ldots \times S_{[\text{n-th}]}$ where now the ordering of K is material and the indices are usually omitted. Any subset $R \subseteq \prod_{k \in K} S_{[k]}$ is a conventional relation. A visualization is shown in Fig.1.

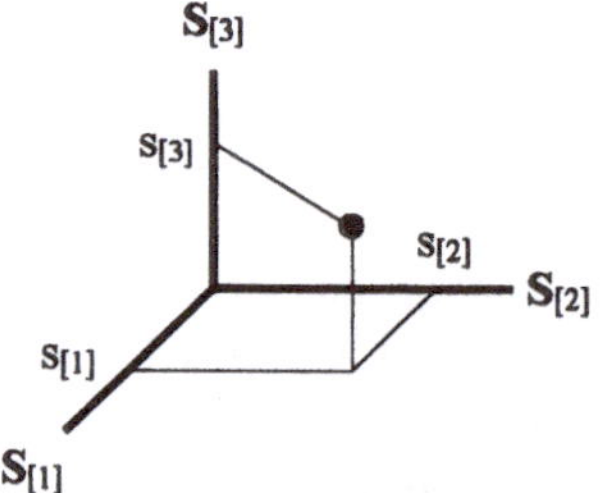

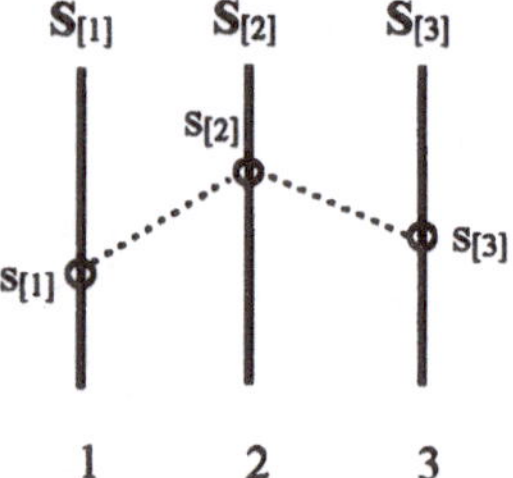

Fig.1

We generalize these concepts by the definition $\bigwedge i \in I\ (s_i =_{def} (i, s_{[i]}))$ and identify $(s_i)_{i \in I}$ with $\{(i, s_{[i]}) \mid i \in I\}$. The family $(s_i)_{i \in I}$ is a "parameterization" of the set $\text{ind}(I) \subseteq S$ by I. With $S_k =_{def} \{k\} \times S_{[k]}$ we have $(s_k)_{k \in K} \in \prod_{k \in K} S_k$ and $\bigwedge k,k' \in K\ (k \neq k' \Rightarrow s_k \neq s_{k'})$ while $s_{[k]}$ may equal $s_{[k']}$. Thus, defining *set of* $(s_k)_{k \in K} =_{def} \text{ind}(K) = \{s_{[k]} \mid k \in K\}$, card *set of* $(s_k)_{k \in K} \leq \text{card}\, K$.

$S = \{s_{[i]} \mid i \in I\}$ and the unparameterized relation $\{(s_{[i]}) \mid i \in I\}$ with one-element families $(s_{[i]})$ are isomorphic.

Object- and index sets are distinguished only by their structural use. A set can be indexed by itself ("canonical indexing"). Interpretations for indices are for example references like names, labels, addresses, space coordinates, time coordinates, or constructive procedures to determine a certain object. Though index-object pairs are extensively used for example in computer science (addressed memory units, identifiers, names for routines, numbering of instructions etc.), they are usually not formally treated in textbooks.

$R \subseteq \prod_{k \in K} S_k$ defines a relation consisting of a set of families $(s_k)_{k \in K}$ which for given parameter set P can be parameterized by a surjective indexing $P \to R$ to give a "parameterized" relation $((s_{pk})_{k \in K})_{p \in P}$. More general, let there be given any family $(R_{[l]})_{l \in L}$ of unparameterized relations $R_{[l]} \subseteq \prod_{k \in K_{[l]}} S_{[l]k}$, then we extend the definition of an unparameterized relation to $R =_{def} \bigcup_{l \in L} R_{[l]}$. A parameterized form of R is written $(((s_{lpk})_{k \in K_{[l]}})_{p \in P_{[l]}})_{l \in L}$. This procedure can be continued to any hierarchical level. Thus in our definition any family of families of families ... is a parameterized relation.

We name an unparameterized/parameterized hierarchical relation with unparameterized/parameterized component relations a "structure". Because a set is isomorphic with its canonical parameterization, this includes set hierarchies by isomorphisms.

Index sets can be structured, for example $\varnothing \neq K \subseteq K_{[1]} \times K_{[2]} \times \ldots K_{[n]}$ is n-dimensional.

Examples:
(1) Hierarchy of power sets: Given a set $S^{(0)}$. For level $n \in \mathbf{N}$ we define $S^{(n)} =_{def}$ (pow $S^{(n-1)}) \cup S^{(n-1)}$.
(2) Hierarchies of product sets: Given a set $S^{(0)}$ of single element families (s). For level $n \in \mathbf{N}$ let there be given an index set $K^{(n)} \neq \varnothing$ and an indexing $K^{(n)} \to$ pow $S^{(n-1)}$. We define $S^{(n)} =_{def} \left(\bigcup_{U \subseteq K^{(n)}} \prod_{k \in U} S_k^{(n-1)} \right) \cup S^{(n-1)}$.

Introducing $K =_{def} \bigcup_{l \in L} K_{[l]}$ and $\bigwedge k \in K \; (L_{[k]} =_{def} \{l \mid l \in L \wedge k \in K_{[l]}\}$, the relation $((s_{kl})_{l \in L_{[k]}})_{k \in K}$ is "conjugate" to $((s_{lk})_{k \in K_{[l]}})_{l \in L}$. This notion can be extended to hierarchical levels > 2.

Given the relation $F = ((s_{lk})_{k \in K_{[l]}})_{l \in L}$ and a decomposition $\boldsymbol{Z}$ by $\bigwedge l \in L \; ((K_{[l]} = D_{[l]} \cup C_{[l]}) \wedge (D_{[l]} \cap C_{[l]} = \varnothing) \wedge (x_l =_{def} (s_{lk})_{k \in D_{[l]}} \wedge y_l =_{def} (s_{lk})_{k \in C_{[l]}})$. The decomposition $\boldsymbol{Z}$ is "functional" if $\bigwedge l,l' \in L \; (y_{[l]} \neq y_{[l']} \Rightarrow x_{[l]} \neq x_{[l']})$. In this case a "function" $f : X =_{def} \{x_{[l]} \mid l \in L\} \to Y =_{def} \{y_{[l]} \mid l \in L\}$ is defined by $\bigwedge l \in L \; (x_{[l]} \mapsto y_{[l]})$, also expressed by $f \subset X \times Y$ and saying f is "functional from domain X to range or codomain Y". According to f, $x_{[l]}$ is "valuated" by $y_{[l]}$. F can have many functional decompositions, each of the resulting functions can have many "arities" card $D_{[l]}$ and card $C_{[l]}$. If for $\bigwedge l \in L \; (K_{[l]} = K \wedge D_{[l]} = D \wedge D \neq \varnothing)$ our definition specializes to the conventional one. Functional decompositions $\boldsymbol{Z}$, $\boldsymbol{Z}'$ of F can be partially ordered by $\boldsymbol{Z} \leq \boldsymbol{Z}' \Leftrightarrow_{def} \bigwedge l \in L\ (D_{[l]} \subseteq D'_{[l]})$.

If X is partitioned into $\{X^{(v)} \mid v \in N \wedge \text{card } N > 1\}$, we can consider $(f^{(v)} =_{def} f \mid X^{(v)}$ [the restriction of f to domain $X^{(v)}$]$)_{v \in N}$. Then any $x \in X$ determines uniquely an $X^{(v)}$ and an $f^{(v)}$. Each $f^{(v)}$ can have its own computational method. Then the constructive computation of f is "polymorphic".

Example (technical "and-gate" $\wedge^*$): For $i \in \{1,2\}$ and $x_{[i]}, y_{[3]} \in \{1,0\}$ the function $\wedge^*$ is given by

$x_{[1]}$	1	1	0	0	0	
$x_{[2]}$	1	0	1	0		0
$y_{[3]}$	1	0	0	0	0	0
			$\wedge$		$\wedge'$	$\wedge''$

Example: $F = ((s_{lk})_{k \in K_{[l]}})_{l \in L}$ with $L = \{1, 2, 3\}$, $K_{[1]} = \{a, b, c\}$, $K_{[2]} = \{a, b, c, d\}$, $K_{[3]} = \{a, b, c\}$, $s_{[1a]} = s_{[2a]}$, $s_{[2b]} = s_{[3b]}$, $s_{[2c]} = s_{[3c]}$, functional decompositions are

	Z		***Z'***	
l	$D_{[l]}$	$C_{[l]}$	$D_{[l]}$	$C_{[l]}$
1	{a, b}	{c}	{a}	{b, c}
2	{a, b, c}	{d}	{a, b}	{c, d}
3	{a, b}	{c}	{a, b}	{c}

with $\boldsymbol{Z}' < \boldsymbol{Z}$.

2.2 Composition of relations

Compositions of sets like the union $\cup$, the intersection $\cap$, the difference $\setminus$, and the symmetrical difference Δ are well known from classical set theory ("algebra of sets"). They apply to unparameterized relations considered as sets. Further, we list some important mappings of families in family notation.

Let **F** be a universal set of families and let $r =_{def} (r_k)_{k \in K} \in \mathbf{F}$. We consider the following mappings $\Gamma: r \mapsto s,\ s \in \mathbf{F}$:

(1) "projections" for the selection of objects by their indices:
Given an index set J, the projection of r with respect to J is $\text{pr}(J)\ r =_{def} (r_k)_{k \in K \cap J}$, the coprojection $\text{cpr}(J)\ r =_{def} (r_k)_{k \in K \setminus (K \cap J)}$. For a cartesian indexing one writes for example simply $\text{pr}_n\ (r_{[1st]}, .. r_{[n\text{-th}]}, ..) = r_{[n\text{-th}]}$ for the n-th projection.

(2) "cuts" for the selection of objects by their values:

Given $c = (c_j)_{j \in J} \in \mathbf{F}$, $\text{cut}(c)\ r =_{def} (r_k)_{k \in K^*}$ with $K^* =_{def} \{k \mid k \in K \wedge \bigvee j \in J\ (c_j = r_k$, i.e. $j = k \wedge c_{[j]} = r_{[k]}\}$.

(3) transformations of objects and indices:

Given a non-empty index set L, a surjective mapping $\tau\colon K \to L$ and a non-empty set S. Then we define $\wedge l{\in}L\ (K_{[l]} =_{def} \overset{-1}{\tau}(\{l\}))$, $\overset{-1}{\tau}$ the reciprocal image of τ, which is a partition of K. Further, let there be given $\wedge l{\in}L\ (\varphi_{[l]}\colon \{r_{[k]} \mid k \in K_{[l]}\} \to \{s_{[l]}\} \subseteq S)$. Then a transformation $\wedge l{\in}L\ ((r_k)_{k\in K[l]} \mapsto (s_l))$ is defined. This applies in particular to the case τ bijective and $\wedge k{\in}K\ (s_{[\tau(k)]} = r_{[k]})$ (transformation of indices only), and to the case $\tau = id$ (transformation of objects only).

Mappings Γ can be applied in case of structured index sets and can be extended to families of families, $(r_q)_{q\in Q}$, by applying Γ to each $r_{[q]}$.

Examples:
(1) Analytical geometry in euklidean $\mathbf{R}^n$, transformation of the coordinates of an object, e.g. displacements, rotations, symmetry operations, affine mappings.
(2) Order preserving relocation of a program in computer memory.

For $\mathbf{R}$ being a universal set of relations we consider assignments $\Gamma\colon (R_j)_{j\in J} \mapsto S$ for $\text{card}\, J > 1$ with $\wedge j{\in}J\ (R_j = ((r_{jqk})_{k\in K_{[jq]}})_{q\in Q_{[j]}} \in \mathbf{R})$ and $S \in \mathbf{R}$, which we name "concatenations":

We use the notations $K_{jq} =_{def} \{j\}\times\{q\}\times K_{[jq]}$, $Q_j =_{def} \{j\}\times Q_{[j]}$, $Q_J =_{def} \prod_{j\in J} Q_j$, and for $\wedge u{\in}Q_J\ (K_u =_{def} \bigcup_{jq\in u} K_{jq} \wedge K_{[u]} =_{def} \bigcup_{jq\in u} K_{[jq]} \wedge \wedge k{\in}K_{[u]}\ (D_k(u) =_{def} \{r_{[jq]k} \mid jq \in u\} \wedge (\varphi_{[u]}$ assumed to be a given function such that $\varphi_{[u]}\colon D_k(u) \mapsto s_{[u]k} \wedge ((\text{card}\, D_k(u) = 1 \wedge r_{[jq]k} \in D_k(u)) \Rightarrow s_{[u]k} = r_{[jq]k}))) \wedge K^*_{[u]} =_{def} \{k \mid k \in K_{[u]} \wedge \wedge jq,j'q'{\in}u\ ((r_{[jq]k}$ and $r_{[j'q']k}$ exist$) \wedge r_{[jq]k} = r_{[j'q']k})\})$. $Q^*_J =_{def} \{u \mid u \in Q_J \wedge \wedge k{\in}K_{[u]}\ (\text{card}\, D_k(u) = 1)\}$, $Q_{J\cap} =_{def} \{u \mid u \in Q_J \wedge K^*_{[u]} \neq \varnothing\}$. We consider the concatenations

(1) $\mathbf{K}((R_j)_{j\in J}) =_{def} ((r_{jqk})_{jqk\in K_u})_{u\in Q_J}$,

(2) $\mathbf{K}_\varphi((R_j)_{j\in J}) =_{def} ((s_{uk})_{k\in K_{[u]}})_{u\in Q_J}$, $\varphi = (\varphi_u)_{u\in Q_J}$,

(3) $\mathbf{K}_{join}((R_j)_{j\in J}) =_{def} ((s_{uk})_{k\in K_{[u]}})_{u\in Q^*_J}$, $\wedge u{\in}Q^*_J\ (\varphi_{[u]} = id)$,

(4) $\mathbf{K}_\cap((R_j)_{j\in J}) =_{def} (pr(K^*_{[u]})((s_{uk})_{k\in K_{[u]}}))_{u\in Q_{J\cap}}$, $\wedge u{\in}Q_{J\cap}\ (\varphi_{[u]} = id)$.

$\mathbf{K}$, $\mathbf{K}_{join}$, $\mathbf{K}_\cap$ are commutative and associative.

Visualization (Fig. 2): $J = \{j, j', j''\}$, $u = (jq, j'q', j''q'')$, $K_{[u]} = \{1, 2, 3, 4, 5\}$.
$\mathbf{K}$ yields $(a_{jq1}, b_{jq2}, d_{jq3}, c_{j'q'2}, \ldots e_{j''q''5})$; $\mathbf{K}_\varphi$ yields $(a_{u1}, \varphi_{u2}(b,c), d_{u3}, a_{u4}, e_{u5})$; $\mathbf{K}_{join}$ is empty; $\mathbf{K}_\cap$ yields (d_{u3}, a_{u4}).

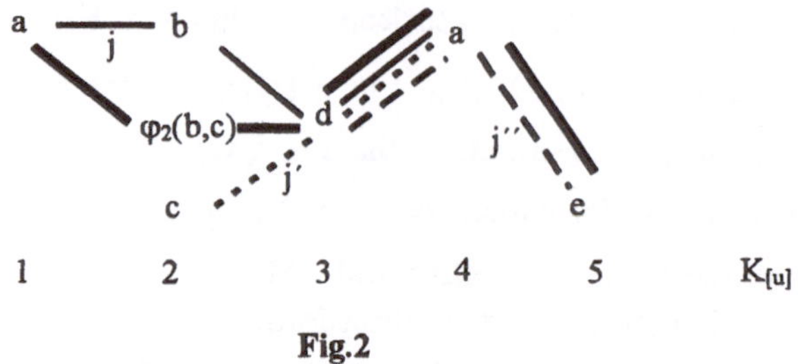

Fig.2

Example: Let be $R = \{(r_{kji})_{kji \in U \subseteq K \times J \times I}\}$ and let $pr(K)U = K$, $pr(J)U = J$, $pr(I)U = I$. We define $U_{[k]} =_{def} cut(\{k\} \times J \times I)\ U$, $R_k =_{def} pr(U_{[k]})R$, $J_{[k]} =_{def} pr(J)\ U_{[k]}$, $U_{[kj]} =_{def} cut(\{k\} \times \{j\} \times I)\ U_{[k]}$, $R_{kj} =_{def} pr(U_{[kj]})R_k$, $I_{[kj]} =_{def} pr(I)\ U_{[kj]}$, $U_{[kji]} =_{def} cut(\{k\} \times \{j\} \times \{i\})\ U_{[kj]} = \{(k,j,i)\}$, $R_{[kji]} =_{def} pr(U_{[kji]})R_{[kj]} = \{(r_{kji})\}$. The decomposition can be reversed by concatenations $\mathbf{K}$: $\mathbf{K}((R_{kji})_{i \in I_{[kj]}}) = R_{kj}$, $\mathbf{K}((R_{kj})_{j \in J_{[k]}}) = R_k$, $\mathbf{K}((R_k)_{k \in K}) = R$. The constructs can be geometrically interpreted. The conjugate representations are obtainable by permuting the sequence of I, J, K. They are all equivalent in the sense that their concatenations result in R. For $(x,y) \in R_{[k]}$ and $(y, z) \in R_{[i]}$, $\mathbf{K}_{join}((x, y), (y, z)) = (x, y, z)$ which is not necessarily an element of R. Because of the ordering of the (non explicitly written) indices, $pr_{13}((x, y, z)) = (x, z)$ expresses a transitivity $(x) \mapsto (z)$ via (y).

For conventional relations $\mathbf{K}_{join}$ is a standard construct in the theory of data bases and in geometry. An example for $\mathbf{K}_\varphi$ is the following: We assume that $R = \{r_q =_{def} (r_{qk})_{k \in K_{[q]}} \in \prod_{k \in K_{[q]}} R_{qk} \mid q \in Q = \{u,v\}\}$ and $\bigwedge k \in \bigcup_{q \in Q} K_{[q]}\ (M_{[k]} = R_{[qk]} \wedge$ $M_{[k]}$ being a module of linear forms with an addition $+ : M_{[k]} \times M_{[k]} \to M_{[k]}$ and with a ring A for scalar multiplication $* : A \times M_{[k]} \to M_{[k]}$). We extend + to $M_{[k]} \to M_{[k]}$ by $+ =_{def} id$ on $M_{[k]}$. Then we define $\bigwedge \alpha =_{def} (\alpha_{[k]})_{k \in K_{[q]}} \in A^{[K_{[q]}]}$ $(\alpha * r_{[q]} =_{def} ((\alpha_{[k]} * r_{[qk]})_k)_{k \in K_{[q]}})$ and $s =_{def} r_{[u]} + r_{[v]}$ by $s =_{def} ((r_{[uk]} + r_{[vk]})_k)_{k \in K_{[u]} \cup K_{[v]}}$ (for an application to manufacturing systems we refer to the paper of R. F. Albrecht (1996c) and the article of E. Canuto, F. Donati, M. Vallauri in this volume).

Example: $K_{[u]} = \{1,2,3\}$, $K_{[v]} = \{2,3,4\}$, $\bigwedge k \in \{1,2,3,4\}$ ($M_{[k]}$ = A = set of integers), $r_{[u]} = ((1,5), (2,0), (3,-1))$, $r_{[v]} = ((2,6), (3,4), (4,7))$, $-2 * r_{[u]} = ((1,-10), (2,0), (3,2))$, $s = ((1,5), (2,6), (3,3), (4,7))$.

We consider again $(R_j)_{j \in J}$ with $R_j = ((r_{jqk})_{k \in K_{[jq]}})_{q \in Q_{[j]}}$. A relation between the index sets K_{jq}, which we name a "connector" C, a relation with index sets given by the elements of C, which we name a "property relation" E(C) on C, and a function φ with dom φ = E(C) can be used to define a general class of concatenations.

For sake of simplicity we formulate this for $R_j = (r_{jk})_{k\in K_{[j]}}$, suppressing q. Let there be given a connector $C \subseteq \bigcup_{U\subseteq J} \prod_{j\in U} K_j$ and a relation $E(C) \subseteq \bigcup_{c\in C} \prod_{jk\in c} M_{jk}$ with non-empty $M_{[jk]}$. We introduce the notations: $U =_{def} \bigcup_{j\in J} K_j$, $U(\varnothing) =_{def} \{jk \mid jk \in U \wedge \neg \bigvee c\in C\ (jk \in c)\}$. We assume $\bigwedge c\in C\ ((r_{jk})_{jk\in c} \in E(C)) \wedge r_c =_{def} \varphi((r_{jk})_{jk\in c}))$. Then with $P =_{def} U(\varnothing) \cup C$ the concatenation

(5) $\mathbf{K}(((r_{jk})_{k\in K_{[j]}})_{j\in J}, E(C), C, \varphi) =_{def} (r_p)_{p\in P}$ is defined.

Projection on $U(\varnothing)$ yields $(r_{jk})_{jk\in U(\varnothing)}$, a relation between the unconnected r_{jk}.

If $E(C)$ is an equality relation, i.e. $\bigwedge c\in C\ ((m_{jk})_{jk\in c} \in E(C) \Rightarrow \bigwedge jk,j'k'\in c\ (m_{[jk]} = m_{[j'k']}))$, then we set $\bigwedge jk,j'k'\in c\ (r_{[c]} =_{def} r_{[jk]} = r_{[j'k']})$ and write

(5′) $\mathbf{K}((r_{jk})_{k\in K_{[j]}})_{j\in J}, E(C), C) = (r_p)_{p\in P}$.

Now let there be $\bigwedge jk\in U\backslash U(\varnothing)\ ((r_{[jk]} \in (G_{[jk]}, +) \wedge (G_{[jk]}, +)$ an additive group) $\wedge\ C(jk) =_{def} \{c \mid c \in C \wedge jk \in c\} \wedge r_{[jk]} = \sum_{c\in C(jk)} r_{[jkc]} \wedge \bigwedge c\in C(jk)\ (r_{[jkc]} \in G_{[jk]}))$. This implies a decision on the decomposition of $r_{[jk]}$. We introduce $U(c) =_{def} \{jkc \mid jk \in c\}$, $U(C) =_{def} \bigcup_{c\in C} U(c)$, $P =_{def} U(C) \cup U(\varnothing)$. We assume, a relation $\widetilde{E}(C) \subseteq \bigcup_{c\subseteq C} \prod_{jkc\in U(c)} M_{jkc}$ with non-empty sets $M_{[jkc]}$ and a function $\widetilde{\varphi}$ with dom $\widetilde{\varphi} = \widetilde{E}(C)$ are defined. If $\bigwedge c\in C\ ((r_{jkc})_{jk\in c} \in \widetilde{E}(C) \wedge r_c =_{def} \widetilde{\varphi}((r_{jkc})_{jk\in c}))$ then

(6) $\mathbf{K}((r_{jk})_{k\in K_{[j]}})_{j\in J}, \widetilde{E}(C), C, \widetilde{\varphi}) =_{def} (r_p)_{p\in P}$ is defined.

If $\widetilde{E}(C)$ is an equality relation, then we set $\bigwedge c\in C\ \bigwedge jkc,j'k'c\in U(c)\ (r_{[c]} =_{def} r_{[jkc]} = r_{[j'k'c]})$ and write

(6′) $\mathbf{K}((r_{jk})_{k\in K_{[j]}})_{j\in J}, \widetilde{E}(C), C) = (r_p)_{p\in P}$.

These concatenations cover for example functional composition if $E(C)$ is an equality relation. For illustration see the directed structure of Fig.3.

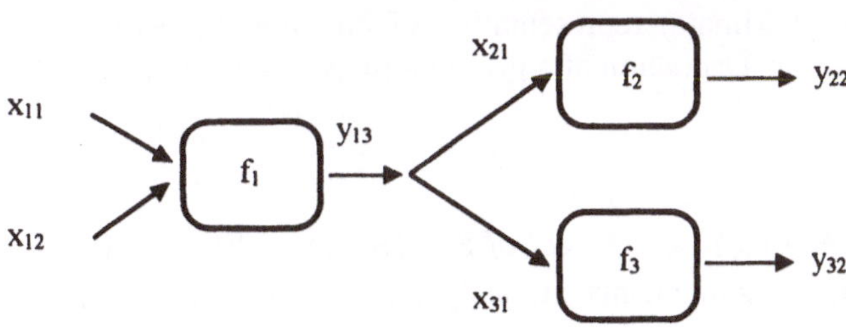

Fig.3

Example: We assume f_1, f_2, f_3 are functions with integers as domain and range, and for given x-values $y_{13} = f_1(x_{11},x_{12})$, $y_{22} = f_2(x_{21})$, $y_{32} = f_3(x_{31})$, $C = \{(13,21), (13,31)\}$,

$E(C) = \{(y_{13}, x_{21}), (y_{13}, x_{31})\}$, $r_{[13,21]} = y_{[13]} = x_{[21]}$, $r_{[13,31]} = y_{[13]} = x_{[31]}$. $S = \mathbf{K}((x_{11}, x_{12}, y_{13})_1, (x_{21}, y_{22})_2, (x_{31}, y_{32})_3), E(C), C) = (x_{11}, x_{12}, r_{13,21}, r_{13,31}, y_{22}, y_{32})$, $U(\varnothing) = \{11, 12, 22, 32\}$, $pr(U(\varnothing)\ S = (x_{11}, x_{12}, y_{22}, y_{32})$ (unconnected values) is functional $(x_{11}, x_{12}) \mapsto (y_{22}, y_{32})$. For $E^*(C) = \{(y_{13}, z_{21}), (y_{13}, x_{31})\}$, $z_{[21]} \neq x_{[21]}$, $\mathbf{K}$ is not defined. For $C' = \{(13,31)\}$, $E'(C') = \{(y_{13}, x_{31})\}$, $r_{[13,31]} = y_{[13]} = x_{[31]}$, $\mathbf{K}((x_{11}, x_{12}, y_{13})_1, (x_{21}, y_{22})_2, (x_{31}, y_{32})_3), E'(C'), C') = (x_{11}, x_{12}, r_{13,31}, x_{21}, y_{22}, y_{32})$ (application in mathematics, computer science).

Example: In the previous example let $y_{[13]}$ be decomposed into $y_{[13]} = y_{[13,(13,21)]} + y_{[13,(13,31)]}$ and let $y_{22} =_{def} f_2(x_{21,(13,21)})$, $y_{32} =_{def} f_3(x_{31,(13,31)})$ and $\tilde{E}(C) = \{(y_{13,(13,21)}, x_{21,(13,21)}), (y_{13,(13,31)}, x_{31,(13,31)})\}$ be an equality relation, $r_{[13,21]} = y_{[13,(13,21)]} = x_{[21,(13,21)]}$, $r_{[13,31]} = y_{[13,(13,31)]} = x_{[31,(13,31)]}$. Then $\mathbf{K}((x_{11}, x_{12}, y_{13})_1, (x_{21}, y_{22})_2, (x_{31}, y_{32})_3), \tilde{E}(C), C) = (x_{11}, x_{12}, r_{13,21}, r_{13,31}, y_{22}, y_{32})$ (application in production systems).

Considering now the general case $(R_j)_{j\in J}$ with $R_j = ((r_{jqk})_{k\in K_{[jq]}})_{q\in Q_{[j]}}$, then those $(q_j)_{j\in J} \in \prod_{j\in J} Q_j$ have to be selected for which $((r_{jq_{[j]}k})_{k\in K_{[j]}})_{j\in J}$ admits concatenation with respect to J.

Combinations of the compositions considered give further compositions.

Example (usual concatenation of cartesian families): $\mathbf{K}_\times((x'_{[i]})_{i=1,2,\ldots n}, (x''_{[i]})_{i=1,2,\ldots m}) = ((x_{[i]})_{i=1,2,\ldots n+m})$ with $x_{[i]} =_{def} x'_{[i]}$ for $i = 1,\ldots n$ and $x_{[i]} =_{def} x''_{[i-n]}$ for $i = n+1,\ldots m$.

Concatenations $\Gamma: (R_j)_{j\in J} \mapsto S$ are algebraic compositions. Introducing $\mathscr{X} \subseteq \mathbf{R}$, $\mathscr{K}$ a set of admitted (composite) concatenations, we consider the equation

$$\text{(E)}\quad S = \text{var}\ \Gamma((\text{var}\ X_j)_{j\in \text{var}J}) \text{ for given } S.$$

General problem classes are: Has (E) a solution $\Gamma \in \mathscr{K}$, $X_{[j]} \in \mathscr{X}$? If so, find one/all solutions (representation, decomposition problem). In particular, part of the var X_j may have been assigned, $(\text{var}\ X_j := R_j)_{j\in J^*}$, $\varnothing \neq J^* \subset \text{var}\ J$ (deduction of S from $(R_j)_{j\in J^*}$).

Examples: (Approximate) representation of functions by arithmetic expressions in numerical analysis. Deduction of a given word S from axioms in a formal language.

2.3 Variables

We consider $S = (S_p)_{p\in P}$, $\mathscr{S} =$ *set of* $S = \{S_{[p]} \mid p \in P\}$, $\bigwedge p\in P\ (S_{[p]} \neq \varnothing \wedge (S_{[p]}$ a structure with all components uniquely parameterized by $I_{[p]}) \wedge i \subset I_{[p]})$. i is assumed to be independent of p and may be composite. Let $\bigwedge p\in P\ (V_{[p]} =_{def} pr(I_{[p]}\backslash i)S_{[p]} \wedge (C =_{def} pr(i)S_{[p]}$ being independent of p (constant part))) and let $\mathscr{V} =_{def} \{V_{[p]} \mid p \in P\}$. Then $S_{[p]} = \mathbf{K}(V_{[p]}, C)$. To facilitate the representation of $\mathscr{S}$ we introduce objects var V and $S(\text{var}\ V) = \mathbf{K}(\text{var}\ V, C)$, a function val: $P \times \{\text{var}\ V\} \times$

$\{S(\text{var } V)\} \to \mathscr{V}\times\mathscr{S}$ with $(p, \text{var } V, S(\text{var } V)) \mapsto (V_{[p]}, S_{[p]})$, and a function val^{-1}: $\{(V_{[p]}, S_{[p]}) \mid p \in P\} \to \{(\text{var } V, S(\text{var } V))\}$. The terminology used is: var V is a controlled "variable" on "variability domain" dom var V = $\mathscr{V}$ with respect to $\mathscr{S}$, S(var V) is a "structure variable" on "variability domain" $\mathscr{S}$, val is an "assignment" or "control" function with "assignment" or "control" parameter (or "program") $p \in P$, $V_{[p]}$ in $S(V_{[p]})$ is "substitutable" by any $V_{[q]}$ (application of val^{-1} and re-assignment) to give $S(V_{[q]})$. The notations used are: var V : $\mathscr{V}$, S(var V) :$\mathscr{S}$, and for assignments according parameter p var V :=(p) $V_{[p]}$, S(var V) :=(p) $S_{[p]}$. However, in some applications there may be constraints prohibiting unlimited application of val^{-1} and re-assignment (for example the number of writings into a computer memory can be limited).

Example for physical realizations of control functions: A universal computer processor is a physical variable on a domain of arithmetic and logic processors, it is controlled by the control unit, the control parameters are the instructions.

The structures and operations considered in **2.1** and **2.2** can be extended to structures and operations with variables, yielding results with variables. Variables can also be defined on structured indices.

Example ("last in - first out" queue): Given var n : $\mathbf{N}_0$, variable integer interval [1, var n], $[1,0] =_{def} \varnothing$, $\varnothing \neq X$ any set, variable $q =_{def} (x_i)_{i\in[1,\,\text{var } n]}$, variables x, $x_{[i]}$: X. Operations are:
"pop": for var n > 0: $(x_{\text{var } n}) = \text{pr}(\text{var } n)q$, $x = x_{[\text{var } n]}$, q := cpr(var n)q; var n := var n – 1; for var n = 0: q := q;
"push": given x, $x_{\text{var } n+1}$:= (var n + 1, x), q := $\mathbf{K}(q, (x_{\text{var } n+1}))$, var n := var n + 1.

If a function f maps X onto $f(X) \subseteq Y$ and var x : X, var y : f(X), we write var y := f(var x). Assignment to var x results in assignment to var y by f: var y depends functionally on var x.

We assume var V is a composite structure with variables var w_j, collected in the family $(\text{var } w_j)_{j\in J}$ with $\bigwedge j\in J$ (var w_j : $W_j = \{w_{j[q]} \mid q \in Q_{[j]}\}$). Further, we assume $P \subseteq \prod_{j\in J} Q_j$, $p = (q_{[p]j})_{j\in J}$ for $p \in P$, and var V :=(p) $V_{[p]}$ corresponds to $(\text{var } w_j)_{j\in J}$:=$(q_{[p]j})_{j\in J}$ $(w_{[p]j})_{j\in J}$.

Notice: An assignment to var V is not defined on $\prod_{j\in J} Q_j \setminus P$ (in practice for this an "error message" can be provided).

In our representation a variable var V : $\mathscr{V}$, card $\mathscr{V}$ > 1 expresses an undeterminacy in the sense that the selection which $V\in \mathscr{V}$ is to be instanciated is not yet made, but the variability domain is determined.

Assignments to $(\text{var } w_j)_{j\in J}$ can be "partial" which means to a subfamily $(\text{var } w_j)_{j\in J^*}$ with $\varnothing \subset J^* \subset J$. Then $(\text{var } w_j)_{j\in J^*}$ has the assignment parameter set pr(J*)P, and if $(q_j')_{j\in J^*}$ is chosen, the supplement $(q_j')_{j\in J\setminus J^*}$ is restricted to parameter set $\{(q_j)_{j\in J\setminus J^*} \mid (q_j)_{j\in J} \in P \wedge pr(J^*)((q_j)_{j\in J}) = (q_j')_{j\in J^*}\}$. This domain specification is for example one part of the "firing conditions" for a processor in a data flow computer architecture (R. F. Albrecht, 1996b).

The variability domain of a variable can contain variables of lower level ("variables on variables" hierarchy). By definition, if var b is on lower level with respect to a variable var a, then var a cannot be assigned to var b (parameterization by hierarchical level). It is assumed that in constructive applications for any variable a sequence of assignments with decreasing level ends after a finite number of assignment steps at a non-variable "terminal" object. The hierarchy of variables in a structure is partial ordered $\prec$. var b $\prec$ var a means, it exists a finite sequence (var b = var $x^{(n)}$, var $x^{(n-1)}$,... var $x^{(1)}$ = var a) with var $x^{(\nu+1)}$ $\in$ domain var $x^{(\nu)}$. In general, assignments need not be in the order "top down". However, when assigning first to a lower level variable var b of a higher level variable var a, var b $\prec$ var a, assignments to var a can be restricted depending on the assignment to var b.

Example: var object : {var (geometrical object)$_1$, var (arithmetical object)$_2$},
var (geometrical object)$_1$: {point$_{11}$, triangle$_{12}$, rectangle$_{13}$},
var (arithmetical object)$_2$: {integer$_{21}$, real$_{22}$},
var object :=(1) var (geometrical object)$_1$, var (geometrical object)$_1$:=(2) triangle$_{12}$, var object :=(1,2) triangle$_{12}$. The hierarchical level is expressed by the order of the indices ("path").

Considering the set **A** of all admissible (partial, total) assignments **a** to var V (the non-assignment included) and denoting an assignment **a** to var V by var$_{\mathbf{a}}$ V (even in case it is a terminal object), we can introduce $\mathbf{V} =_{def}$ (var$_{\mathbf{a}}$ V)$_{\mathbf{a}\in\mathbf{A}}$, $\mathbf{S} =_{def}$ (S(var$_{\mathbf{a}}$ V))$_{\mathbf{a}\in\mathbf{A}}$, the "state spaces" (or configuration spaces) of var V, S(var V), with "states" var$_{\mathbf{a}}$ V and S(var$_{\mathbf{a}}$ V), respectively.

Example: For illustration see Fig.4.
var f : $\{f_p \mid p \in \{1,2\}\}$ with $f_{[1]} = (.)^2 : \mathbf{R} \to \mathbf{R}_+$, var y_1 := f_1(var x_1), $f_{[2]} = (.)^{1/2} : \mathbf{R}_+ \to \mathbf{R}_+$, var y_2 := f_2(var x_2), + : $\mathbf{R} \times \mathbf{R} \to \mathbf{R}$, (var x_{+1}, var x_{+2}) : {(var x_{+1p}, var x_{+2p}) | p $\in$ {1,2}}, (var x_{+11}, var x_{+21}) : $\mathbf{R} \times \mathbf{R}$, (var x_{+12}, var x_{+22}) : $\overset{-1}{+}(\mathbf{R}_+)$, var y_{+3} := var x_{+1} + var x_{+2} . Concatenated by C = {((+,3),(var f,1)}, E(C) = {var y_{+3} = var x_{p1} = var z_p} to var y := var f(var z), var z := (var x_{+1} + var x_{+2}) with the constant part + of the structure, we have:
V = ((var x_{+1}, var x_{+2}, var z, var f, var y), (var x_{+11}, var x_{+21}, var z_1, f_1, var y_1), (x_{+11}, x_{+21}, var z_1, f_1, var y_1), ...(var x_{+12}, var x_{+22}, var z_2, f_2, var y_2),....).
S = ((var y := var f(var z), var z := var x_{+1} + var x_{+2}), (var y_1 := (var z_1)^2 , var z_1 := var x_{+11} + var x_{+21}),...).
Free "input" parameters:

2cd control level: p $\in$ {1,2},

1st control level, "data" level: $\{(p, (x_{+1p}, x_{+2p})) \mid p \in \{1,2\}, (x_{+11}, x_{+21}) \in \mathbf{R}\times\mathbf{R}, (x_{+12}, x_{+22}) \in \overset{-1}{+}(\mathbf{R}_+)\}$
"intermediate" variable: var z, can also be output,
"output" variable: var y, depending on input parameters.
Let be assigned var x_{+1} := var x_{+2} := −1. Consequently, only var f := $f_{[1]}$ is admitted.

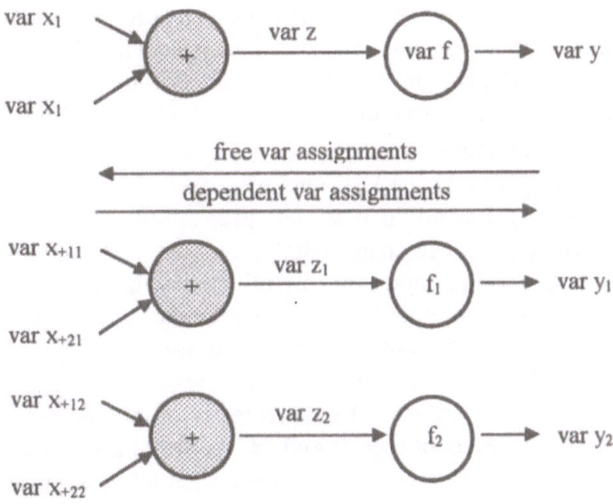

Fig.4

A transition $\text{var}_a\ V \mapsto \text{var}_b\ V$ from one state, with $a \in \mathbf{A}$, to another state, with $b \in \mathbf{A}$, is named an "event" $(\text{var}_b\ V, \text{var}_a\ V)$. In particular, $\text{var}_b\ V$ can be the result of a given functional assignment, the function depending on $\text{var}_a\ V$ and on a parameter $p \in P(\text{var}_a\ V)$. The state changes admitted at $\text{var}_a\ V$ are expressed by the parameter set $P(\text{var}_a\ V)$.
Notice: $a \in \mathbf{A}$ is a state-, $p \in P(\text{var}_a\ V)$ a local state change-parameter. For example, the change can be a proceeding to a lower, a substitution on the same, a return to a higher var-level.

We consider a simple or composite variable var V : $\{V_{[p]} \mid p \in P\}$ with simple or composite control parameters $p \in P$ and control function $\text{val}^{(1)}$. So far, for any assignment var V :=(p) $V_{[p]}$ we assumed p to be "given". However, introducing var p : $\{p_{[q]} \mid p_{[q]} \in P \wedge q \in Q\}$, Q a parameter set to parameterize P, the selection of a parameter p is the result of application of a control function $\text{val}^{(2)} : Q\times\{\text{var } p\} \to P$ on control level 2. The reasoning can be continued to any hierarchical control level ("higher order control").

Example: Q the statements of a higher order programming language, $\text{val}^{(2)}$ simulated by a compiler, producing for $q \in Q$ a sequence of instructions p as input to the processor control unit which realizes $\text{val}^{(1)}$.

2.4 Selection by properties

The basic types of selection of indexed objects from a given structure are by properties of indices or properties of values (**2.2**). Mathematically, "properties" are expressed by membership of the object in (hierarchical) sets or relations. In physical systems, properties can be physical qualities and quantities of an object, interpreted, the object belongs to the set of all objects, of which we can recognize these physical properties. Sets or relations, selected by the basic selection types or combinations of these, can then be subject to set or relation constructors. Valuations of objects, expressing their "importance" or "weight", can be an additional selection tool. For example, let there be given a structure $R = (R_u)_{u \in U}$, $U \neq \varnothing$, of non-empty relations R_u, for all $u \in U$ a valuation ψ: $R_u \mapsto v_u$, $v_{[u]} \in V$, and a non-empty $W \subseteq (\text{pow } U) \backslash \varnothing$. For all $w \in W$ let to $(R_u, v_u)_{u \in w}$ cuts with respect to V be applied, resulting in a subrelation $(R_u)_{u \in w'}$. For all $u \in w'$ let projections of components of R_u be applied resulting in R_u'. Concatenate $((R_u')_{u \in w'})_{w' \in W'}$, W' being the set of all w'. We apply this principle to the important case of valuated binary relations R_u:

Let there be given a non-empty set Y of elements y, which we name "objects", and a non-empty set X of elements x, which we name "properties", bijective parameterizations ind: $J \leftrightarrow Y$, ind′: $I \leftrightarrow X$, and a relation $R = (y_j, x_i)_{(j,i) \in U}$, $U \subseteq J \times I$, with $pr_1 U = J$, $pr_2 U = I$. Because of the symmetry of our representation, we could as well have named X the set of objects and Y the set of properties. If V is a non-empty set and an indexing ind: $U \rightarrow V$ is given, then a bijective valuation ψ: $R \rightarrow \text{ind}(U)$ is defined by $v_{ji} = \psi((y_j, x_i))$ and R can be represented by $M =_{def} (v_{ji})_{(j,i) \in U}$. We consider $\bigwedge j \in J \; (\text{pr}(\{j\} \times I) \; M = (v_{ji})_{i \in I_{[j]}})$, $\bigwedge i \in I \; (\text{pr}(J \times \{i\}) \; M = (v_{ji})_{j \in J_{[i]}})$, which define $I_{[j]}$, $J_{[i]}$. We assume $\bigwedge j,j' \in J \; ((j \neq j') \Rightarrow ((v_{[j]i})_{i \in I_{[j]}} \neq (v_{[j']i})_{i \in I_{[j']}}))$, $\bigwedge i,i' \in I \; ((i \neq i') \Rightarrow ((v_{j[i]})_{j \in J_{[i]}} \neq (v_{j[i']})_{j \in J_{[i']}}))$, which requires $\text{card } Y \leq (\text{card } V)^{\text{card } X}$ and $\text{card } X \leq (\text{card } V)^{\text{card } Y}$. We name $((y_j, x_i), v_{ji})_{ji \in U}$ and also $(y_j, x_i, v_{ji})_{ji \in U}$ a "binary valuated relation" (or a "knowledge module") **K**.

We introduce η: $V \times V \rightarrow \mathbf{B} = (\{\text{"t"}, \text{"f"}\}, \sqcap, \sqcup)$ (as a boolean lattice) with $\eta(\text{diag } (V \times V)) = \{\text{"t"}\}$, $\eta((V \times V) \setminus \text{diag } (V \times V)) = \{\text{"f"}\}$. For any finite set B with $\text{card } B \leq \text{card } I, \text{card } J$, we consider functions $\varphi_{\text{card } B} \in (\mathbf{B}^{\text{card } B} \rightarrow \mathbf{B})$. For $\varnothing \neq \tilde{I} \subseteq I$ and $\varnothing \neq \tilde{J} \subseteq J$ and $\tilde{I}$ and $\tilde{J}$ finite let be $\bigwedge i \in \tilde{I} \; ((\tilde{v}_{[i]} \in V) \wedge \bigwedge j \in J_{[i]} \; (\eta_{[ji]} =_{def} \eta(\tilde{v}_{[i]}, v_{[ji]})))$, $\bigwedge j \in \tilde{J} \; ((\tilde{v}_{[j]} \in V) \wedge \bigwedge i \in I_{[j]} \; (\eta_{[ji]} =_{def} \eta(\tilde{v}_{[j]}, v_{[ji]})))$. We define for $\beta_{[\tilde{I}]}, \beta_{[\tilde{J}]} \in \mathbf{B}$:

$Y((x_i, \tilde{v}_i)_{i \in \tilde{I}}, \varphi_{\text{card } \tilde{I}}, \beta_{[\tilde{I}]}) =_{def} (y_j)_{j \in J^*}$, $J^* =_{def} \{j \mid j \in J \wedge \tilde{I} \subseteq I_{[j]} \wedge \varphi_{\text{card } \tilde{I}}((\eta_{[ji]})_{i \in \tilde{I}}) = \beta_{[\tilde{I}]}\}$,

$X((y_j, \tilde{v}_j)_{j \in \tilde{J}}, \varphi_{\text{card } \tilde{J}}, \beta_{[\tilde{J}]}) =_{def} (x_i)_{i \in I^*}$, $I^* =_{def} \{i \mid i \in I \wedge \tilde{J} \subseteq J_{[i]} \wedge \varphi_{\text{card } \tilde{J}}((\eta_{[ji]})_{j \in \tilde{J}}) = \beta_{[\tilde{J}]}\}$.

This defines functions f: $(x_i, \tilde{v}_i)_{i\in\tilde{I}} \mapsto (y_j)_{j\in J^*}$ and g: $(y_j, \tilde{v}_j)_{j\in\tilde{J}} \mapsto (x_i)_{i\in I^*}$, mapping families on families.

Especially, for $\tilde{I} = \{i\}$, $\tilde{J} = \{j\}$, $\varphi_1 = \text{id}$,
$Y(x_i, \tilde{v}_i) =_{def} Y((x_i, \tilde{v}_i), \eta(\tilde{v}_{[i]}, v_{[ji]}) = \text{"t"})$, and conjugate,
$X(y_j, \tilde{v}_j) =_{def} X((y_j, \tilde{v}_j), \eta(\tilde{v}_{[j]}, v_{[ji]}) = \text{"t"})$, are non-empty for
$\tilde{v}_{[i]} \in \{v_{[ji]} \mid j \in J_{[i]}\}$, $\tilde{v}_{[j]} \in \{v_{[ji]} \mid i \in I_{[j]}\}$, and $\{$*set of* $Y(x_i, \tilde{v}_i) \mid i \in I\}$ and $\{$*set of* $X(y_j, \tilde{v}_j) \mid j \in J\}$ are then sets of generators to construct by set operations subsets of Y and X, respectively. We have in this case
set of $Y((x_i, \tilde{v}_i)_{i\in\tilde{I}}, \varphi_{\text{card}\,\tilde{I}}, \text{"t"}) = \Phi_{\text{card}\,\tilde{I}}((\textit{set of}\, Y(x_i, \tilde{v}_i))_{i\in\tilde{I}})$, and conjugate,
set of $X((y_j, \tilde{v}_j)_{j\in\tilde{J}}, \varphi_{\text{card}\,\tilde{J}}, \text{"t"}) = \Phi_{\text{card}\,\tilde{J}}((\textit{set of}\, X(y_j, \tilde{v}_j))_{j\in\tilde{J}})$, $\Phi_{\text{card}\,\tilde{I}}$, $\Phi_{\text{card}\,\tilde{J}}$ the set functions corresponding to the boolean functions $\varphi_{\text{card}\,\tilde{I}}$, $\varphi_{\text{card}\,\tilde{J}}$ respectively.
Notice: *set of* $Y((x_i, \tilde{v}_i)_{i\in\tilde{I}}, \varphi_{\text{card}\,\tilde{I}}, \text{"f"}) \subseteq Y \setminus$ *set of* $Y((x_i, \tilde{v}_i)_{i\in\tilde{I}}, \varphi_{\text{card}\,\tilde{I}}, \text{"t"})$.

For $\wedge i\in I\, (v_{[i]} \in \{v_{[ji]} \mid j \in J_{[i]}\})$, $\wedge j\in J\, (v_{[j]} \in \{v_{[ji]} \mid i \in I_{[j]}\})$ let there be given
$(1_\cap)$: $I^{(1)}, I^{(2)}$, both $\neq \varnothing$ and $\subseteq I$, $\tilde{I} =_{def} I^{(1)} \cup I^{(2)}$, $(v_i)_{i\in\tilde{I}}$. Then we have
set of $Y((x_i, v_i)_{i\in I^{(1)}}, \sqcap_{\text{card}\,I^{(1)}}, \text{"t"})$ and *set of* $Y((x_i, v_i)_{i\in I^{(2)}}, \sqcap_{\text{card}\,I^{(2)}}, \text{"t"}) \supseteq$ *set of* $Y((x_i, v_i)_{i\in\tilde{I}}, \sqcap_{\text{card}\,\tilde{I}}, \text{"t"})$.
If $Y((x_i, v_i)_{i\in I^{(1)}}, \sqcap_{\text{card}\,I^{(1)}}, \text{"t"}) = Y((x_i, v_i)_{i\in I^{(2)}}, \sqcap_{\text{card}\,I^{(2)}}, \text{"t"})$, then $(x_i, v_i)_{i\in I^{(1)}}$ and $(x_i, v_i)_{i\in I^{(2)}}$ are in an equivalence relation $\equiv_\cap$, expressing that they specify the same subset of Y. For the conjugate case let be given analogously
$(1'_\cap)$: $J^{(1)}, J^{(2)}$, both $\neq \varnothing$ and $\subseteq J$, $\tilde{J} =_{def} J^{(1)} \cup J^{(2)}$, $(v_j)_{j\in\tilde{J}}$. Then we have
set of $X((y_j, v_j)_{j\in J^{(1)}}, \sqcap_{\text{card}\,J^{(1)}}, \text{"t"})$ and *set of* $X((y_j, v_j)_{j\in J^{(2)}}, \sqcap_{\text{card}\,J^{(2)}}, \text{"t"}) \supseteq$ *set of* $X((y_j, v_j)_{j\in\tilde{J}}, \sqcap_{\text{card}\,\tilde{J}}, \text{"t"})$.

This expresses the "inheritance" principle: the larger the set of common properties/objects, the smaller the set of objects/properties possessing these common properties/objects.

We consider

$\wedge j\in J\, (\mathscr{Y}(j) =_{def} \{\textit{set of}\, Y((x_i, v_{[j]i})_{i\in\tilde{I}}, \sqcap_{\text{card}\,\tilde{I}}, \text{"t"}) \mid \tilde{I} \in (\text{pow}\, I_{[j]}) \setminus \varnothing\})$. $\mathscr{Y}(j)$ is a filter base with $\lim \mathscr{Y}(j) = \bigcap_{i\in I_{[j]}}$ *set of* $Y(x_i, v_{[j]i})$.

$\wedge i\in I\, (\mathscr{X}(i) =_{def} \{\textit{set of}\, X((y_j, v_{j[i]})_{j\in\tilde{J}}, \sqcap_{\text{card}\,\tilde{J}}, \text{"t"}) \mid \tilde{J} \in (\text{pow}\, J_{[i]}) \setminus \varnothing\}$. $\mathscr{X}(i)$ is a filter base with $\lim \mathscr{X}(i) = \bigcap_{j\in J_{[i]}}$ *set of* $X(y_j, v_{j[i]})$.

To distinguish $(v_{[j]i})_{i\in I_{[j]}}$ from $(v_{[j']i})_{i\in I_{[j']}}$ in case $(v_{[j]i})_{i\in I_{[j]}} = \text{pr}(I_{[j]})\, (v_{[j']i})_{i\in I_{[j']}}$, we can introduce a property expressing the difference of the index sets

$I_{[j]}$, $I_{[j']}$, for example $I_{[.]}$ with valuation $I_{[j]}$, and similar, an object expressing the difference of index sets $J_{[i]}$.

If $\wedge j \in J$ ($\lim \mathscr{Y}(j)$ = $\{y_{[j]}\}$) and $\wedge i \in I$ ($\lim \mathscr{X}(i)$ = $\{x_{[i]}\}$), then we say $((v_{ji})_{i \in I_{[j]}})_{j \in J}$ and $((v_{ji})_{j \in J_{[i]}})_{i \in I}$ "characterize" or "identify" Y and X, respectively. Characterizations can be partial ordered by set inclusion. There may exist filter bases $\mathscr{Y}(j)^*$ and $\mathscr{X}(i)^*$ with less elements, also converging to $\{y_{[j]}\}$ and $\{x_{[i]}\}$, respectively. It can be of practical interest to find minimal characterizations $\mathscr{Y}(j)^*$ and $\mathscr{X}(i)^*$.

For the dual case let be given

$(2_\cup)$: $I^{(1)}$, $I^{(2)}$, both $\neq \varnothing$ and $\subseteq I$, $\tilde{I} =_{def} I^{(1)} \cup I^{(2)}$, $(v_i)_{i \in \tilde{I}}$. Then we have

set of $Y((x_i, v_i)_{i \in I^{(1)}}, \sqcup_{card\, I^{(1)}}, "t")$ and *set of* $Y((x_i, v_i)_{i \in I^{(2)}}, \sqcup_{card\, I^{(2)}}, "t") \subseteq$

set of $Y((x_i, v_i)_{i \in \tilde{I}}, \sqcup_{card\, \tilde{I}}, "t")$.

If $Y((x_i, v_i)_{i \in I^{(1)}}, \sqcup_{card\, I^{(1)}}, "t") = Y((x_i, v_i)_{i \in I^{(2)}}, \sqcup_{card\, I^{(2)}}, "t")$ then $(x_i, v_i)_{i \in I^{(1)}}$ and $(x_i, v_i)_{i \in I^{(2)}}$ are in an equivalence relation $\cong_\cup$, expressing that they specify the same subset of Y. For the conjugate case let be given analogously

$(2'_\cup)$: $J^{(1)}$, $J^{(2)}$, both $\neq \varnothing$ and $\subseteq J$, $\tilde{J} =_{def} J^{(1)} \cup J^{(2)}$, $(v_j)_{j \in \tilde{J}}$. Then we have

set of $X((y_j, v_j)_{j \in J^{(1)}}, \sqcup_{card\, J^{(1)}}, "t")$ and *set of* $X((y_j, v_j)_{j \in J^{(2)}}, \sqcup_{card\, J^{(2)}}, "t") \subseteq$

set of $X((y_j, v_j)_{j \in \tilde{J}}, \sqcup_{card\, \tilde{J}}, "t")$.

This expresses the dual to the inheritance principle: the larger the set of alternative properties/objects, the larger the set of objects/properties possessing these alternative properties/objects.

We consider

$\wedge j \in J(\mathscr{Y}(j) =_{def} \{$*set of* $Y((x_i, v_{[j]i})_{i \in \tilde{I}}, \sqcup_{card\, \tilde{I}}, "t") \mid \tilde{I} \in (pow\, I_{[j]}) \setminus \varnothing\}$. $\mathscr{Y}(j)$ is an ideal base with $\lim \mathscr{Y}(j) = \bigcup_{i \in I_{[j]}}$ *set of* $Y(x_i, v_{[j]i})$.

$\wedge i \in I(\mathscr{X}(i) =_{def} \{$*set of* $X((y_j, v_{j[i]})_{j \in \tilde{J}}, \sqcup_{card\, \tilde{J}}, "t") \mid \tilde{J} \in (pow\, J_{[i]}) \setminus \varnothing\}$. $\mathscr{X}(i)$ is an ideal base with $\lim \mathscr{X}(i) = \bigcup_{j \in J_{[i]}}$ *set of* $X(y_j, v_{j[i]})$.

Example: V = {a,b,c,d,e}, I = {1,2,3,4,5,6}, J = {1,2,3,4}, relation M shown in Fig.5. M is characterizing for Y and for X. As illustrated in Fig.6, minimal characterizations are for example $\{a_{12}, I_{[2]}, e_{32}\}$ or $\{c_{15}, I_{[2]}, e_{32}\}$ for Y, and $\{a_{12}, e_{24}, c_{31}, d_{25}, b_{33}\}$ for X.

We have $Y((x_1, c_1), id, "t") = (y_1, y_3)$, $Y((x_5, c_5), id, "t") = (y_1)$, *set of* $Y(((x_1, c_1), (x_5, c_5)), \sqcap_2, "t") = \{y_{[1]}, y_{[3]}\} \cap \{y_{[1]}\} = \{y_{[1]}\}$, $Y(((x_3, b_3), (I_{[.]}, \{3,4,5\})), \sqcap_2, "t") = (y_2)$. $Y(((x_1, c_1), (x_2, b_2)), \sqcup_2, "t") = (y_1, y_3)$, $Y((x_2, a_2), id, "f") = (y_3)$.

$(x_5, c_5)\} \cong \{(x_1, c_1), (x_5, c_5)\}$ with respect to characterizing y_1, $\{(x_5, c_5), (I_{[.]}, \{5\})\}$ is coarser characterizing than $\{(x_1, c_1), (x_5, c_5), (I_{[.]}, \{1,5\})\}$, finest characterization of y_1 in M is $\{(x_1, c_1), (x_2, a_2), (x_4, e_4), (x_5, c_5), (I_{[.]}, \{1,2,4,5\})\}$.

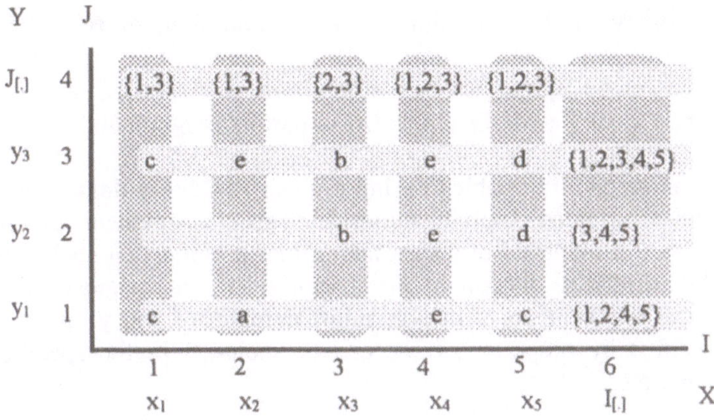

Fig.5. Relation M

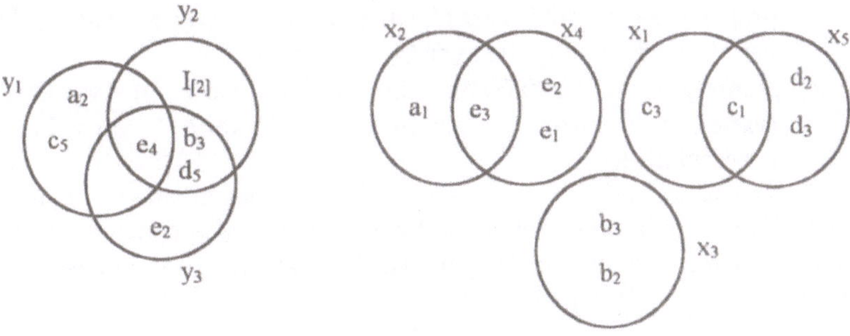

Fig.6

As an example for hierarchical selection we consider $Y(((x_3, b_3), ((x_1, c_1), (x_5, c_5), \sqcap_2), \sqcup_2, \text{"t"}) = (y_1, y_2, y_3)$, according $\eta(b_{[3]}, v_{[j3]}) \sqcup_2 (\eta(c_{[1]}, v_{[j1]}) \sqcap_2 \eta(c_{[5]}, v_{[j5]})) = \text{"t"}$ for $j = 1, 2, 3$.

To show the utilization of variables, let var v_{23}: $\{a, b\}$ in structure M, which results in var M: {M with $v_{[23]} = a$, M with $v_{[23]} = b$}. Thus var $Y((x_3, b), id, \text{"t"})$: $\{(y_3), (y_2, y_3)\}$.

Objects y, properties x, and also valuations v in our module **K** can of course be composite (see **6.2**).

Examples:

(1) Y a set of mathematical functions, X a set of computer programs, if function $y \in Y$ can be computed by program $x \in X$, (y,x) is valuated "t", else "f".

(2) Y = set of proper or improper triangles in euklidean $\mathbf{R}^2$, X = $\{((x_{n1}, x_{n2})_{n=1,2,3} \mid$ point coordinates $(x_{[n1]}, x_{[n2]}) \in \mathbf{R}^2$ for $n = 1,2,3\}$, v = "f" if the points are linearly dependent, else v = "t".
(3) (y, x) a proposition in propositional calculus, $(\mathbf{R}, <)$ used as model time, v = $(\text{"t"}(t))_{t\in[b,e]}$ means (y, x) is "true" during time interval [b,e] of $\mathbf{R}$, $b \le e$.

Modules **K** can be represented by a variable var **K** = ((var y_j, var x_i), var $v_{ji})_{ji\in \text{var U}}$, var U $\subseteq$ var J $\times$ var I, with the possibility of partial assignments.

So far we assumed all variables having a given variability domain which may contain lower level variables. Applying admissible partial assignments to some of the variables, we obtain a manifold of states, taking into account that each assignment may reduce the variability domain of the instantly not yet assigned variables, as outlined in **2.3**. If the elements of the variability domains of var y and var x are themselves specified by properties, these can be included in the specification of objects and properties.

Example: var V := {"t", "f"}, to var y "any" object can be assigned,
property x_1: var y is a non-empty set,
property x_2: an algebraic composition var © : var y $\times$ var y $\rightarrow$ var y is defined,
property x_3: property x_2 $\wedge$ var © (var e′, var e″) = var © (var e″, var e′) for var e′, var e″ : var y, i.e. var © is commutative,
property x_4: property x_2 $\wedge$ var © (var e′, var © (var e″, var e‴)) = var © ((var © (var e′, var e″), var e‴) for var e′, var e″, var e‴ : var y, i.e. var © is associative,
(((var y, x_i), var $v_i)_{i=1,2,3,4}$, $\sqcap_4$ (var $v_i)_{i=1,2,3,4}$ = "t") specifies an abstract commutative semi-group. An interpretation is obtained by assignment of a concrete object y to var y, and a concrete © to var ©, for which (var $v_i)_{i=1,2,3,4}$ are decidable, for example var y := {"t", "f"}, var © := $\sqcap$.

3 Hierarchies of valuated objects, logics

Applying the concepts of chapter **2** we define recursively the following hierarchical structure of valuated objects:

Given the non-empty sets $Z^{(0)}$, $V^{(0)}$, and $\wedge z\in Z^{(0)}$ $(\varnothing \ne V^{(0)}_{[z]} \subseteq V^{(0)})$.

Using the notation $V_z = \{z\} \times V_{[z]}$, we set $\mathbf{L}^{(0)} =_{def} \bigcup_{z\in Z^{(0)}} V^{(0)}_z$.

For n, N $\in \mathbf{N}$, $1 \le n \le N$, let there be given $\varnothing \ne W^{(n)} \subseteq (\text{pow } Z^{(n-1)}) \setminus \varnothing$ and $V^{(n)} \ne \varnothing$. We define

$Z^{(n)} =_{def} \bigcup_{w\in W^{(n)}} \prod_{z\in w} V^{(n-1)}_z$, and if $\wedge z\in Z^{(n)}$ $(\varnothing \ne V^{(n)}_{[z]} \subseteq V^{(n)})$ are given, $\mathbf{L}^{(n)}$ $=_{def} \bigcup_{z\in Z^{(n)}} V^{(n)}_z$. The structure $\mathbf{L}(N) =_{def} \bigcup_{n=1}^{N} \mathbf{L}^{(n)}$ has hierarchical level N over $\mathbf{L}^{(0)}$.

The components of $\mathbf{L}(N)$ can be subject to relations, e.g. functional dependencies, orderings, index transformations, cardinality attributes (a property may hold for "all", "infinite many", "finite many", "exactly m", $0 \leq m$, elements of a component), etc..

The relationship with binary modules $\mathbf{K}$ as described in **2.4** is the following: Let be $V = V^{(0)} = V^{(1)}$ and let K be a bijective parameterization of $Z^{(0)}$. If K is partitioned into $K = I \cup J$, $J \cap I = \varnothing$, then we define $X =_{def} \{x_{[i]} =_{def} z_{[i]} \mid i \in I\}$, $Y =_{def} \{y_{[j]} =_{def} z_{[j]} \mid j \in J\}$. For any valuation, $\{(z_{[k]}, v_{[k]}) \mid k \in K\} \subseteq \mathbf{L}^{(0)}$, $W^{(1)} \leftrightarrow U \subseteq J \times I\}$, $\{((y_{[j]}, v_{[j]}), (x_{[i]}, v_{[i]})) \mid (j,i) \in U\} \subseteq Z^{(1)}$, $\{(((y_{[j]}, v_{[j]}), (x_{[i]}, v_{[i]})), v_{[ji]}) \mid (j,i) \in U\} \subseteq \mathbf{L}^{(1)}$, projections yield $\{((y_{[j]}), (x_{[i]})), v_{[ji]}) \mid (j,i) \in U\}$, concatenation yields $\{(y_{[j]}, x_{[i]}), v_{[ji]}) \mid (j,i) \in U\}$ isomorphic $((y_j, x_i), v_{ji})_{(j,i)\in U}$. The reasoning can be extended to higher hierarchical levels and to n-ary modules, $n > 2$.

We make further assumptions, reducing the generality of the structure: Let $\bigwedge 0{\leq}n{\leq}N$ $(((V^{(n)}, \leq^{(n)}, \sqcap^{(n)}, \sqcup^{(n)})$ be a lattice) $\wedge$ $W^{(n)}$ be a set of finite sets). For all n, $1{\leq}n{\leq}N$, and for any $w{\in}W^{(n)}$ let the valuation $v^{(n)}_{[w]}$ of $(z, v^{(n-1)}_{[z]})_{z\in w}$ be given by a homo- or antimorphism $\varphi^{(n)} : (v^{(n-1)}_{[z]})_{z\in w} \mapsto v^{(n)}_{[w]}$ with respect to a transitive relation $\prec^{(n-1)}$ on $\bigcup\limits_{w\in W^{(n)}} \prod\limits_{z\in w} V^{(n-1)}_{[z]}$ which is an extension of $\leq^{(n-1)}$, i.e. $\prec^{(n-1)}$ reduces to $\leq^{(n-1)}$ on $V^{(n-1)}$, $\varphi^{(n)}$ is independent of $z{\in}w$ and commutative. In case on each level n a set $\Phi^{(n)}$ of morphisms $\varphi^{(n)}$ is given and only these functions were applied for valuations, we say $\mathbf{L}(N)$ is a "logic" and the $\varphi^{(n)}$ are "logic functions".

Given $A = (a_{[z]})_{z\in w}$, $B = (b_{[z]})_{z\in w'}$ on the same level, examples for $\prec$ are:

$A \prec_{\wedge\vee} B \Leftrightarrow_{def} \bigwedge a{\in}A \bigvee b{\in}B\ (a \leq b)$,

$A \prec_{\vee\wedge} B \Leftrightarrow_{def} \bigvee a{\in}A \bigwedge b{\in}B\ (a \leq b)$,

$A \prec_{\wedge\vee\wedge} B \Leftrightarrow_{def} A \prec_{\wedge\vee} B \wedge A \prec_{\vee\wedge} B$,

$A \prec_{\beta} B \Leftrightarrow_{def} \bigvee$bijection $\beta: w \leftrightarrow w'\ (\bigwedge a_{[z]}{\in}A\ (a_{[z]} \leq b_{[\beta(z)]}))$.

We have $A \prec_{\beta} B \Rightarrow A \prec_{\wedge\vee} B$, $(B = pr(w')B^*) \wedge (A \prec_{\wedge\vee} B) \Rightarrow A \prec_{\wedge\vee} B^*$, $(A = pr(w)A^*) \wedge (A \prec_{\vee\wedge} B) \Rightarrow A^* \prec_{\vee\wedge} B$.

Example: $Z^{(0)}$ a set of propositions z, $V^{(0)} = V^{(1)} = V = (\{"t", "f"\}, \neg, \wedge, \vee)$, $"f" < "t"$, is a boolean lattice, $\mathbf{L}^{(0)} = Z^{(0)} \times V^{(0)}$, $W^{(1)} =$ set of finite subsets of $Z^{(0)}$, $Z^{(1)} = \bigcup\limits_{w\in W^{(1)}} \prod\limits_{z\in w} V^{(0)}_{z}$ with elements $z^{(1)} = (z, v_{[z]})_{z\in w}$, $\Phi^{(1)} = \{$unary $\wedge_1 = \vee_1 = id$ is a $\leq$-homomorphism, unary $\neg$ is a $\leq$-antimorphism, $\vee_{card\ w}$ for card $w > 1$ is a $\prec_{\wedge\vee}$ - homomorphism $\prod\limits_{z\in w} V_{[z]} \rightarrow V$, $\wedge_{card\ w}$ for card $w > 1$ is a $\prec_{\vee\wedge}$ - homomorphism $\prod\limits_{z\in w} V_{[z]} \rightarrow V$, $\mathbf{L}^{(1)} = \{((z, v_{[z]})_{z\in w}, \varphi^{(1)}((v_{[z]})_{z\in w})) \mid (z, v_{[z]})_{z\in w} \in Z^{(1)} \wedge \varphi^{(1)} \in \Phi^{(1)}\}$ (example for boolean algebra).

Example: ($\mathbf{Z}$, $\leq$, +) integers, $\mathbf{R}$ reals, $Z^{(0)} = \{[n, n+1) \mid n \in \mathbf{Z} \wedge [n, n+1)$ interval on $\mathbf{R}\}$, $V^{(0)} = (\mathbf{Z}, \leq)$, $V^{(1)} = (10 * \mathbf{Z}, \leq) \subset V^{(0)}$ are lattices, $\mathbf{L}^{(0)} = Z^{(0)} \times V^{(0)}$, $W^{(1)}$ = set of finite subsets of $Z^{(0)}$, $+_1$ = id, $+_{\text{card } w}$ = card w - ary addition, is a $\prec_\beta$ - homomorphism, $\bigwedge v \in V^{(0)} ((i \in \mathbf{Z} \wedge 10i < v \leq 10(i+1)) \Rightarrow \rho v =_{def} 10(i+1))$, ρ is a rounding topology: $v \leq v' \Rightarrow \rho v \leq \rho v'$, $\rho v = \rho\rho v$, $v \leq \rho v$, $\Phi^{(1)} = \{\rho +_{\text{card } w} \mid w \in W^{(1)}\}$, $\mathbf{L}^{(1)} = \{((z, v_{[z]})_{z \in w}, \rho +_{\text{card } w} ((v_{[z]})_{z \in w}) \mid w \in W^{(1)}\}$ (example for rounded summation).

4 Topological structures

Classical topology is the well developed mathematical discipline dealing with neighborhoods and approximations and it is to be expected that all "fuzziness", "vagueness", "softness" theories are covered by it.

4.1 Filters and ideals

Given a complete lattice $(\mathscr{L}, \leq, \sqcap, \sqcup, \mathbf{o}, \mathbf{e})$ and a subset $\mathscr{B} = \{B_{[k]} \mid k \in K\} \subset \mathscr{L}$, the indexing bijective, with the following properties: $\bigwedge k \in K (B_{[k]} \neq \mathbf{o}) \wedge \bigwedge k', k'' \in K (\bigvee k''' \in K ((B_{[k''']} \leq B_{[k']}) \wedge (B_{[k''']} \leq B_{[k'']})))$. Then $\mathscr{B}$ is a "filter base" on $\mathscr{L}$, attributed "proper" for $\sqcap \mathscr{B} \neq \mathbf{o}$. If in addition $\bigwedge k \in K \bigwedge L \leq \mathbf{e} (B_{[k]} \leq L \Rightarrow L \in \mathscr{B})$ then $\mathscr{B}$ is a "filter". The dual notions to filter base and filter are "ideal base" and "ideal". If S is a non-empty set, then this applies to the complete, atomic, boolean lattice (pow S, $\subseteq$, $\cap$, $\cup$, $\varnothing$, S) which we consider in the following. For filter base $\mathscr{B} = \{B_{[k]} \mid k \in K\}$, $B^* =_{def} \lim \mathscr{B} = \bigcap_{k \in K} B_{[k]}$ (because S itself is assumed to have the discrete topology).

The neighborhood of any $s \in B$, $B =_{def} \bigcup_{k \in K} B_{[k]}$, to the elements of B^* can be expressed by membership or non-membership of s in certain $B_{[k]}$: Let $\bigwedge s \in B ((K(s) =_{def} \{k \mid k \in K \wedge s \in B_{[k]}\}) \wedge \overline{K}(s) =_{def} K \setminus K(s))$, $\mathscr{B}_\cap(s) =_{def} \{B_{[k]} \mid k \in K(s)\}$, $\mathscr{B}_\cup(s) =_{def} \{B_{[k]} \mid k \in \overline{K}(s)\}$. We have $s \in \bigcap_{k \in K(s)} B_{[k]} \cap \bigcap_{k \in \overline{K}(s)} \mathbf{C}B_{[k]}$, $\mathbf{C}$ the complement with respect to B. Let $K_{min}(s) =_{def} \{k \mid k \in K(s) \wedge \neg\bigvee k' \in K(s) (B_{[k']} \subset B_{[k]})\}$, $\overline{K}_{max}(s) =_{def} \{k \mid k \in \overline{K}(s) \wedge \neg\bigvee k' \in \overline{K}(s) (B_{[k']} \supset B_{[k]})\}$, then $\mathscr{B}_{\cap min}(s) =_{def} \{B_{[k]} \mid k \in K_{min}(s)\}$, $\mathscr{B}_{\cup max}(s) =_{def} \{B_{[k]} \mid k \in \overline{K}_{max}(s)\}$. General "distance" / "similarity" *relations* of s from / with $s^* \in B^*$ are then given by $\bigwedge s \in B (D_\cap(s^*, s) =_{def} \mathscr{B}_{\cap min}(s) \wedge D_\cup(s^*, s) =_{def} \mathscr{B}_{\cup max}(s))$. $D_\cap(s^*, s) = D_\cap(s^*, s')$ and $D_\cup(s^*, s) = D_\cup(s^*, s'')$ define equivalence relations $s \sim_\cap s'$ and $s \sim_\cup s''$. In particular, if $\mathscr{B}$ is itself a complete lattice then $d_\cap(s^*, s) =_{def} \bigcap D_\cap(s^*, s) \in \mathscr{B}$ and $d_\cup(s^*, s) =_{def} \bigcup D_\cup(s^*, s) \in \mathscr{B}$ are *functional* in s and $d_\cup(s^*, s) \subset d_\cap(s^*, s)$. Dual results hold for $\mathscr{B}$ an ideal base. For illustration see Fig.7.

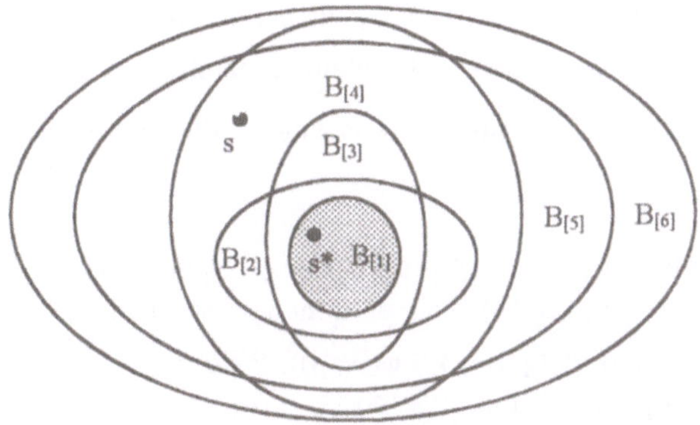

Fig.7. $\mathscr{B} = \{B_{[1]}, B_{[2]}, B_{[3]}, B_{[4]}, B_{[5]}, B_{[6]}\}$, $\lim\mathscr{B} = B_{[1]}$, $D_\cap(s^*,s) = \{B_{[4]}, B_{[5]}\}$, $D_\cup(s^*,s) = \{B_{[2]}, B_{[3]}\}$

4.2 Comparison and composition of bases

Given a non-empty set S and two filter bases $\mathscr{B} = \{B_{[k]} \mid k \in K\} \subset \text{pow } S$ with $B =_{def} \bigcup_{k\in K} B_{[k]}$, $B^* =_{def} \bigcap_{k\in K} B_{[k]} \neq \varnothing$, $\mathscr{C} = \{C_{[l]} \mid l \in L\} \subset \text{pow } S$ with $C =_{def} \bigcup_{l\in L} C_{[l]}$, $C^* =_{def} \bigcap_{l\in L} C_{[l]} \neq \varnothing$, all indexings bijective. We say $\mathscr{B}$ is "finer" than $\mathscr{C}$, $\mathscr{B} \prec \mathscr{C} \Leftrightarrow_{def} \bigwedge l\in L \bigvee k\in K(B_{[k]} \subseteq C_{[l]})$, $\mathscr{B}$ is "equivalent" $\mathscr{C}$, $\mathscr{B} \sim \mathscr{C} \Leftrightarrow_{def} \mathscr{B} \prec \mathscr{C} \wedge \mathscr{C} \prec \mathscr{B}$, and for finite cardinalities, $\mathscr{B}$ "finer granulated" than $\mathscr{C}$ if card $\mathscr{B}$ > card $\mathscr{C}$.

$\mathscr{S} =_{def} \{B_{[k]} \cup C_{[l]} \mid (k,l) \in K \times L\}$ is a filter base on pow S with $S^* =_{def} \bigcap_{kl\in K\times L} (B_{[k]} \cup C_{[l]}) = B^* \cup C^*$, $\mathscr{D} =_{def} \{B_{[k]} \cap C_{[l]} \mid (k,l) \in K \times L\}$ is a filter base on pow S only if $\bigwedge(k,l)\in K\times L\ ((B_{[k]} \cap C_{[l]}) \neq \varnothing)$, then $D^* =_{def} \bigcap_{kl\in K\times L} (B_{[k]} \cap C_{[l]}) = B^* \cap C^*$.

Let $\mathbf{B} = \{\mathscr{B}_{[m]} \mid m \in M\}$ be a set of either all being proper filter bases or all being proper ideal bases on pow S. To compare the bases $\mathscr{B}_{[m]}$ by a uniform neighborhood / similarity measure we introduce a filter base $\mathbf{D} =_{def} \{D_{[q]} \mid q \in Q\} \subset$ pow $(\mathbf{B} \times \mathbf{B})$ with the following properties: $\mathbf{B} \times \mathbf{B} \in \mathbf{D}$, diag $(\mathbf{B} \times \mathbf{B}) =_{def} \{(\mathscr{B}_{[m]}, \mathscr{B}_{[m]}) \mid m \in M\} \subseteq \bigcap_{q\in Q} D_{[q]}$, and $\bigwedge q\in Q\ (D_{[q]} = \overset{-1}{D_{[q]}})$. Then for $F_{[mq]} =_{def} \text{cut}\ (\mathscr{B}_{[m]})\ D_{[q]} = \{\mathscr{B}_{[m']} \mid (\mathscr{B}_{[m]}, \mathscr{B}_{[m']}) \in D_{[q]}\}$, $\mathscr{F}_{[m]} =_{def} \{F_{[mq]} \mid q \in Q\}$ is a filter base with $\mathscr{B}_{[m]} \in \bigcap \mathscr{F}_{[m]}$. Consequently, $\mathbf{D}$ defines for all $\mathscr{B}_{[m]}$ a uniform and symmetric neighborhood system. Thus according **4.1**, for any pair $(\mathscr{B}_{[m]}, \mathscr{B}_{[m']})$ the neighborhood / similarity measures $D_\cap(\mathscr{B}_{[m]}, \mathscr{B}_{[m']})$ and $D_\cup(\mathscr{B}_{[m]}, \mathscr{B}_{[m']})$ can be applied.

Example: $B = (\{0,1\}, \leq)$, $S = B\times B\times B$ can be partial ordered by component-wise $\leq$, $\bigwedge s', s''\in S$ ($d(s', s'') =_{def} s' \oplus s''$, $\oplus$ addition mod 2) is a uniform generalized distance, $h(s', s'') =_{def} \sum_{n=1}^{3} pr_n(d(s', s''))$ is a metric distance (Hamming).

4.3 Valuated bases

Given a proper filter base $\mathscr{B} = \{B_{[k]} \mid k \in K\}$ on (pow B, $\subseteq, \cap, \cup$), a non-empty complete lattice $(V, \leq, \sqcap, \sqcup)$, and a $\leq$–homomorphism φ: pow B $\to$ V, i.e. $\bigwedge k,k'\in K\ ((B_{[k]} \subseteq B_{[k']}) \Rightarrow (\varphi(B_{[k]}) \leq \varphi(B_{[k']})))$. With $v_{[k]} = \varphi(B_{[k]})$ it follows from $(B_{[k]} \subseteq B_{[k']}) \wedge (B_{[k]} \subseteq B_{[k'']})$ that $(v_{[k]} \leq v_{[k']}) \wedge (v_{[k]} \leq v_{[k'']})$, hence $\varphi(\mathscr{B})$ is a filter base on V if all $v_{[k]} \neq \mathbf{o}$, and $\varphi(\lim \mathscr{B}) \leq \lim \varphi(\mathscr{B})$.

We consider $\overset{-1}{\varphi}: V \to \text{pow } B$ defined by $\bigwedge v\in V\ (\overset{-1}{\varphi}(v) =_{def} \bigcup_{\varphi(U)=v} U)$. Then $\overset{-1}{\varphi}$ is a homomorphism. If $\mathscr{V} = \{v_{[l]} \mid l \in L\}$ is a filter base on V and $\bigwedge v\in\mathscr{V}\ (\overset{-1}{\varphi}(v) \neq \varnothing)$, then $\overset{-1}{\varphi}(\mathscr{V})$ is a filter base on pow B. This for example is the case if φ is the set extension of a function f: B $\to$ C and V = pow C. Then we have $\mathscr{B} \prec \overset{-1}{\varphi}(\varphi(\mathscr{B}))$.

φ being a homomorphism corresponds to the "neighborhood to lim $\mathscr{B}$" interpretation. Choosing φ as antimorphism, $\varphi(\mathscr{B})$ is an ideal base, which corresponds to the "similarity to lim$\mathscr{B}$" interpretation.

In the function $(B_{[k]}, v_{[k]})_{k\in K}$ the elements $B_{[k]}$ of a (filter / ideal) base $\mathscr{B}$ with "support" $B = \bigcup \mathscr{B}$ are valuated by $v_{[k]}$ with $v_{[k]} = \varphi(B_{[k]})$. Now we consider a family $(\varphi_k)_{k\in K}$ of homomorphisms $\varphi_{[k]}$: pow B $\to$ V, a filter base $\mathscr{V}$ on V and the family $(\mathscr{B}_{[k]} =_{def} \overset{-1}{\varphi}_{[k]}(\mathscr{V}))_{k\in K}$ of bases $\mathscr{B}_{[k]} = \{B_{[kv]} \mid v \in \mathscr{V}\}$, assuming all $B_{[kv]} \neq \varnothing$. On the set of filter bases $\mathscr{B}_{[k]}$ we define an ordering $\leq_{\mathscr{V}}$ by $\bigwedge k,k'\in K$ $((\mathscr{B}_{[k]} \leq_{\mathscr{V}} \mathscr{B}_{[k']} \Leftrightarrow_{def} \bigwedge v\in\mathscr{V}(B_{[kv]} \subseteq B_{[k'v]}))$. On $(\{\mathscr{B}_{[k]} \mid k \in K\}, \leq_{\mathscr{V}})$ filter bases can be considered and for $\mathscr{B}_{[k]} \leq_{\mathscr{V}} \mathscr{B}_{[k']}$ a generalized distance is defined by $d(\mathscr{B}_{[k]}, \mathscr{B}_{[k']}) = (B_{[k'v]} \setminus B_{[kv]})_{v\in\mathscr{V}}$ (in general $\Delta(B_{[k'v]}, B_{[kv]}) = (B_{[k'v]} \setminus B_{[kv]}) \cup (B_{[kv]} \setminus B_{[k'v]})$). A similar reasoning can be applied to the case where a fixed filter base $\mathscr{B}$ is mapped onto filter bases $(\varphi_{[k]}(\mathscr{B}))_{k\in K}$.

Example: The general case for any f: B $\to$ C, φ the set extension of f, any filter base $\mathscr{B}$ on pow B mapped by $\varphi(\mathscr{B})$, and any filter base $\mathscr{C}$ on pow C mapped by $\overset{-1}{\varphi}(\mathscr{C})$, can be found in text books (e.g. N. Bourbaki, 1951). The particular case $B = [a,b] \subset \mathbf{R}$, $a < b$, $C = [0,1] \subset \mathbf{R}$, f: B $\to$ C, $\bigvee x\in B\ (f(x) = 1)$, $\mathscr{C} = \{[\alpha,1] \mid \alpha\in[0,1]\}$, $B_{[\alpha]} =_{def} \overset{-1}{\varphi}([\alpha,1])$ an "α - cut", $\mathscr{B} = \{B_{[\alpha]} \mid \alpha \in [0,1]\}$, was introduced by L.A.

Zadeh (1965), B named a "fuzzy set" with "membership function" f. Let f': [a,b] → [0,1] be another function with $\bigvee x \in B\ (f'(x) = 1)$ with corresponding quantities primed. If $\bigwedge x \in B\ (f(x) \leq f'(x))$ then $B_\alpha \subseteq B'_\alpha$ and $\mathscr{B} \leq_{\mathscr{C}} \mathscr{B}'$. $\lim \mathscr{B} \subseteq \lim \mathscr{B}'$. If f is integrable, then for example the functional $\int_{x \in B_{[\alpha]}} f(x)dx$ corresponds to a valuation of $(f(x))_{x \in B_{[\alpha]}}$, the functional $\int_\alpha^1 B_{[\beta]} d\beta$ corresponds to a valuation of $(B_{[\beta]})_{\beta \in [\alpha,1]}$, both have equal values, $\int_{x \in B} |f(x) - f'(x)| dx$ is an example of a neighborhood measure for $\mathscr{B}, \mathscr{B}'$.

Example (interval arithmetic on **R**): Let **R** be the real numbers and **IR** be the set of all closed intervals on **R**. An arithmetic function f: $\mathbf{R}^2 \to \mathbf{R}$ (division by zero excluded), extended to a set function F, maps $(\mathbf{IR})^2 \to \mathbf{IR}$. For i = 1,2 let be given $\mathscr{I}^{(i)} = \{\emptyset \neq I_{[1]}^{(i)} \subset I_{[2]}^{(i)} \subset \dots I_{[v]}^{(i)} \subset \dots I_{[n]}^{(i)} \subset \mathbf{R}\}$, $I_{[v]}^{(i)} \in \mathbf{IR}$, and $I_{[v]}^{(i)}$ valuated by 1/v. F maps the monotone filter bases $\mathscr{I}^{(1)}, \mathscr{I}^{(2)}$ on filter base $\mathscr{I} = \{\emptyset \neq F(I_{[1]}^{(1)}, I_{[1]}^{(1)}) \subset \dots F(I_{[v]}^{(1)}, I_{[v]}^{(2)}) \subset \dots F(I_{[n]}^{(1)}, I_{[n]}^{(2)}) \subset \mathbf{R}\}$, $F(I_{[v]}^{(1)}, I_{[v]}^{(2)}) \in \mathbf{IR}$ and valuated by 1/v, **R** valuated by 0. Usually n = 1.

Example (rounded computations): For constructive computation of a function f: X → Y, idempotent mappings (roundings) $\rho_X : X \to \overline{X}$ and $\rho_Y : Y \to \overline{Y}$ onto finite subsets $\overline{X} \subset X$, $\overline{Y} \subset Y$ are applied to reduce X and Y to representable subsets. For example, roundings of **R** to machine numbers. The set extensions of roundings are idempotent $\subseteq$-homomorphisms: for $X' \subseteq X'' \subseteq X$ we have $\rho_X(X') \subseteq \rho_X(X'')$, $\rho_X \circ \rho_X(X') = \rho_X(X')$, and $\rho_X(X') \subseteq X'$, similar for ρ_Y. f is approximated by a constructive function $\bar{f} : \overline{X} \to \overline{Y}$. Resulting functional assignments are shown in Fig.8.

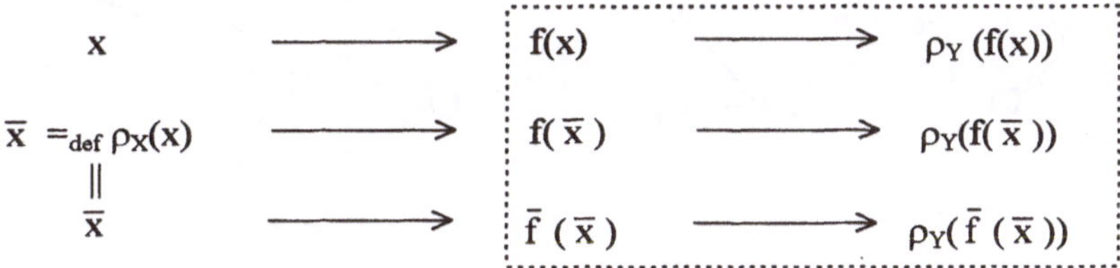

Fig.8. Rounded computations

The study of topological distance relations between the framed quantities in Fig.8, which can all be distinct, in dependency of distance relations between the given x, $\overline{x}$, f, $\bar{f}$ is a fundamental part of numerical analysis.

In the special case of rounded floating point number arithmetic, basic results were obtained by the "Karlsruhe Group" (U. Kulisch 1975, 1996, U. Kulisch and W. L. Miranker 1981, and others).

5 Time and processes

5.1 Time sets

To model the time behavior of systems we introduce sets of "times". A set of times $(\mathbf{T}, <, \sqcap, \sqcup, \mathbf{O}, \mathbf{E})$ is a complete, atomic, lattice, with $\mathbf{O}$ denoting the zero and $\mathbf{E}$ denoting the unit element, and two structures:

1. lattice operations join $\sqcup$ and meet $\sqcap$, by which the partial ordering $\sqsubseteq$ on $\mathbf{T}$ is defined,
2. an irreflexive (acyclic) partial ordering $<$ on the set $A(\mathbf{T}) \subset \mathbf{T}$ of atoms of $\mathbf{T}$.

For each $U \in \mathbf{T}$, $\alpha(U) \subseteq A(\mathbf{T})$ denotes the set of all atoms of U, $U = \bigsqcup \alpha(U)$.
$<$ on $A(\mathbf{T})$ induces relations on $\mathbf{T}$. For $A, B \in \mathbf{T}$ we can define for example:

$$\bigvee a \in \alpha(A) \bigwedge b \in \alpha(B)\,(a < b) \Leftrightarrow_{def} A <_{\vee\wedge} B,$$

$$\bigwedge a \in \alpha(A) \bigvee b \in \alpha(B)\,(a < b) \Leftrightarrow_{def} A <_{\wedge\vee} B,$$

$$A <_{\vee\wedge\vee} B \Leftrightarrow_{def} A <_{\vee\wedge} B \wedge A <_{\wedge\vee} B,$$

$$\bigwedge a \in \alpha(A) \bigwedge b \in \alpha(B)\,(a < b) \Leftrightarrow_{def} A <_{\wedge\wedge} B,$$

$$\bigvee a \in \alpha(A) \bigvee b \in \alpha(B)\,(a < b) \Leftrightarrow_{def} A <_{\vee\vee} B,$$

$$A <_{\vee\vee} B \wedge B <_{\vee\vee} A \Leftrightarrow_{def} A <_{\vee\vee} >_{\vee\vee} B,$$

$$A <_{\vee\vee} B \wedge \neg(B <_{\vee\vee} A) \Leftrightarrow_{def} A <_{\vee\vee} \neg >_{\vee\vee} B.$$

Not all of these relations are independent, for example

$A <_{\wedge\wedge} B \Rightarrow A \sqcap B = \mathbf{O}$ and $A <_{\wedge\wedge} B \Rightarrow A <_{\vee\wedge\vee} B \Rightarrow A <_{\vee\vee} \neg >_{\vee\vee} B$,

and not all are transitive, for example $A <_{\vee\vee} B$. The transitive ones can be extended to the transitive hulls. Restricted to atoms, they reduce to $<$. For illustration see Fig.9 ($<$ is symbolized by $\rightarrow$).

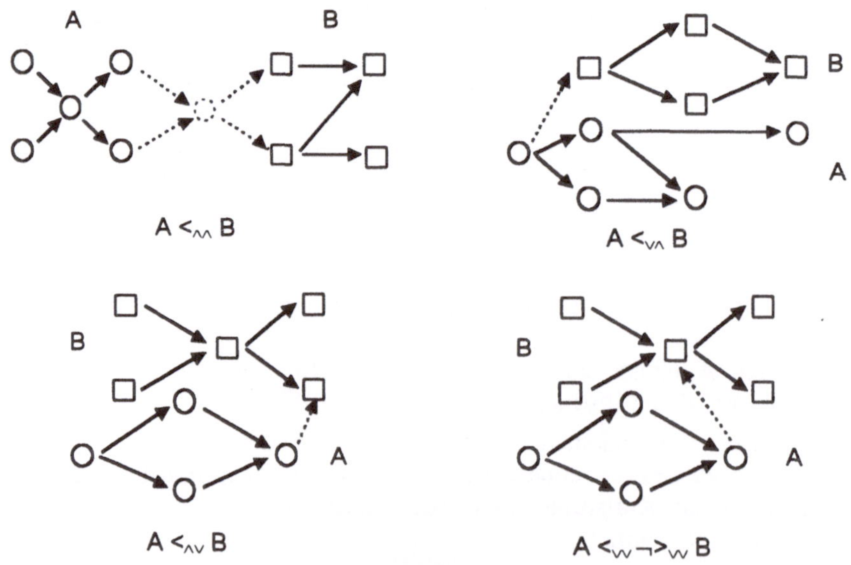

Fig.9

Considerations related to our concept were first made by J. F. Allan 1984, using times intuitively given by intervals on linearly ordered $\mathbf{R}$, modeling physical time, and without the mathematical structures presented here.

Further relations can be defined by combinations, distinguishing cases $A \sqcap B = \mathbf{O}$ and $A \sqcap B \neq \mathbf{O}$, by admitting $\leq$ in the above definitions, by taking topological properties into account if a topological closure τ is defined on $\mathbf{T}$, for example, $\tau(A) = \tau(B)$.

We define an "interval" I on $A(\mathbf{T})$ as a subset $\varnothing \neq I \subseteq A(\mathbf{T})$ such, that for $\bigwedge x,z \in I \bigwedge y \in A(\mathbf{T})$ $((x<y<z) \Rightarrow y \in I)$. Let $\Im(A(\mathbf{T}))$ denote the set of all intervals on $A(\mathbf{T})$. For all $C \in \mathbf{T}$ let be defined a topological closure $\overline{C} \in \mathbf{T}$ with $\alpha(\overline{C}) \in \Im(A(\mathbf{T}))$. The closure properties are: $C \sqsubseteq \overline{C}$, $D \sqsubseteq C \Rightarrow \overline{D} \sqsubseteq \overline{C}$, $\overline{C} = \overline{\overline{C}}$.
We use $\overline{C}$ to define a "length" of C.

Example: $\mathbf{T} = \text{pow}\,\overline{\mathbf{R}}$, $A(\mathbf{T}) = \{\{r\} \mid r \in (\overline{\mathbf{R}}, <)\}$, $\sqcap = \cap$, $\sqcup = \cup$, $\sqsubseteq = \subseteq$, $\mathbf{O} = \varnothing$, $\mathbf{E} = \overline{\mathbf{R}}$. For $\varnothing \neq U \subseteq \overline{\mathbf{R}}$, let be $\overline{U} =_{def} \{r \mid r \in \overline{\mathbf{R}} \wedge (z_1 \leq r \leq z_2) \wedge (z_1 = \text{glb}\, U \text{ in } \overline{\mathbf{Z}} \wedge z_2 = \text{lub}\, U \text{ in } \overline{\mathbf{Z}})\}$, $\overline{\mathbf{Z}}$ being the completion of the integers.

From a given $(\mathbf{T}^{(0)}, <^{(0)}, \sqcap, \sqcup, \mathbf{O}, \mathbf{E}^{(0)})$ with atoms $A^{(0)}(\mathbf{T}^{(0)})$ we can obtain hierarchies of "coarser" time sets $(\mathbf{T}^{(n)}, <^{(n)}, \sqcap, \sqcup, \mathbf{O}, \mathbf{E}^{(n)}))$ over $\mathbf{T}^{(0)}$ with atoms $A^{(n)}(\mathbf{T}^{(n)})$ by the following induction:
Let be given $\mathbf{T}^{(n-1)}$ for $0 < n$. If $\mathbf{T}^{(n-1)} \neq \{\mathbf{O}, \mathbf{E}^{(n-1)}\}$, select a set $\mathbf{U}$, $\varnothing \neq \mathbf{U} \subset \mathbf{T}^{(n-1)}$, with $\bigwedge U,V \in \mathbf{U}$ $(U \neq \mathbf{O} \wedge U \neq V \Rightarrow U \sqcap V = \mathbf{O})$, and define $A^{(n)}(\mathbf{T}^{(n)}) =_{def} \mathbf{U}$. If one of the above transitive relations on level $(n-1)$ is applicable, it is taken as $<^{(n)}$. Then $\mathbf{T}^{(n)} =_{def} \{\sqcup W \mid W \subseteq A^{(n)}(\mathbf{T}^{(n)})\}$. $\sqcap$, $\sqcup$ on higher level are restrictions of those on lower level.

For $\mathbf{T}$ boolean and $U \in \mathbf{T}$, the complement of U with respect to $\mathbf{T}$ is given by $\text{compl}(\mathbf{T})(U) = \sqcup\, \mathit{compl}(A(\mathbf{T}))(\alpha(U))$, join of the set theoretical complement of $\alpha(U)$ with respect to $A(\mathbf{T})$.

5.2 Processes

We consider :
Z_0 a non-empty set of objects z, named "primitive process states", $Z = \text{pow}\, Z_0$,
$(\mathbf{T}, <, \sqcap, \sqcup)$ a set of times, $U \in \mathbf{T}$, $\alpha(U)$ the set of atoms of U, and an indexing: $\alpha(U) \rightarrow Z$.
$P = (z_t)_{t \in \alpha(U)}$ defines a "process" ("time function") of "length" $\overline{U}$ with "process state" $z_{[t]}$ at time instant $t \in \alpha(U)$. If $<$ is linear on $\alpha(U)$, the process is named "sequential". If $<$ is non-linear and $\neg(t < t') \wedge \neg(t' < t)$, then (z_t) and $(z_{t'})$ are "parallel" with respect to $<$ on $\alpha(U)$. Parallelism is a *structural* property with respect to the underlying $<$. U is the reference time, $(\alpha(U), <)$ is the reference time process to P with canonical indexing. $<$ on $\alpha(U)$ induces an irreflexive ordering $\prec$ on P. Processes can be states of other processes.

If $U = P \sqcup F$ with $P <_{\wedge\wedge} F$, then $(z_t)_{t\in\alpha(P)}$ is the past process to the future process $(z_t)_{t\in\alpha(F)}$.

If $P = (P_l)_{l\in L}$ is a family of processes $P_l = (z_{lt})_{t\in\alpha(U(l))}$, $\bigwedge l\in L\ (U(l) \in \mathbf{T})$, then $U(l)$ is the "local" time of P_l, $U = \bigsqcup \{U(l) \mid l\in L\}$ is the "global" time of P. If for P_l, P_m, $l \neq m$, $V =_{def} \bigsqcap \{U(l), U(m)\} \neq \mathbf{O}$, then P_l, P_m are "concurrent" during V. P can be concatenated $\mathbf{K}_\varphi$ by the connector $\alpha(U)$ and $\varphi = \bigcup$ to $(z_t = \bigcup_{l\in L} pr(t)\ P_l)_{t\in\alpha(U)}$.

For two processes
$P' = (z'_t)_{t\in\alpha(U)}, U \in (\mathbf{T}', <', \sqcap', \sqcup')$, $P'' = (z''_t)_{t\in\alpha(V)}, V \in (\mathbf{T}'', <'', \sqcap'', \sqcup'')$
there is in general no relationship between $\mathbf{T}'$ and $\mathbf{T}''$, i.e. the local times U and V are independent, the processes P', P'' are parallel, i.e. each $(z'_{t'})$ is parallel to any $(z''_{t''})$. However, if $C \subseteq \alpha(U)\times\alpha(V) \cup \alpha(V)\times\alpha(U)$ is an acyclic non-empty connector, we can extend the partial orderings to an ordering $<$ on $\alpha(U) \cup \alpha(V)$ by $< \mid \alpha(U) = <'$, $< \mid \alpha(V) = <''$ (| means "restricted to"), $\bigwedge c\in C$ $((c = (t_i, t_j) \in \alpha(U)\times\alpha(V) \Rightarrow_{def} t_i < t_j) \wedge (c = (t_{j'}, t_{i'}) \in \alpha(V)\times\alpha(U) \Rightarrow_{def} t_{i'} < t_{j'}))$. Taking the transitive hull, a global time W for P' and P'' can be defined by $\alpha(W) = \alpha(U) \cup \alpha(V)$ with ordering $<$.

Given a process $P = (s_t)_{t\in\alpha(U)}$, $U\in\mathbf{T}$, $U \neq \mathbf{O}$, $\alpha(U)$ finite, $(\mathbf{T}, <, \sqcap, \sqcup)$. P can (in general many ways) be partitioned into parts which form process states of a sequential process S. We apply the following algorithm:

($\mathscr{A}$): $n := 0$, $A := \alpha(U)$,
while $A \neq \varnothing$ do
$n := n+1$,
select an A_n, $A_n \neq \varnothing$, $A_n \subseteq A_n^* =_{def} \{t \mid t\in A \wedge \neg\bigvee\tau\in A\ (\tau<t)\}$,
$A := A\backslash A_n$,
end while, end ($\mathscr{A}$).

Because $(\alpha(U),<)$ is finite and acyclic, ($\mathscr{A}$) terminates at $n = N$, card (maximal $<$-chain in $\alpha(U)$) $\leq N \leq$ card $\alpha(U)$. The process states s_t, $t\in A_n$, are parallel, maximal parallelism is achieved if for all n holds $A_n = A_n^*$. A compatible extension $\mathbf{<}$ to $<$ on $\alpha(U)$ is given by setting $\bigwedge 1\leq n<N\ \bigwedge t \in A_n\ \bigwedge t' \in A_{n+1}\ (t \mathbf{<} t')$, i.e. $A_n \mathbf{<}_{\wedge\wedge} A_{n+1}$, and taking the transitive hull of $\mathbf{<}$. The irreflexive orderings $<, \mathbf{<}$ on $\alpha(U)$ induce irreflexive orderings $\prec, \mathbf{\prec}$ on P. $(S = (S_n =_{def} \bigcup_{t\in A_n} s_t)_{n=1,2,..N}, \mathbf{\prec}_{\wedge\wedge})$ is a partition of P, isomorphic with $(\{1,2,...N\}, 1<2<...<N)$ and isomorphic with any maximal $\prec$-chain in $(P, \prec)$. A mapping $h: (P, \prec) \rightarrow (\{1,2,...N\}, 1<2<...<N)$ defined by $\bigwedge s_t\in P\ (t \in A_n \Leftrightarrow_{def} h: s_t \mapsto n)$ is a homomorphism. $S = (S_n)_{n=1,2,..N}$ defines an "algorithm" with "algorithmic steps" $S_n = \overset{-1}{h}(n)$. While the original ordering $<$ on $\alpha(U)$ can express functional or causal dependency $\prec$ on P, $\mathbf{<}$ corresponds to linear "algorithmic" or "logical" time $\{1,2,...N\}$, $1<2<...<N$. The

algorithm ($\mathscr{A}$) is for example fundamental for data flow computer architectures: A_n* marks executable operations, A_n marks those chosen for execution on available resources.

Example: Process $P' = (x_i)_{i\in I}$, I = {1,2,3,4,5,6,7; < '} is local time for P', process $P'' = (y_k)_{k\in K}$, K = {a,b,c,d,e,f,g; < ''} is local time for P''. I and K connected by the acyclic connector C = {(2, d), (e, 5)} gives global time W = {1,2,...7, a,b,...g} with < as shown in Fig.10. H(I) = {(1↦b), (2↦c), (3↦d), (4↦e), (5↦5), (6↦6), (7↦7)} and H(K) = {(a↦a), (b↦b), (c↦c), (d↦d), (e↦e), (f↦6), (g↦7)} are homomorphisms on the maximal chain w = (a≺b≺c≺d≺e≺5≺6≺7) of W. w taken as linear global time for P', P'' yields process $P = (\{y_a\}_a, \{x_1, y_b\}_b, \{x_2, y_c\}_c, \{x_3, y_d\}_d, \{x_4, y_e\}_e, \{x_5\}_5, \{x_6, y_f\}_6, \{x_7, y_g\}_7)$ (see Fig.10). P is in 1:1 correspondence with $S = (\{y_a\}_1, \{x_1, y_b\}_2, \{x_2, y_c\}_3, \{x_3, y_d\}_4, \{x_4, y_e\}_5, \{x_5\}_6, \{x_6, y_f\}_7, \{x_7, y_g\}_8)$, which is one possible result of an algorithm ($\mathscr{A}$).

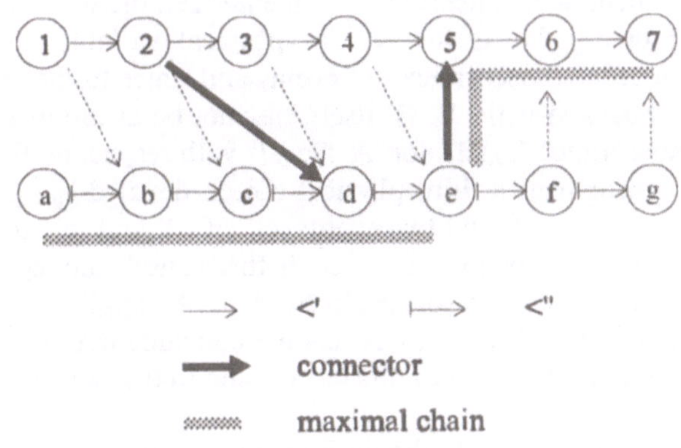

Fig. 10

Projections, coarsenings and refinements of processes:
Let be U,V ∈ **T**, **T** a set of times, $P = (z_t)_{t\in\alpha(U)}$ a process. We define the "projection of P onto V" by $pr(V)P =_{def} (z_t)_{t\in\alpha(W)}$ with $W =_{def} V \sqcap U$.

Given two time sets **T**', **T**'' over time set **T**, **T**'' coarser than **T**', with sets of atoms $A'(\mathbf{T}')$, $A''(\mathbf{T}'')$. Let be V ∈ **T**'', then V can be represented in **T**'' by $\alpha''(V)$, each element t'' of $\alpha''(V)$ has a representation $\alpha'(t'')$ in **T**'. Then $P'' = (z''_{t''})_{t''\in\alpha''(V)}$ with $z''_{t''} =_{def} (z'_{t'})_{t'\in\alpha'(t'')}$ is a "coarsening" of $P' = (z'_{t'})_{t'\in\alpha'(V)}$ and P' is a "refinement" of P''. P'' is a process of processes.

5.3 Physical objects and processes

Assume, A and B denote physical objects which need not be distinct. If A has the capability to recognize the existence of B and to subjectively sense, observe, perceive "properties" of B or to reason on these because of memorized previous experiences, A may gain an "impression" or an "imagination" of B. Further, A may have gradual capabilities

- to recognize non altering properties identifying B for A,
- to recognize changes of further properties, "events",
- to recognize the "state" resulting from a change,
- to subjectively "understand", "interpret" the event and resulting state,
- to bring the perceptions in spatial, temporal, qualitative, causal relationships,
- to memorize and retrieve memorized perceptions (by a memory with limited capacity and storage duration),
- to form classes and "concepts",
- to find causal dependencies and to make "conclusions".

Of course, from the subjective perceptions, their interpretation and "processing", A cannot draw a conclusion on "absolute" reality, spatial and temporal structuring and completeness of properties of B.

If, from the point of view of A, object B changes its properties (states and/or behavior (i.e. a sequences of states)) and A memorized the previous and recognizes the subsequent property, B is for A a "physical variable", the observed and memorized partial ordered occurrences of events and states forms for A a "physical process" $P_A(B)$, associated with B. B itself may not be aware of this process. This process can define a "time" $T_A(B)$ for A, "local" with respect to B.

Assume, A recognizes a third physical object, denoted by C, not necessarily being distinct from A or B, and two events, e_B of B and e_C of C. A for some reason may conclude e_B depends on e_C, e_C is the "cause" and e_B is the "effect" of this cause. For example, one reason could be that A applies a reference time in which e_B, e_C can be related $e_B < e_C$. A cannot conclude that another observer A′ of B and C recognizes the same events e_B, e_C and in the same order.

Example: A hunter shoots a duck, the duck changes from being alive to being dead. For the hunter, a bang and the death of the duck are both caused by his shot. A far distanced observer, assumed not to see the hunter, sees first the death of the duck and later hears the bang. He may conclude the bang is caused by the death of the duck.

Confirmed by our multiple and repeatable physical experiences, classical paradigms are:

- the reaction, an effect cannot be "earlier" than the generating action or cause, independent of the observer′s time scale,
- physical observing, sensing, processing is subject to this time condition, the event observed cannot be later then the event of observation,
- every physical effect has a cause ("causality principle").

Let both, A and B, observe themselves and being carriers of processes $P_A(A)$, $P_B(B)$, defining local times $T_A(A)$, $T_B(B)$. If an event e_A for A is caused by an event e_B of B, then by the above paradigms, the occurrence of e_A cannot be earlier than the occurrence of e_B in the local time $T_A(A)$ and the occurrence of e_B cannot be later than the occurrence of e_A in the local time $T_B(B)$. This justifies the introduction of a directed connector to join the (in general partial) ordered local times to one partial ordered time global to A and B, as outlined in **5.2**. However, this does not include that A and B "know" about the structure of the global time. To enable A to recognize and interpret the (partial) order with respect to the local time $T_B(B)$ of those events e_B which cause events e_A, techniques are available under suitable assumptions on the abilities ("intelligence") of A. For example, the implementation of a priory knowledge in A about the time of occurrence of e_B, or, in case e_B is the sending of a load of things and e_A is the receiving of this load, the attachment of identifiers for classification and labels for temporal ordering of e_B with respect to $T_B(B)$ to be interpreted by A.

Example (see Fig.11): $e_B \prec_B e'_B \prec e'_A \prec_A e_A$, order information to A: $e_B \prec_B e'_B$.

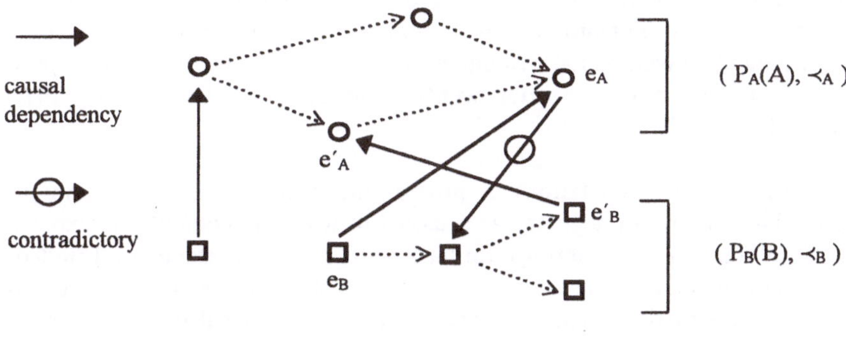

Fig.11

Example (packet switching in communication networks): A message is to be sent from a sender to certain addressees. The message is partitioned into packets, the packets may go on different paths through the network. Each packet has a message identification, sender identification, addressee identification, numbering label to link received packets in the correct order, begin and end label or length information, local time of sending it off if "updates" are possible, etc..

Example (consistency of data in distributed data bases): An information I should be made available to many users. For this it is duplicated, I′, I″, ..., and stored in different memories. Users may read and overscribe the content of these different memories in course of time, this results in processes $(I'_{t'})_{t' \in U'}$, $(I''_{t''})_{t'' \in U''}$, ... Reading and storing consumes time and, at least writing, blocks multiple simultaneous access. A frequent assumption is, the "last" version is the wanted one. The problems are: which one is the last and where is it stored? The search for it consumes time and may block access to updates intended by other users. A solution by imposing an enforced global time and access constraints goes on cost of efficiency. On the other hand, a user may be satisfied with a version "not too old", or even needs the version of a particular earlier date, which then may be already lost.

The occurrences of events e_A of a process $P(A)$ can depend on events e'_A of $P(A)$ caused by events e'_B ("signals") of a process $P(B)$. Then $P(B)$ "triggers" the e_A, the process of the e_A is "synchronized" with the "clock" process $P(B)$. Synchronization is widely applied to match time constraints.

If the event sequences of $P_C(A)$ and $P_C(B)$ can be compared by an observing object C ($C = A$ or $C = B$ permitted), and if for C $P_C(B)$ is finer $P_C(A)$, then $P_C(B)$ can serve as "reference time" process for $P_C(A)$. A hypothetical linearly ordered "finest" universal reference process is mathematically modeled by the reals $\mathbf{R}$. Partial orderings can be considered as superposition of the linear chains contained in them. Thus a hypothetical finest partially ordered universal reference process is modeled by $\mathbf{R}^{\mathbf{R}}$.

A physical process $P(A)$ carried by object A may have been determined up to an instant t in its local time. Then a possible process continuation into a not yet existing local time future of t is a process variable with a possibly not yet determined variability domain. The determination of the domain for a subsequent process fragment as well as the assignment to this variable can depend on previous states of the process and/or exterior events influencing A. The process $P(A)$ is named "evolutionary".

Example (interactive computation): A programmer inputs some instructions to his computer. This program (-segment) is a model of what the computer is expected to do and simultaneously the control information if decoded by the control unit of the computer. The model time is given by the sequence of the instructions. The computer executes these instructions. This is a physical process. Its local time is measured by us in our time scale, e.g. in msecs. After termination, the programmer decides on a continuation of the process, the decision depending on results obtained so far, on environmental influences, and on his own intention. The execution process may be interrupted, unexpected and unforeseen by the programmer, for example by hardware or software failures. The programmer may analyze the interrupt and then input instructions for process continuation depending on already computed results, the results of his error analysis and his own intention at this moment. In any case the computer performs an evolutionary process. We do not reserve the name "process" for the one we were initially planning to be performed by the computer.

In our notation, from the point of view of observer B, in $A_p = (p, A_{[p]})$ $A_{[p]}$ denotes the identified object A considered with respect to observable property p, which can be a state of a variable var p: P, for example a location in space or an instant in time.

6 Directed systems

6.1 Induced structures

We introduced structures as hierarchies of relations parameterized by index sets. The index sets can be structured, then indexing induces a structure on the indexed objects. Important for many applications are index sets with an irreflexive partial ordering (acyclic) as already employed in chapter 5.

Let us suppose $R = (r_i)_{i \in I}$ is given with I structured, (I, <), < an irreflexive partial ordering, and *set of* R = $S = \{r_{[i]} \mid i \in I\}$. Then induced structures are

$\bigwedge r_i, r_{i'}' \in R\ (i < i' \Leftrightarrow_{def} r_i \prec r_{i'}')$, $\prec$ is an irreflexive partial ordering on R,

$\bigwedge r_{[i]}, r_{[i']}' \in S\ (i < i' \Leftrightarrow_{def} r_{[i]} \to r_{[i']}')$, $\to$ is a "direction" on S, transitive but not irreflexive.

The mapping $h: (R, \prec) \to (S, \to)$ induced by the indexing is a homomorphism. In addition, an undirected relation $\sim$ can be defineded on $(R, \prec)$, mapped on $\cong$.

If card I is finite, then the structures can be represented by partially directed and undirected graphs $\hat{R}$, $\hat{S}$, containing edges which can be undirected or directed.

Visualization (Fig.12.): S = {a, b, c}, R = (a_1, b_2, c_3, a_4, a_5), with I = {1,2,3,4,5} and partial irreflexive ordering 1<2<4<5. $\hat{S}$ has cycles, $\hat{R}$ cannot uniquely be reconstructed from $\hat{S}$.

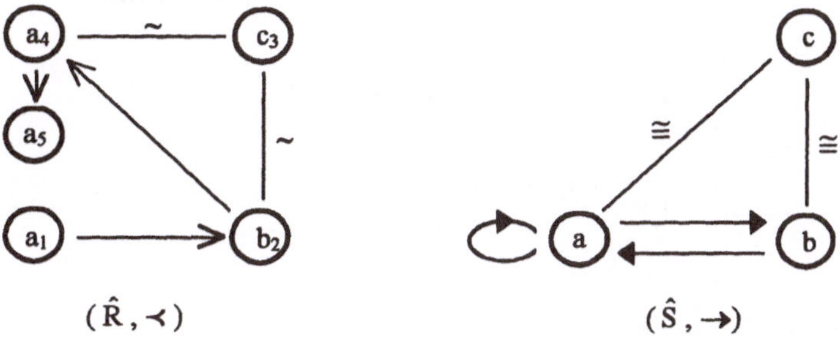

Fig.12

We consider a set $M = \{m_p \mid p \in P\}$, an index set $(J, <_J, \sim_J)$, $\bigwedge j \in J$ (an index set $(I_{[j]}, <_j, \sim_j) \wedge$ indexings $I_j \to M$ with $ji \mapsto m_{p\,ji} \wedge r_j =_{def} (m_{p\,ji})_{i \in I_{[j]}}$), $R =_{def} (r_j)_{j \in J}$. All orderings < are assumed to be partial and irreflexive. Then the induced structures are hierarchical: $(R, \prec_J, \sim_J)$, $\bigwedge j \in J\ ((r_j, \prec_j, \sim_j))$. Homomorphisms $h: (R, \prec_J, \sim_J) \to (S, \to, \cong)$ and $\bigwedge j \in J\ (h_j: (r_j, \prec_j, \sim_j) \to (s_j, \to_j, \cong))$ can be applied. Structures $(S, \to, \cong)$ and $(s_j, \to_j, \cong_j)$ can of course be defined without reference to h, h_j, and several directed and undirected relations $\to$, $\cong$ can exist on S, s_j. This construction can be extended to any hierarchical level.

Example ("entity relationship" model): In the literature on data base modeling usually three model types are considered, the "hierarchical", the "relational" and the "entity relationship model". All three (and even generalizations) are subsumed under our concept of hierarchical relations. Fig.13 illustrates the constructs $(S, \to)$ and $(s_j, \to_j)$. The elements $s_{[ji]}$ are to interpret as "entities", aggregated in relation s_j, the $s_{[j]}$ are entities on next higher hierarchical level, aggregated in relation S. In

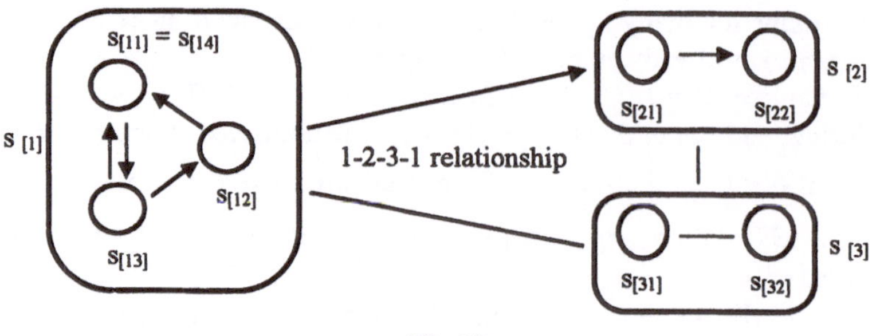

Fig.13

the corresponding graph, entities are represented by nods, "relationships" by attributed edges and arrows.

6.2 Input-output systems

We return to **2.4** and consider the binary relation $R = (y_j, x_i)_{(j,i)\in U}$ with $U \subseteq J \times I$, bijective indexings $I \leftrightarrow X$, $J \leftrightarrow Y$, $pr_1\, U = J$, $pr_2\, U = I$, and with

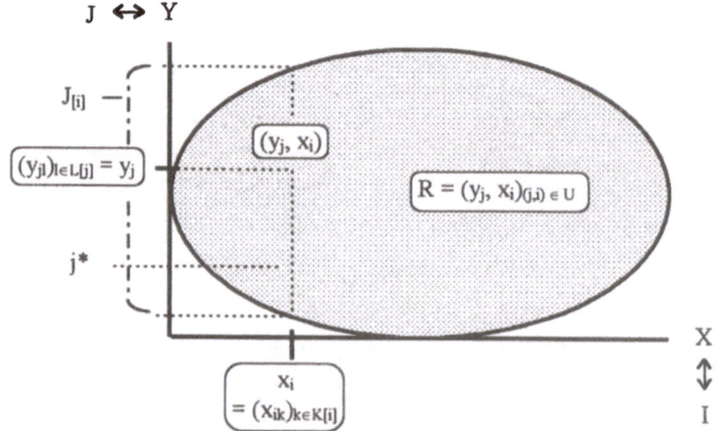

Fig.14

$\wedge i{\in}I\ (x_i = (x_{ik})_{k\in K_{[i]}})$, $\wedge j{\in}J\ (y_j = (y_{jl})_{l\in L_{[j]}})$ given, $K = \bigcup_{i\in I} K_{[i]}$, $L = \bigcup_{j\in J} L_{[j]}$. For illustration see Fig.14. The assignment $(x_i) \mapsto pr(Y)\ (cut((y_j, x_i)_{j\in J})\ R) = (y_j)_{j\in J_{[i]}}$ is for card $J_{[i]} = 1$ functional, otherwise relational, i.e. functional into $\{Y^C \mid C{\in}pow\ J \wedge \text{card } C > 1\}$. A physical system modeled by R and employing this assignment direction is an "input-output system", the input is modeled by X, the output by Y. If R is functional on X, the system is "functional". If $(x_i) \mapsto (y_j)_{j\in J_{[i]}}$ is not functional, there are two ways to reach a functional assignment:

(1) To choose a parameter $j^* \in J_{[i]}$ and to assign $(x_i) \mapsto (y_{j^*})$. Expressing this by a variable we have var $y_j : \{y_j \mid j \in J_{[i]}\}$ and the assignment var $y_j := (j^*)\ y_{j^*}$.

Notice, the variability domain of var y_j depends on x_i.

(2) to concatenate $(y_j)_{j\in J_{[i]}}$ to one value $y_{[i]}$.

In the functional case, introducing $\wedge(ji)\in U$ $((W_{ji} =_{def} L_j \cup K_i) \wedge z_{ji} =_{def} (z_w)_{w\in W_{ji}}$ with $z_w =_{def} x_{ik}$ for $w = ik$ and $z_w =_{def} y_{jl}$ for $w = jl)$, we have a relation $S =_{def} (z_w)_{w\in W}$ with $W =_{def} \bigcup_{ji\in U} W_{ji}$, which has the functional decomposition $\boldsymbol{Z}$ given by $z_w =_{def} x_{ik}$ for $w = ik$ and $z_w =_{def} y_{jl}$ for $w = jl$. This way we started in **2.1**. S may have more functional decompositions.

As mentioned in **2.4**, conjugate assignments $(y_j) \mapsto pr(X)\ (cut((y_j, x_i)_{i\in I})\ R) = (x_i)_{i\in I_{[j]}}$ can be considered as well.

6.3 Input-output systems with valuations

Next we introduce valuations by elements of sets **V**, **W** and consider $S = ((y_j, w_{ji}), (x_i, v_{ji}))_{ji\in U}$, with all $v_{[ji]} \in \mathbf{V}$ and all $w_{[ji]} \in \mathbf{W}$. We want to find functional relations with respect to S between families $(x_i, \tilde{v}_i)_{i\in\tilde{I}}$, $\tilde{I} \subseteq I$, x_i valuated by $\tilde{v}_{[i]} \in \mathbf{V}$, and families $(y_j, w_j)_{j\in J(\tilde{I})}$, $J(\tilde{I}) \subseteq J$, y_j valuated by $\tilde{w}_{[j]} \in \mathbf{W}$. Because of the isomorphism induced by the indexing and the assumption that valuations are independent of the valuated elements, we can split S into $R = (y_j, x_i)_{ji\in U}$, $M = (w_{ji}, v_{ji})_{ji\in U}$, and M into $W = (w_{ji})_{ji\in U}$ and $V = (v_{ji})_{ji\in U}$.

To cover a wide variety of applications, we make the following assumptions: $(\mathbf{W}, +, *)$ is an algebraic structure with associative binary compositions $+, *$; $+$ commutative, $*$ distributive with respect to $+$. Further, let be $\wedge w, w_{[1]},\dots w_{[n]} \in \mathbf{W}$ $(+w = *w = w \wedge +(w_{[1]},\dots w_{[n]}) = w_{[1]}+\dots+w_{[n]} \wedge *(w_{[1]},\dots w_{[n]}) = w_{[1]}*\dots*w_{[n]})$. Then for any given non-empty index set C, we consider $\mathscr{W} =_{def} (\{\mathbf{W}^c \mid \varnothing \neq c \subseteq C\}, "+", "*")$, "+", "*" defined by component-wise $+, *$, as already employed earlier, i.e. for "∘" $\in$ {"+", "*"}, $w' = (w'_q)_{q\in c'}$, $w'' = (w''_q)_{q\in c''}$, $c', c'' \subseteq C$, w' "∘" $w'' =_{def} (w_q)_{q\in c'\cup c''}$ with $w_q = w'_q$ for $q \in c'\setminus c''$, $w_q = w''_q$ for $q \in c''\setminus c'$, $w_{[q]} = w'_{[q]} \circ w''_{[q]}$ for $q \in c'\cap c''$, in other words, w is the concatenation by ∘ of w' and w'', $w = \mathbf{K}_\circ(w', w'')$. "+", "*" are associative, "+" commutative, "*" distributive. $\mathscr{W}$ is an algebraic module over **W**. If it is clear from context, we write ∘ for "∘".

Examples:

(1) $(\mathbf{B}, \sqcap, \sqcup)$ a boolean lattice with $* = \sqcap$, $+ = \sqcup$, which is no ring. With $* = \sqcap$, $+ = \Delta$ (the symmetrical difference), it is a ring. According to our definition, $\sqcap$, $\sqcup$ are (primitive) logic functions with respect to $\leq$ as defined by $\sqcap, \sqcup$. Neutral elements are **e**, **o**.

(2) $(\mathbf{R}_0, *, +, \leq)$ the non-negative real numbers with 1 and 0 as neutral elements, $*, +$ are logic functions.

(3) $(\{f \mid f\colon [0, 1] \to (\mathbf{R}, *, +, \leq) \wedge f$ continuous on $[0, 1] \subset \mathbf{R}\}, *, +)$ is a ring

with divisors of zero. Neutral elements are $f(x) = const = 1$, $f(x) = const = 0$. $w = f(x)$, $x \in [0, 1]$, writes in our notation $(w_x)_{x\in[0,1]}$.

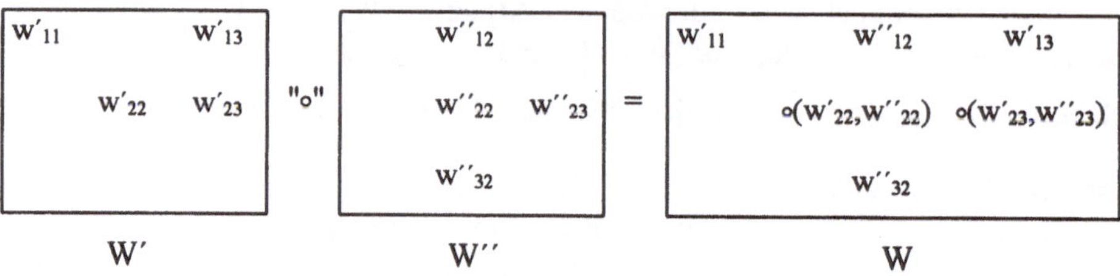

Fig.15. Visualization of an operation "∘"

Let there be given $\varnothing \neq \tilde{I} \subseteq I$, $\tilde{v}_{\tilde{I}} =_{def} (\tilde{v}_i)_{i\in\tilde{I}}$, $\tilde{v}_{[i]} \in \mathbf{V}$, (an "input pattern"), and $\eta{:}\mathbf{V}\times\mathbf{V} \to \mathbf{B} = \{"t", "f"\}$ with $\eta(v',v'') = "t"$ for $v' = v''$, $\eta(v',v'') = "f"$ for $v' \neq v''$. We apply selection with respect to relation V as described in **2.4**: $\wedge j\in J\ (\tilde{I}_{[j]} =_{def} \{i \mid i \in \tilde{I} \cap I_{[j]} \wedge \eta(\tilde{v}_{[i]}, v_{[ji]}) = "t"\})$, thus $\tilde{I}_{[j]} \subseteq \tilde{I} \cap I_{[j]}$, $J(\tilde{I}) =_{def} \{j \mid j \in J \wedge \tilde{I}_{[j]} \neq \varnothing\}$. Selected is then $\tilde{V} = ((v_{ji})_{i\in\tilde{I}_{[j]}})_{j\in J(\tilde{I})}$ with the conjugate $((v_{ij})_{i\in\tilde{J}_{[i]}})_{i\in I(\tilde{J})}$.

Depending on the application field, other selection modes may be chosen, for example $\tilde{I} \subseteq I_{[j]}$ and $\tilde{I}_{[j]} =_{def} \{i \mid i \in \tilde{I} \wedge (\eta(\tilde{v}_{[i]}, v_{[ji]}) = "t")\}$.

Thus, in general a relational input-output assignment $\tilde{v}_{\tilde{I}} = (\tilde{v}_i)_{i\in\tilde{I}} \mapsto ((w_{ji})_{i\in\tilde{I}_{[j]}})_{j\in J(\tilde{I})}$ with conjugate $((w_{ij})_{j\in\tilde{J}_{[i]}})_{i\in I(\tilde{J})}$ is defined.

To $((w_{ji})_{i\in\tilde{I}_{[j]}})_{j\in J(\tilde{I})}$ module operations can be applied. We choose a family of operations $(\circ_j)_{j\in J(\tilde{I})}$ (often $\wedge j,j'\in J(\tilde{I})\ (\circ_{[j]} = \circ_{[j']})$) and as connectors $(\tilde{I}_{[j]})_{j\in J(\tilde{I})}$. Concatenation yields $(w^{(1)}_{j\tilde{I}_{[j]}})_{j\in J(\tilde{I})} =_{def} (\circ_j\ (w_{ji})_{i\in\tilde{I}_{[j]}})_{j\in J(\tilde{I})}$. In the conjugate case the corresponding concatenations result in $(w^{(1)}_{i\tilde{J}_{[i]}})_{i\in I(\tilde{J})} =_{def} (\circ_i\ (w_{ij})_{j\in\tilde{J}_{[i]}})_{i\in I(\tilde{J})}$. Because of the 1-1 indexing, we have now a functional input-output assignment $(x_i, \tilde{v}_i)_{i\in\tilde{I}} \mapsto (y_j, w^{(1)}_{j\tilde{I}_{[j]}})_{j\in J(\tilde{I})}$, and, if $(y_j)_{j\in\tilde{J}_{[i]}}$ is concatenated to an object $\bar{y}_{[i]}$, $(x_i, \tilde{v}_i)_{i\in\tilde{I}} \mapsto (\bar{y}_i, w^{(1)}_{i\tilde{J}_{[i]}})_{i\in I(\tilde{J})}$.

Let a family $((x_i, \tilde{v}_{i[p]})_{i\in\tilde{I}_{[p]}})_{p\in P}$ be given to which for each p by concatenations like above (possibly with different ∘'s for each p) the family $((y_j, w^{(1)}_{j\tilde{I}_{[jp]}})_{j\in J(\tilde{I}_{[p]})})_{p\in P}$ is assigned. Then with $J(P) =_{def} \bigcup_{p\in P} J(\tilde{I}_{[p]})$, $\tilde{I}_{[jP]} =_{def} \{\tilde{I}_{[jp]} \mid p$

$\in$ P}, and for given $(\circ_j)_{j\in J(P)}$, concatenations $(w^{(2)}_{j\tilde{I}[jP]})_{j\in J(P)} =_{def} (\circ_j (w^{(1)}_{j\tilde{I}[jp]})_{\tilde{I}[jp]\in\tilde{I}[jP]})_{j\in J(P)}$ on next higher concatenation level can be performed resulting in a valuation $(y_j, w^{(2)}_{j\tilde{I}[jP]})$ of y_j for $j \in J(P)$. Similar results hold for the conjugate case. The reasoning can be continued to any concatenation level.

Generalizing, we can consider variables var $\tilde{I}$: pow I, var $\tilde{v}_i$: **V**, var $\circ$: {+ , $*$}, and the relation variable var M = (var w_{ji}, var $v_{ji})_{ji\in U}$. An important case is var w_{ji} = var ϕ_{ji}(var v_{ji}) with adjustable assignments to var ϕ_{ji} to achieve a certain input-output behavior.

We assume, **V** has an analogue structure (**V**, +, $*$) with module $\mathscr{V} = \{\mathbf{V}^c \mid c \subseteq J\}$, "+", "$*$"). Then a similar reasoning can be applied to the conjugate case with $\tilde{J}$, $\tilde{w}_{\tilde{J}}$ being given, and with operations from module ($\mathscr{V} = \{\mathbf{V}^c \mid c \subseteq I\}$, "+", "$*$"). We distinguish the operations of **V**, **W** by indices, $(+_V, *_V)$, $(+_W, *_W)$.

For any non-empty $\tilde{v}_{\tilde{I}} = (\tilde{v}_i)_{i\in\tilde{I}}$ we consider the relational assignment by M (with given selection mode): $\tilde{v}_{\tilde{I}} = (\tilde{v}_i)_{i\in\tilde{I}} \mapsto ((w_{ji})_{i\in\tilde{I}[j]})_{j\in J(\tilde{I})}$. We agree upon

$$+_V (\tilde{v}_i)_{i\in\tilde{I}} \mapsto (+_W (w_{ji})_{i\in\tilde{I}[j]})_{j\in J(\tilde{I})} \text{ and } *_V (\tilde{v}_i)_{i\in\tilde{I}} \mapsto (*_W (w_{ji})_{i\in\tilde{I}[j]})_{j\in J(\tilde{I})},$$

and for the conjugate case, for any non-empty $\tilde{w}_{\tilde{J}} = (\tilde{w}_j)_{j\in\tilde{J}}$,

$$+_W (\tilde{w}_j)_{j\in\tilde{J}} \mapsto (+_V (v_{ji})_{j\in\tilde{J}[i]})_{i\in I(\tilde{J})} \text{ and } *_W (\tilde{w}_j)_{j\in\tilde{J}} \mapsto (*_V (v_{ji})_{j\in\tilde{J}[i]})_{i\in I(\tilde{J})}.$$

Hierarchical expressions of $+_{V/W}$-terms and $*_{V/W}$-terms are homomorphically mapped onto hierarchical families of $+_{W/V}$-terms and $*_{W/V}$-terms, respectively.

The valuated relations of the type considered have many applications, for example in the theory of concepts, classification, associative memories, pattern recognition, data bases, formal logic, knowledge bases, inference rules, adaptive networks like artificial neural networks.

Example (knowledge module, data base, "inference rules" in 2-valued logic):
V = **W** = ({0, 1}, $\wedge$, $\vee$) is boolean. The relation M is shown in Fig.16. v's, w's are variables on {0,1}, letters for indices are variables, ψ is a variable on $\{\wedge,\vee,\neg\}$. x, y are propositions. For all j, j′ $(\circ_{[j]} = \circ_{[j']})$ and selection criteria of type $\tilde{I} \subseteq I_{[j]}$ are chosen.
Let there be given on level 1:
$p = 1$, $\tilde{I} := \{1, 2\}$, $\tilde{v}^{(1)}_{[\tilde{I}]} := (\tilde{v}_{[1]} := 1, \tilde{v}_{[2]} := 0)$, $\tilde{v}^{(1)}_{[12]} = \vee(\tilde{v}_{[1]}, \tilde{v}_{[2]})$,
selected are: $\tilde{J}(\tilde{I}) := \{1, 3\}$, $w_{[11]} = 0$, $w_{[12]} = 1$, $w_{[31]} = 1$, $w_{[32]} = 1$,
assigned are according $w^{(1)}_{[j,[1,2]]} = \vee(w_{[j,[1]]}, w_{[j,[2]]})$: $w^{(1)}_{[1,[12]]} := 1$, $w^{(1)}_{[3,[12]]} := 1$;
$p = 2$, $\tilde{I} := \{3, 4\}$, $\tilde{v}^{(1)}_{[\tilde{I}]} := (\tilde{v}_{[3]} := 1, \tilde{v}_{[4]} := 1)$, $\tilde{v}^{(1)}_{[34]} := \wedge(\tilde{v}_{[3]}, \tilde{v}_{[4]})$,
selected are: $\tilde{J}(\tilde{I}) := \{2, 3\}$, $w_{[23]} = 1$, $w_{[24]} = 0$, $w_{[33]} = 1$, $w_{[34]} = 1$,

assigned are according $w^{(1)}_{[j,[34]]} = \wedge (w_{[j,[3]]}, w_{[j,[4]]})$: $w^{(1)}_{[2,[34]]} := 0$, $w^{(1)}_{[3,[34]]} := 1$.
Let there be given on level 2:
$\tilde{I} := \{\{1,2\}, \{3,4\}\}$, $\tilde{v}^{(2)}_{[[12],[34]]} = \wedge (\tilde{v}^{(1)}_{[12]}, \tilde{v}^{(1)}_{[34]})$
selected are: $\tilde{J}(\tilde{I}) := \{3\}$, $w^{(1)}_{[j,[12]]} = 1$, $w^{(1)}_{[j,[34]]} = 1$,
assigned is according $\tilde{w}^{(2)}_{[j,[\tilde{I}]]} = \wedge (w^{(1)}_{[j,[12]]}, w^{(1)}_{[j,[34]]})$: $w^{(2)}_{[3,[[12],[34]]]} := 1$.

$w^{(2)}$	$w'^{(1)}$	$w^{(1)}$	J					
1	1	1	3	1,1	1,0	1,1	1,1	
	0		2		1,1	1,1	0,1	$M=(w_{ji},v_{ji})_{ji \in U}$
		1	1	0,1	1,0			
				1	2	3	4	I
		$\vee$		1	0			$\tilde{v}^{(1)}$
	$\wedge'$					1	1	$\tilde{v}'^{(1)}$
$\wedge$				(1	0)	(1	1)	$\tilde{v}^{(2)}$

Fig.16

Let us assume, the x and y are composite and valuated, for example
$(x_1, v_1) = (((x_{11}, v_{11}), (x_{12},v_{12})), v_{[1]} = \wedge(v_{[11]}, v_{[12]}))$,
$(x_2, v_2) = (((x_{21}, v_{21}), (x_{22},v_{22})), v_{[2]} = \vee(v_{[21]}, v_{[22]}))$,
$(x_3, 1) = (((x_{31}, v_{31}), (x_{32},v_{32})), 1 = \wedge(v_{[31]}, v_{[32]}))$,
$(x_4, v_4) = ((x_{41}, v_{41}), v_{[4]} = id(v_{[41]}))$,
$(y_2, w_2) = ((y_{21}, 1), w_{[2]} = \psi(1))$,
$(y_3, w_3) = ((y_{31}, w_{31}), (y_{32}, w_{32}), w_{[3]} = \vee(w_{[31]}, w_{[32]})$.
According M, we have for example:
"if" $(((v_{[11]} \wedge v_{[12]}) = 1 \vee (v_{[21]} \vee v_{[22]})) = 0 \wedge ((v_{[31]} \wedge v_{[32]}) = 1 \wedge v_{[41]} = 1))$
"then" $(w_{[31]} \vee w_{[32]}) = 1$. Utilizing the conjugate relation, we have for example:
"if" $w_{[2]} = 0$ (i.e. $\psi = \neg$) "then" $v_{[4]} = 1$.

Under suitable assumptions on their domains and codomains, relations $S = ((y_j, w_{ji}), (x_i, v_{ji}))_{ji \in U}$ with algebraic modules $\mathscr{W}$, $\mathscr{V}$ can be hierarchically concatenated.

Example (simplified neural network model): $V = W = \mathbf{R}$, input $\tilde{v} = (v_1, v_2, v_3)$, $I = \tilde{I} = \{1,2,3\}$, $v_{[i]} \in V$, $J = \{1,2\}$. Relation M is given by Fig.17, $w_{ji} := *(p_{ji}, v_i^n)$. Classically $n = 1$ or $n = 2$. p_{ji} is a variable, p_{ji}: P, $P \subseteq \mathbf{R}$, P is the domain of "weights". Concatenation + on connector (j,I) applied for j = 1,2 results in $w_j^{(1)}$. On the next step $w_j^{(2)} := f_j(w_j^{(1)})$ ("excitation" and "output" function). Usually all $f_{[j]}$ are the same and not variable. The two steps make one layer. The next layer has the same structure, possibly other I, J, f. In the "learning phase" the parameters p_{ji} are systematically altered such that their sequence $p_{ji}^{(v)}$ (hopefully) converges to limits which adjust the network to the behavior wanted in the subsequent application phase.

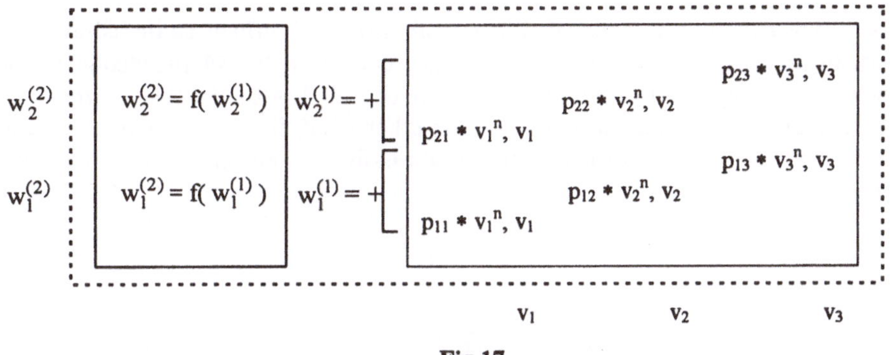

Fig.17

In case the module **W** has neutral elements **o** for + and **e** for *, one can introduce $\eta: \mathbf{V} \times \mathbf{V} \rightarrow \{\mathbf{o}, \mathbf{e}\}$ with $\eta(v', v'') = \mathbf{e}$ for $v' = v''$, $\eta(v', v'') = \mathbf{o}$ for $v' \neq v''$. Instead of canceling indices i with $(\tilde{v}_{[i]} \neq v_{[ji]})$, one can use $\eta(\tilde{v}_{[i]}, v_{[ji]}) = \mathbf{o}$ as scalar factor to w_{ji}, $(\eta(\tilde{v}_{[i]}, v_{[ji]}) * w_{[ji]})_{ji}$. However, the result with respect to a subsequent concatenation by * is not the same as before, we have another selection mode.

6.4 Time dependent systems

Mathematically, any structure referenced by a partial ordered set represents a process. For example, such a reference can be a model time, modeling a physical time given by observable events, a logical (or algorithmic) time given by hierarchies of concatenations or assignments to variables.

Describing the structure by variables with variability domains, the domain of a variable may contain lower level variables. As pointed out in **2.3** there may be dependencies between and constraints on these domains. In addition, from the parameterized hierarchical structure one cannot conclude that the unparameterized structure is acyclic. If the structure models physical objects and for two structure elements s_i, s_k holds $s = s_{[i]} = s_{[k]}$, then the use of a physical object named s for s_i is alternative with the use of s for s_k. The use of s for s_i may "consume" or change s. Otherwise s may be later reused for s_k. To instanciate a particular state in a structure with variables, the applied (partial) assignments have to be tested for their global admissibility.

For example, let there be given a relation var $R = (\text{var } r_l)_{l \in L}$ with domains dom var R = D, dom var $r_l = D_l$. Then in general $D \subseteq \prod_{l \in L} D_l$. If the projection $pr_l\, D = D_l$, then for an element $(r_{l'})_{l' \in L} \in D$ the choice of $r_l \in D_l$ is "free", i.e. independent of the other $r_{l'}$. If var R is combined with var S to var W = (var S, var R) with dom var S = E, dom var W = F, then $F \subseteq \prod(D,E)$. An assignment $(r_l)_{l \in L} \in D$ may no more be admissible for assignments to var W. It follows $pr(E)F \subseteq E$, $pr(D)F \subseteq D$, $pr_l(pr(D)F \subseteq D_l$. Thus, $(r_l)_{l \in L} \in \prod_{l \in L} pr_l\,(pr(D)F$ is necessary, but in general not sufficient to reach F. The overall structure influences the admissible part structures. Further, a change of a D_l to D'_l can cause $F = \varnothing$, the change is not consistent with the goal to reach at least part of F. A visualization is given in Fig.18. See also the Example to Fig.4. On the other hand, if one insists in dom var R, var S has to be adapted to it to find an admissible extension. This need not be possible.

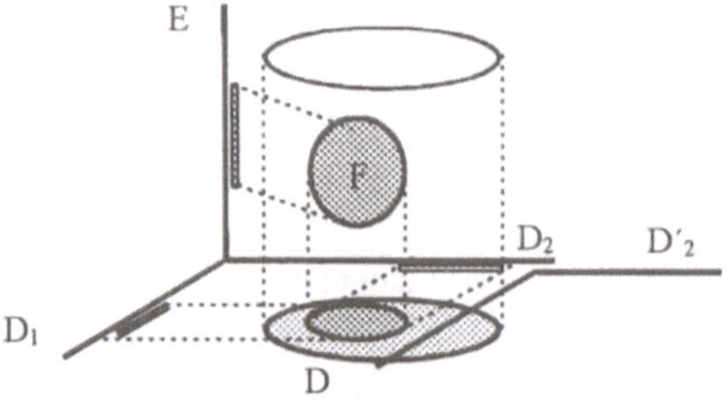

Fig.18

Example (construction process of a building): var building: {var church, var factory, var house}, var house = (var $\text{wall}_{\text{var } n}$, var roof, var basement), var $\text{wall}_{\text{var } n}$ = (var $\text{window}_{\text{var } n,\, \text{var } m}$, var $\text{door}_{\text{var } n,\, \text{var } k}$, var concrete $\text{part}_{\text{var } n}$) and so on. Constraints are for example: The height of the walls has to be the same, a window and a door cannot be at the same place, number, size and position of doors cannot be such that a wall consists of doors only. There is a tendency to specify objects formally by their properties. Let us try "wall": upright structure made of stone, brick, or similar material, serving for enclosure, division, support of a building, etc. (College Dictionary). The specification is not formal. Moreover, usually not wanted curved, not rectangular, not plane walls are in conformance with this specification. Supposing the construction process is well defined, to realize it physically, passive and active "resources" have to be allocated at the right time to the work to be done. This resource allocation and process scheduling causes another problem. If the bricks provided are used up for wall_1, the process has no continuation to wall_2, installing doors and windows in wall_1 would be senseless. If bricks are provided for wall_2, the workers employed in building wall_1 can at later time at another place build wall_2, they are "reusable" active resources. There are also passive reusable resources, e.g. hardware tools. However, both might be reusable only for a limited number of times.

The Example shows the difficulties in finding complete, consistent, appropriate formal specifications, resource allocations and schedules.

In the mathematical representation of time dependent systems we use a set of times $(\mathbf{T}, <, \sqcap, \sqcup)$ as introduced, and processes as structural objects. For given $K \neq \varnothing$ we consider $\bigwedge k \in K\ (U_{[k]} \in \mathbf{T})$, $U =_{def} \bigsqcup (U_{[k]})_{k \in K}$, and the process $R_U = ((r_{kt})_{t \in \alpha(U_{[k]})})_{k \in K}$. A continuation of R_U into a future of U is representable by a variable var $\Delta R_{var\ V} =_{def} ((var\ r_{lt})_{t \in \alpha(var\ V_{[l]})})_{l \in var\ L}$ and a concatenation $\mathbf{K}(R_U$, var $\Delta R_{var\ V}$, var C). Herein var $\Delta R_{var\ V} : \{((r_{[p]lt})_{t \in \alpha(V_{[pl]})})_{l \in L_{[p]}} \mid p \in P(R_U, U)\}$, var V $=_{def} \bigsqcup (var\ V_{[l]})_{l \in var\ L}$, var $V \sqcap U = \mathbf{o}$, var $V \neg<_{\vee\vee} U$. An assignment to var C determines, which partprocesses of R_U continue into which partprocesses of ΔR_V when it is assigned to var $\Delta R_{var\ V}$. Possible cases are:

(1) An evolution law (a state function) f is given, var $\Delta R_{var\ V} := f(U, R_U, p)$, $p \in P(U, R_U)$, the set of control parameters, $V \in \mathbf{T}$.

(2) var $\Delta R_{var\ V}$ is assigned by observation of new events, either $V \in \mathbf{T}$, or $V \notin \mathbf{T}$, then U can be enlarged to $\bigsqcup \alpha(U) \cup \alpha(V)$.

A physical system depending on a physical time is named "dynamic".

6.5 Modeling of physical systems

When applying mathematical constructs to represent physical systems one has to be aware of the fundamental conceptual differences between both. In mathematics, concepts exist, or are defined by existing ones, or are related to define structures. The constructive computation by physical means of implicitly defined mathematical objects is not seen as part of pure mathematics. In the physical world we have objects which from existing objects generate, produce objects not existing for us or not known to us before this activity. This involves time, the event "objects given" is earlier than the event "other objects produced from the given ones".

If in mathematics a function $f: X \rightarrow Y$ is defined, this includes the definition of X and Y. To a value $x \in X$ the value $y = f(x) \in Y$ is defined, but not always explicitly given. In a physical interpretation (realization), x symbolizes a physical object x, f symbolizes a physical object f, y symbolizes a physical object y, which is produced by f. To ensure the correct correspondence, f needs a "specification". The activity of f is a physical process in physical time, however, mathematically we have a functional assignment $x \mapsto y$, no time is involved. This does not mean that we are not able to model the physical process mathematically using a model time.

Examples:

(1) If in mathematics the family $(f^n(x))_{n \in (\mathbf{N}_0, <)}$, $f: \mathbf{R} \rightarrow \mathbf{R}$, is considered, it is the definition of infinite many, not necessarily distinct elements of $\mathbf{R}$. $(\mathbf{N}_0, <)$ is a logical time. In the physical counterpart, for given objects x, f, we have a physical process $(x_{t1}, x_{t2} = f(x_{t1}), x_{t3} = f(x_{t2}), \ldots)$ in physical time $t1 < t2 < t3 \ldots$, corresponding to $(\mathbf{N}_0, <)$. From the point of view of object f, the output x_{tn} is fed

back as input to give output $x_{t(n+1)}$, thus forming a "loop". However, parameterizing (labeling) the activities of f by the physical time they are happening, we have a cycle free process. If input x and output y are processes in physical times which are refinements of tn and t(n+1), then tn $<_{\wedge\wedge}$ t(n+1) is assumed. y = f(x) denotes a function, x = f(x) is an equation which can have no, one or many solutions.
(2) For n = 1,2,3,...let f be a time shift by 1, $x_{[n]} = x_{[n+1]}$, input is the process (x_n, x_{n+1}), output (x_{n+1}, x_{n+2}) if there is no feed back. With feed back and overlapping values added, we have as n-th output $(x_{n+1}, (n+1)x_{n+2})$, $\{n, n+1\} \neg <_{\wedge\wedge} \{n+1, n+2\}$.
(3) For some computer scientists, a certain quintuple of 3 sets and 2 functions defines an "abstract machine". For a mathematician, this machine "does" nothing, it remains a quintuple forever (unless redefined). The meaning is that, "someone" can formulate algorithms in logical time, using these functions, to model a physical machine behavior.

In our mathematical representation a variable var v was defined on a domain dom var v = V. To express controllability of var v, we introduced a parameterization by a parameter set P, V = $\{v_{[p]} \mid p \in P\}$, and an assignment or control function val: P×{var v} → V, var v :=(p) $v_{[p]}$, specifying var v within a structure. In a physical interpretation var v symbolizes a variable physical object *var v* taking states (properties, configuration, behavior) from a variety of states according a control object *p* symbolized by p. Physical assignment is understood as a physical process carried out by a control object *val* (*p*, *var v*), creating or forming the physical object $v_{[p]}$. The control object *p* can depend on the present and previous states of (partially assigned) *var v*, in this case recognizing, sensing, memorizing these states is another physical activity. Object *p* can also be the result of activities in the physical environment of *var v*. *val* and the domain of *var v* can themselves be variables, depending on current physical time, the memorized history of previous assignments and possible input to *val* by a higher order control object.

Physical objects are located in space and can move or can be transported from one place to another by transportation objects which can be controllable, using a passive object (path, link). Locating and dislocating are processes. In the mathematical model we can use indices as space coordinates, connectors C as relations between locations (paths, links), and relations E(C) to describe placing of objects.

Physically, we have further (mostly) passive objects with the capability to store and release other objects on demand. They have properties like capacity, ordering of stored objects, access regulations and may be dedicated to store only objects of a certain class. They have states like "being loaded with objects so and so", "open to access (load or fetch)", "closed to access". Mathematically, stores are modeled by sets or families changing their members in course of time.

All active objects in their state "active" (or "in operation") perform physical processes. Other states are for example "not in operation", "ready to operate", "operation interrupted, waiting", "operation suspended", "operation regularly terminated". To adapt the local time of a process to global time conditions imposed on a system, one introduces physical objects ("clocks", "timers") performing time reference processes for synchronization and limitation of process lenght. For example, a time controlled transportation process can be equivalent to storing the transported objects for a while, a time controlled output process of an object, the output provided as input to another object, can be delayed or accelerated to match the input time conditions without needing a storage ("just in time transportation or production").

All physical objects and processes under consideration can have valuations (costs, priorities, graded similarities, accuracy, etc.). This corresponds to valuated and topological mathematical structures.

In rough outlines, a variable, controllable system has a control or system level (components for system management and operation control) and an application (or production) level (components for production, computation). We have

on system level:

A composite active system control object, centralized or distributed, with external and internal control input, and eventually control input processing capabilities ("intelligence"), to control utilization, scheduling and operation of objects on application level (the "resources").

Active objects for structuring and configuration of composite (complex) objects on application level. They serve to compose or decompose objects made of primitive application objects (corresponding to defining sets/relations, composition, concatenation of relations, cuts, projections), to allocate system objects to and to release them from a task, to incorporate system and application objects from the environment and to release objects to the environment.

Active objects for recognition and creation of system component properties (e.g. equality, "part of" property, quantification, location, time of occurrence, time of creation, creation of references, recognition of object classes, similarities between objects, processing of properties, etc.).

Active objects for selection and aggregation by properties (corresponding to the mathematical $\{.\,|\,\text{property-list}\}$ construct).

Passive and active control objects (stored programs, status signals, signal links, signal generators, control criteria tester, signal processors, counters, clocks, timers, exception handlers, etc.).

on application level:

Primitive passive application objects (materials, data, fixtures, storage, links, etc.).

Primitive (controllable) active application objects (production modules, processors, agents, transportation facilities, etc.) and their control objects.

The primitive objects can be systems on lower hierarchical level, but are "black boxes" on the level considered.

Basic operating principles are

"reuse" of system components for further tasks after they are freed from one task,

concurrent processing of independent tasks, if sufficient system- and application-objects are available.

The system and its components and operations are subject to logical and physical constraints and conditions, expressed by "syntactic" rules. However, these do not imply that all physically possible and admissible configurations and processes make sense within the application field under consideration. Therefore "semantic" rules have to be added, depending on the semantics of the primitive objects.

6.6 Functional systems

Logical model

Let there be given a finite index set Q and for all $q \in Q$ different functions $g_{[q]}: X_{[q]} \rightarrow Y_{[q]}$ with domain $X_{[q]} \subseteq \bigcup_{u \subseteq I_{[q]}} \prod_{i \in u} X_{[q]i}$ and range $Y_{[q]} \subseteq \bigcup_{v \subseteq J_{[q]}} \prod_{j \in v} Y_{[q]j}$. Let be $x_{[q]u} =_{def} (x_{[q]i})_{i \in u} \in X_{[q]}$ and $y_{[q]v} =_{def} (y_{[q]j})_{j \in v} = g_{[q]}(x_{[q]u}) \in Y_{[q]}$. Then the arities are $1 \leq \text{card } u \leq \text{card } I_{[q]}$ and $1 \leq \text{card } v \leq \text{card } J_{[q]}$. $G = \{g_{[q]} \mid q \in Q\}$ is used as set of "primitive functions" or "generators". A visualization is given in Fig.19.

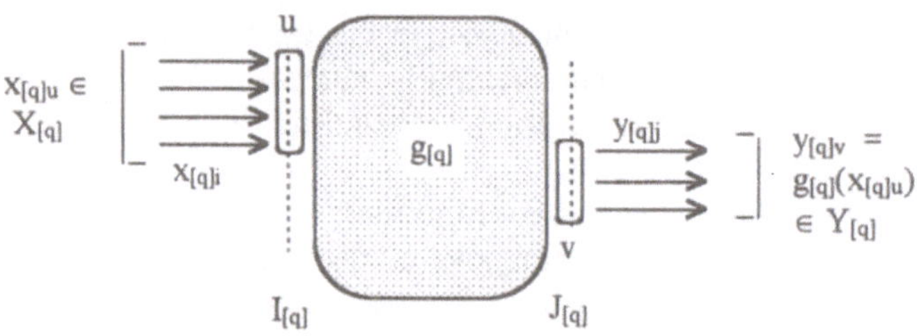

Fig.19

Now we consider a finite index set P, an indexing $ind: P \rightarrow Q$ with assignment $p \mapsto q_{[p]}$. For notational convenience we write in the following p instead of q_p, $f_p =_{def} g_{q_p}$, $f_p: X_p \rightarrow Y_p$. We postulate the following "concatenation conditions":

$C \subseteq \bigcup_{p \in P} J_p \times \bigcup_{p \in P} I_p$ is a *cycle free* connector, functional from right to left, i.e. from $(pj, qi), (p'j', q'i') \in C$ and $pj \neq p'j'$ follows $qi \neq q'i'$,

$E(C)$ is given by $\bigwedge c = (pj, p'i) \in C \; (y_{[pj]} = x_{[p'i]})$ (identity),

$\bigwedge p \in P \; (x_{[p]} \in \text{dom } f_{[p]})$ (the global domain condition).

If $R =_{def} \{(x_p = (x_{pi})_{i \in u_{[p]}}, y_p = (y_{pj})_{j \in v_{[p]}} = f_p(x_p)) \mid p \in P\}$ satisfies the concatenation conditions, then $(\mathbf{K}((x_p, y_p)_{p \in P}, E(C), C)$ exists and with $U(\varnothing) =_{def} \bigcup_{p \in P} u_p \setminus pr_2 C$ and with $V(\varnothing) =_{def} \bigcup_{p \in P} v_p \setminus pr_1 C)$ the assignment $(x_{pi})_{pi \in U(\varnothing)} \mapsto (y_{pj})_{pj \in V(\varnothing)}$ is functional. Thus concatenation of R defines a composite function F which assigns to unconnected x-values unconnected y-values. We say R is concatenable. F can then be a primitive function of a relation on next higher hierarchical level with its own domain condition.

Notice: The concatenation conditions are global and in general they cannot be replaced by any rule local to the f_p. Changing the structure, for example by adding another function f^* to $(f_p(x_p))_{p\in P}$, can violate the concatenation conditions.

We used x_p and y_p already as variables. Now we consider a structure variable over the fixed set G of primitives:

var S = ((var $f_p)_{p\in \text{var } P}$, E(var C), var C) : $\mathscr{S}$, $\mathscr{S}$ a set of concatenable functional structures with E(var C) being an equality relation, yielding a composite function variable var F when concatenated **K**(var S),

var P : $\mathscr{P}$, a set of finite index sets,

var ind(var P) : $\mathscr{Ind}$(var P → G) a set of indexings,

var $f_{[p]}$: G, var f_p : var x_p = (var $x_{pi})_{i\in \text{var } I_{[p]}}$ ↦ var y_p = (var $y_{pj})_{j\in \text{var } J_{[p]}}$,

var x_p : var X_p, var X_p : var $\mathscr{X}$,

var C : pow($\bigcup_{p\in \text{var } P}$ var J_p × $\bigcup_{p\in \text{var } P}$ var I_p).

The component variables of var S are not "free" but subject to the condition var S: $\mathscr{S}$. Each occurring variable is assumed to be controllable by a control function val(.,.). val(., var S) is externally controlled.

Examples: G = {+,−,* : $\mathbf{R}^2$ → **R**, / : **R**×(**R**\{0}) → **R**}, $\mathscr{P}$ a set of finite subsets of integers **I**. $\mathscr{S}$ arithmetic functions over G, e.g. var S = (var f_1 : (var x_{11}, var x_{12}) → var y_{13}, var f_2 : (var x_{21}, var x_{22}) → var y_{23}, E(var C), var C : {∅, (13, 21), (13, 22)}). Visualization in Fig. 20, val . written for val(.,var .). See also Examples in **2.3**.

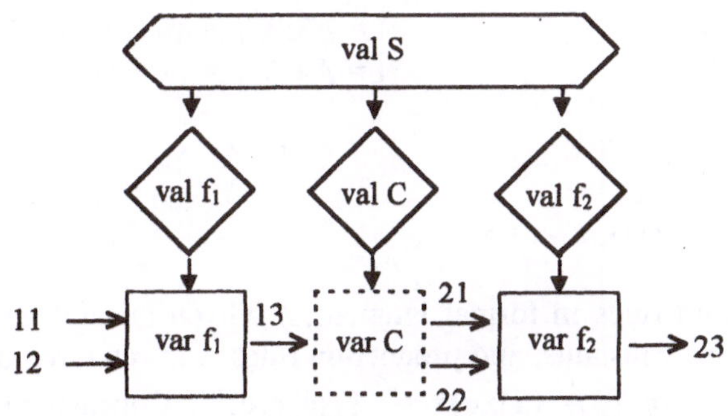

Fig.20

The component objects of var S are in an irreflexive partial order ≺ by functional dependency, for example, if var y_p results from var x_p by var f_p, var x_p ≺ var y_p, if var f_p is connected to var f_{p^*} by a relation E(C), then var f_p ≺ var f_{p^*}, for val(.,var z) we have val(.,var z) ≺ var z. Therefore algorithms according algorithm scheme ($\mathscr{A}$) of **2.3** can be applied as soon as the dependencies are fixed (then we have an "algorithmic structure").

A sequence of partial assignments to the composite variable var S, assumed to be admissible within class $\mathscr{S}$, can be for example:

(S1) var P := P,

(S2) $\wedge p \in P$ (var $f_p := f_p$ with $f_{[p]} \in G$), follows $\wedge p \in P$ (var $X_p := X_p \wedge$ var $I_p := I_p$ $\wedge$ var $Y_p := Y_p \wedge$ var $J_p := J_p$),

(S3) var $C := C \subseteq \bigcup_{p \in P} J_p \times \bigcup_{p \in P} I_p$, follows var $S := S = ((f_p)_{p \in P}, E(C), C)$,

(S4) $\mathbf{K}((f_p)_{p \in P}, E(C), C)$, follows var $F := F$, $F: X_F \rightarrow Y_F$.

This is an algorithm or logical process on "system level" in logical time $T = \{S1,S2,S3,S4\}$. Concurrency holds e.g. at instant (S2).

F according (S4) is a composite function, the components are partially ordered $\prec$. Therefore

(A) starting with a (possibly partial) assignment to var $x_F : X_F$, the functional assignment by F can be partitioned into algorithmic steps of an algorithm on "application level".

Notice: Algorithmic steps on "system" - and on "application" - level can overlap if they are consistent with the ordering $\prec$.

Example (like the previous): Written x, y instead of var x, var y.

$t = S1$: var $P := \{1,2\}$,

$t = S2$: var $f_1 := +_1$, var $f_2 := *_2$,

$t = S3$: var $C := \{(13, 21)\}$,

$t = S4$: $\mathbf{K}((x_{11}, x_{12}, y_{13} := (x_{[11]} + x_{[12]})_{13}$, $(x_{21}, x_{22}, y_{23} := (x_{[21]} * x_{[22]})_{23}$), E(C), C)

follows $\widetilde{F} : (x_{11}, x_{12}, x_{22}) \rightarrow (y_{13,21}, y_{23})$, $F: (x_{11}, x_{12}, x_{22}) \rightarrow (y_{23})$;

$t = A1$: $x_{11} := 3$, $x_{12} := 4$,

$t = A2$: $y_{13} := 3 + 4$,

$t = A3$: $x_{21} := y_{[13]}$, $x_{22} := 5$,

$t = A4$: $y_{23} := (7 * 5)$;

overlapped:

$t = 1$: var $f_1 := +_1$, $x_{11} := 3$, $x_{12} := 4$,

$t = 2$: $y_{13} := 3 + 4$, var $f_2 := *_2$,

$t = 3$: var $C := \{(13, 21)\}$, $x_{22} := 5$,

$t = 4$: $x_{21} := y_{[13]}$,

$t = 5$: $y_{23} := (7 * 5)$.

Example (production rules in formal languages): Given an alphabet $\{s,v,u,a,b,c\}$, s,v,u variables, a,b,c constants, and production rules s ::= ua | uv, u ::= ab | v, v ::= c, and any context $\alpha, \beta \in \{a,b,c\}^* = \bigcup_{n \in N} \{a,b,c\}^n$. Concatenation is cartesian, "variable" ::= "symbol$_1$" | "symbol$_2$" means in our terminology "variable" : {"symbol$_1$", "symbol$_2$"}, rule application to the structure α"variable"β means assignment α"variable"β := α"symbol"β, for example $\alpha s\beta := \alpha ua\beta$, $\alpha ua\beta := \alpha aba\beta$. In this Example the grammar is context free and productions terminate. A case like s ::= ua, u ::= bs, is excluded in **2.3**.

Time parameterized physical model

We base the physical model on the logical model which expresses the functionality of the system. As outlined in **6.5**, the correspondences between objects of the physical and the logical model are:

physical model	logical model
processors, agents, production modules	functions, relations
transportation objects	directed relations, E(C)
transportation links, paths, connections	connectors
storage, memeories	sets, families
variable physical object	mathematical variable
control instance, controller	val-function, assignment function
references, location, time	indices
process in physical time	process in logical time

Physical processes carried out by physical objects proceed in physical times. Dependencies $\prec$ of components in the logical model correspond to irreflexive time orderings $<_{\vee\wedge}$ of the physical processes performed by the physical objects which correspond to the $\prec$ ordered logical objects. With $\prec$ compatible logical (algorithmic) times correspond to extensions of $<_{\vee\wedge}$ in physical times. This holds for processes on system level as well as for processes on application (production) level. All active input-output objects are oriented by their input - output direction.

System control has to provide and to allocate the necessary physical system resources at the right time, has to control their operational processes, and has to release them after a task is finished for possible reassignment to another task. An utilized and released object can be relocated, however, utilization may have changed the object properties. A useful practice is to underlay the logical model to the time parameterized physical model, to keep book which system resource is at which time at which place employed for operation and which others are available to be assigned to a task. The available resources are kept in a (logical) "pool", incoming or preprogrammed demands are collected in an ordered demand list. If a demand matches with an available resource, the system resource management allocates this resource to the demanding task and updates the pool and the demand list.

Example (simplified computer architecture for an arithmetic function, control operators omitted): 2 controllable arithmetic processors and 8 buses/memories are available. The time parameterized logical problem structure represents the program. The time they are allocated, the physical objects superpose the corresponding logical objects. The demand on resources for the next execution step is determined by one program step "look ahead".

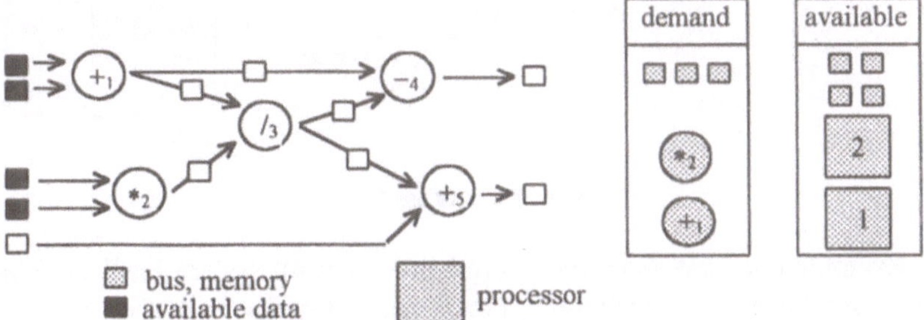

Fig.21a. Initial state

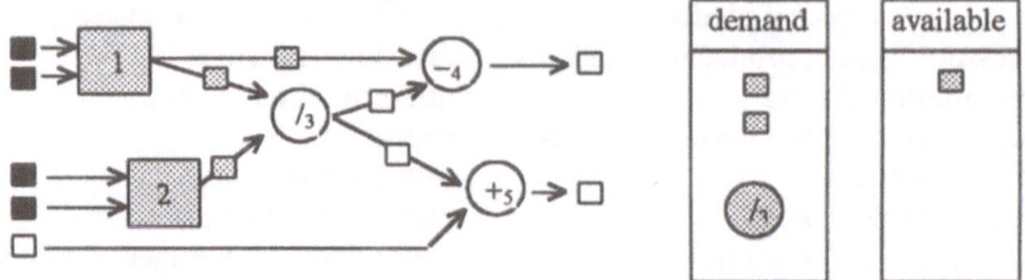

Fig.21b. Resources allocated, 1st process part started

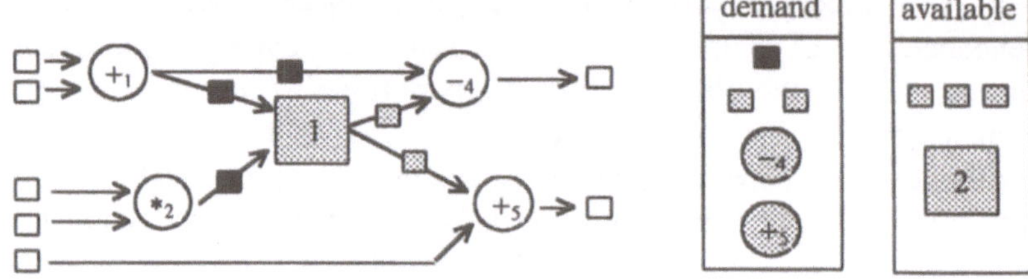

Fig.21c. Resources allocated, 2cd process part started

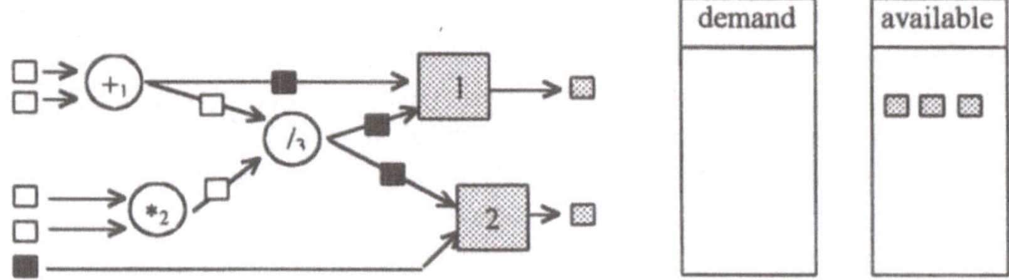

Fig.21d. Resources allocated, 3rd process part started

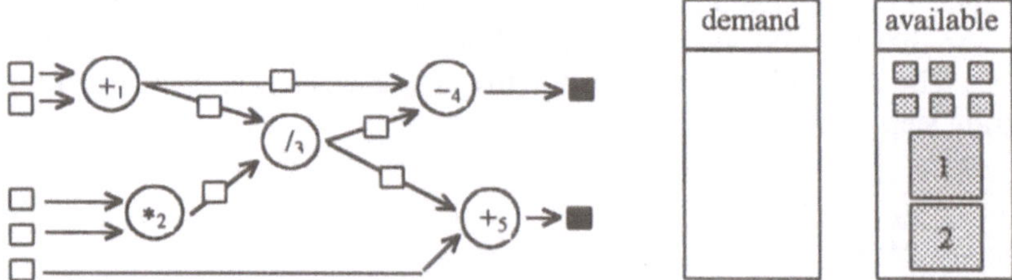

Fig.21e. Process terminated

The Example does not show the control processes on system level in detail nor the control processes of the arithmetic processors. Assumed is that all processes run in finite time and that an operation (arithmetic or transfer) can be executed if operands and operation code and the necessary hardware (processor, bus/memory) are available. Otherwise the execution is suspended until the needed resources are provided. In the 1st and 3rd step we have concurrency. If the overall process has a continuation then possibly *all* memory contents have to be saved for later use. If the process is interrupted within one process step, all so far obtained memory contents have to be saved for possible later restart of this step. Release and later relocation of a resource to the same part process goes on cost of efficiency. In our Example processor 1 and its input and output memories stay on the same "path". A multi-processor architecture can be based on this principle (R.F. Albrecht (1980), L. Bic (1992)). Control - and processing - activities "flow" through the system and at each process step occupy part of the resources. If sufficient resources are available to be permanently assigned to the operations, then in our Example each of the 3 computational steps can belong to a separate computation. At most three computations can go on concurrently, the first already being in step 3, the second in step 2, the third in step 1 ("pipelining" principle). The two outputs can be input to a subsequent system. This applies in particular, if the subsequent system is of the same structural shape and the two outputs are "fed back" to give two of the five inputs, the other three supplied from another source (recurrence of program structure).

Practiced strategies for control of the *production* (*computation*) *execution* are:
Imperative (or *procedural*) *mode*:
The algorithm for production operations and their control operations as well as the allocation of the resources for their execution are predetermined by external control input ("program") to the system control ("do exactly what is ordered at the time").
Autonomous mode:
A system modeled by a concatenable structure variable $((\mathrm{var}\ f_p)_{p \in \mathrm{var}\ P}, E(\mathrm{var}\ C), \mathrm{var}\ C)$ is stepwise and evolutionary supplied by external control and data input and its own intermediate execution results. As soon as the "preconditions" for further control- or production operations are satisfied these operations are executed. Preconditions are: Control-/processing- data are available, are in the domain of the operation, if clock synchronized, clock signal available, the performing physical resources are allocated and ready for execution ("physical modus ponens"). This test is performed by an active system object "precondition tester" , centralized or distributed, which sends the request for resources to another active system object "resource manager", which according to availability allocates the requested resources. Duration of organizational and processing operations and availability of resources determine the (mutually asynchronous) flow of control-/processing- data through the system ("do right now whatever can be done", "data flow" principle). Notice: The "preconditions" or "admittance rule" tester has in general to test data and the control parameters ("operation code") of the next following operation with respect to *global* concatenability. If the requested operation is composite, it may be started with *part* of input data, *part* of control data and *part* of resources allocated. The executability can depend on the values of the input data (R.F. Albrecht, 1996). For illustration see Fig.21f.

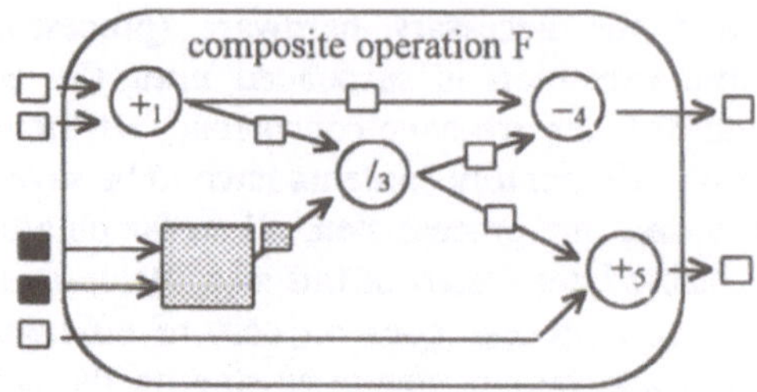

Fig.21f. F has partially assigned variables and resources and can get started

Mixed Mode:
Imperative mode on lower and autonomous mode on higher hierarchical level are in use.

Example: Like the previous, one step in autonomous mode shown in Fig.22. Process control and processor control objects are omitted.

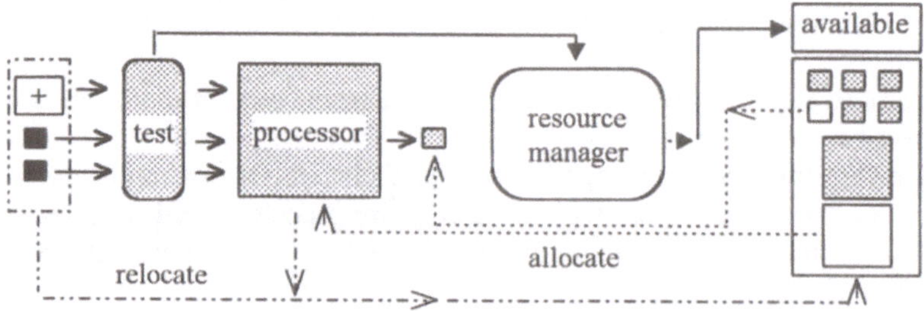

Fig.22

Steps of the system control process are:

(1) search for all functionally independent (assigned) operations in the program file, if found, call of tester,

(2) the tester checks these operations (in parallel) for executability ("admittance rule": operation code and data complete and in the permitted domain), if executable, the tester calls the resource manager for needed resources,

(3) the resource manager allocates the needed resources to as many of the operations as resources are available and updates the list of free resources,

(4) the tester calls the process control to start selected input-, processor-, output-operations,

(5) the process control starts these processes and updates the program file,

(6) resources of terminated processes are released.

If the problem structure is properly posed (no "dead locks", i.e. no cycles), if all processes have a finite duration, and if there are sufficient resources available to solve the problem at all (no "starving to death"), then all process steps can be performed within finite time.

We consider the time behavior of any controllable active object *obj* (e.g. a control object, a processing object, a transportation object). The input process is $x_{uS} =_{def} ((x_{it})_{t\in S_{[i]}})_{i\in u}$. If activated, *obj* generates the output process $y_{vT} =_{def} ((y_{jt})_{t\in T_{[j]}})_{j\in v}$, $S =_{def} \bigsqcup (S_{[i]})_{i\in u}$, $T =_{def} \bigsqcup (T_{[j]})_{j\in v}$, $\bigwedge i\in u, j\in v\ (S_{[i]}, T_{[j]} \in (\mathbf{T}, <, \sqcap, \sqcup))$, S, T are assumed to be bounded and non-empty. A consequence of causality is $S <_{\vee\wedge} T$, often the case is $S <_{\wedge\vee\wedge} T$. A visualization is given in Fig.23.

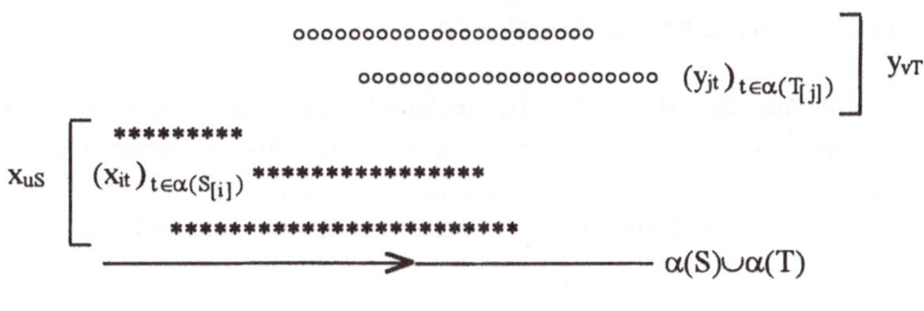

Fig.23

A simplified model of states and events of *obj* is illustrated in Fig.24:

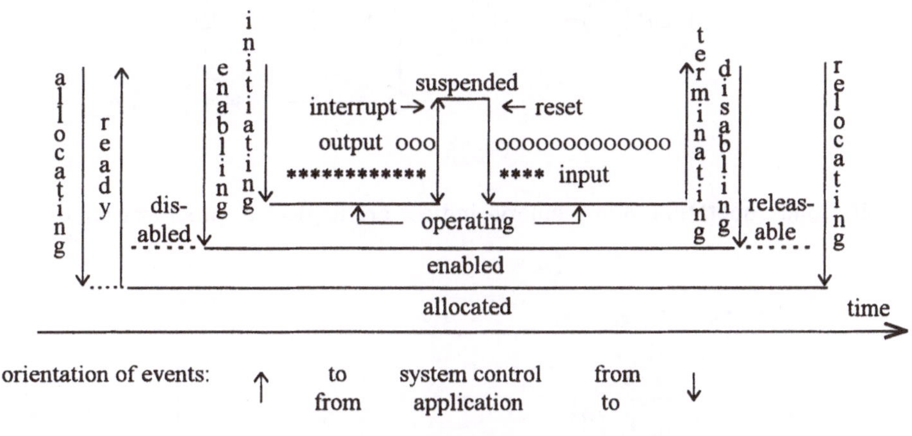

Fig.24

Shown is one interrupt with suspension, there may be more or none. Interrupts may be caused by failures in the operation of *obj* as well as by exceptional events affecting the system. Different from suspension is irregular termination with removing *obj* from the task.

Unparameterized physical model

We introduced parameterization to distinguish objects by their logical, spatial, temporal and other properties. As treated in **6.**1, passing from a parameterized structure S to an unparameterized representation corresponds formally to the application of the *set of* operation. If S is structured by an irreflexive ordering $<$ on its index set, *set of* is a homomorphism with respect to a relation $\rightarrow$, which in general is not acyclic. The homomorphism causes a loss of structural information. An unparameterized physical system model is closer to the engineering practice because it contains the persisting physical objects only once. To give such a model an unambigeous interpretation and to express the dynamic behavior, a parameterization is now attempted by attributes, labels, floating tokens, flags, colors etc.. The usual "firing rules" apply only to simple cases with low granularity. As the difficulties in understanding recursive data flow computations showed, this is no alternative to a sound dynamic system analysis and modeling.

Example (like the one in Fig.21a): The arithmetic function can be computed by a single programmable processor as depicted in Fig.25. Only 3 memories a, b, c are needed, the result is in b, c. Between operations $*_2$, $/_3$, $-_4$ "register transfer" is applied. We have 5 algorithmic steps. Processor control is omitted.

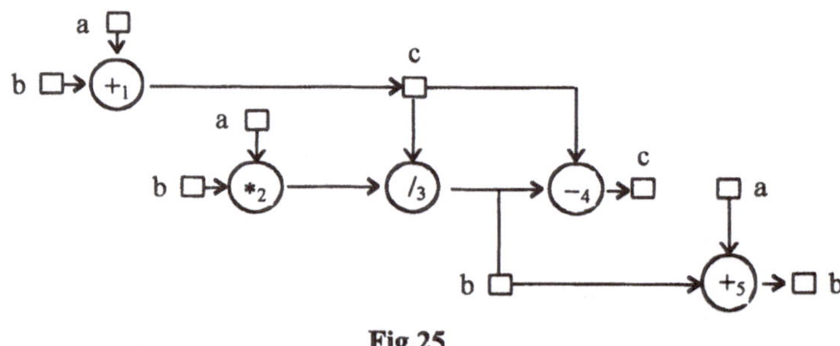

Fig.25

In Fig.26, time-, operation- and memory- indices are neglected, processor control is included.

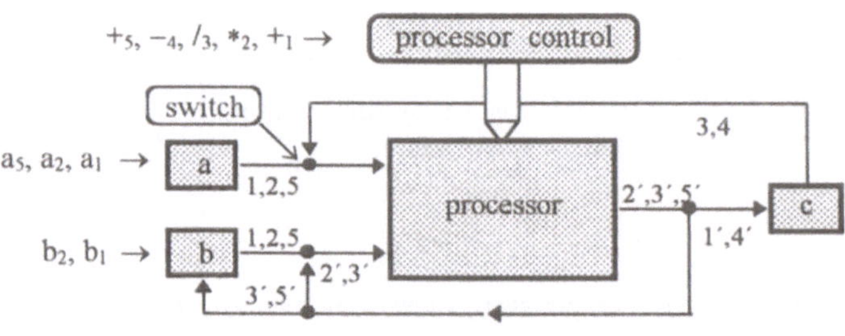

Fig.26

The Figure shows "hardware", "wiring" and necessary "switches" at the marked nods, data- and control-flow are partly labeled, for n the input number, n′ is the corresponding output number, synchronization by a clock and switch control are not depicted.

7 Applications

7.1 Approximation of functions

Let there be given var y = var f(var x) with var f : $\{f_{[p]} : X_{[p]} \rightarrow Y_{[p]} \mid p \in P\}$, var x : $\widetilde{X} =_{def} \{x_{[pq]} \mid x_{[pq]} \in X_{[p]} \wedge (p,q) \in R \subseteq P \times Q\}$, var y : $\widetilde{Y} =_{def} \{f_{[p]}(x_{[pq]}) \mid (p,q) \in R\}$, and let $pr_1 R = P$, $pr_2 R = Q$. We define $\bigwedge p{\in}P$ ((var x_p : $X_{[p]}$) $\wedge$ ($Q_{[p]} =_{def} pr_2$ (cut({p}) R))). We introduce a logical time $T = \{t(n) \mid n \in (\mathbf{N_o}, <)\}$, with an ordering induced by $<$ on $\mathbf{N_o}$, to model the assignment and evaluation process for one function. Starting at t(0), the assignment steps are at

t(0): given $p(t(0)) \in P$,
t(1): assignments $f(t(1)) =_{def}$ val(p(t(0)), var f), $X(t(1)) =_{def} X_{[p(t(0))]}$,
t(2): given $q(t(2)) \in Q_{[p(t(0))]}$,
t(3): assignment $x(t(3)) =_{def}$ val(q(t(2)), var $x_{p(t(0))}$),
t(4): evaluation and assignment $y(t(4)) =_{def} f(t(1))(x(t(3)))$.

To simplify notations we used t(n) for $t_{[n]}$, x(t(n)) for $x_{[p(t(n))]}$, f(t(n)) for $f_{[p(t(n))]}$.

This 5-step cycle can be repeated under agreement that each of the left hand side objects keeps its value constant from assignment until next assignment. Then we consider for coarser logical time $S = \{t(m) \mid m = 5n \wedge n \in [1,2,..N)\} \subset T$, $N \leq \infty$, the processes $(p(t(m)))_{t(m)\in S}$, $(q(t(m)))_{t(m)\in S}$, and the processes $(f(t(m)))_{t(m)\in S}$, $(x(t(m)))_{t(m)\in S}$, $(y(t(m)))_{t(m)\in S}$. For suitable control processes and under further (sufficient) assumptions, given in the sequel, the processes $(f(t(m)))_{t(m)\in S}$, $(x(t(m)))_{t(m)\in S}$, $(y(t(m)))_{t(m)\in S}$ converge with increasing logical time:

Let $X =_{def} \bigcup_{t(m)\in S} X(t(m))$ with $X(t(m)) =_{def} X_{[p(t(m))]}$ be a uniform space, generated by the filter base $D_X \subset pow(X{\times}X)$, which means $diag(X{\times}X) \subseteq \bigcap D_X$, $\bigwedge d{\in}D_X \bigvee c{\in}D_X$ $(c \subseteq \overset{-1}{d})$, $\bigwedge d{\in}D_X \bigvee c{\in}D_X$ $(\mathbf{K}_{join}(c,c) \subseteq d)$, and let $\bigcup_{t(m)\in S} f(t(m))(X(t(m))) \subseteq Y$, Y being a complete uniform space, generated by the filter base $D_Y \subset pow(Y{\times}Y)$. Further, let be $\bigwedge t,t'{\in}S$ $(t < t' \Rightarrow X(t) \subseteq X(t'))$. Then $\bigwedge t,t'{\in}S$ $(t < t' \Rightarrow f(t')(X(t)) \subseteq f(t')(X(t')))$ and $\bigwedge x{\in}X \bigvee t(x){\in}S \bigwedge t{\in}S$ $(t(x) \leq t \Rightarrow x \in X(t))$. We assume

(1) $\bigwedge d_Y{\in}D_Y \bigvee t(d_Y){\in}S \bigwedge t(d_Y) \leq t{\in}S$ $(x{\in}X(t) \Rightarrow (f(t(d_Y))(x), f(t)(x)) \in d_Y)$ (uniform Cauchy filter base),

(2) $\bigwedge d_Y{\in}D_Y \bigvee t(d_Y){\in}S \bigwedge t(d_Y) \leq t{\in}S \bigvee d_X(d_Y){\in}D_X \bigwedge x,x'{\in}X(t)$ $((x,x') \in d_X(d_Y) \Rightarrow (f(t)(x), f(t)(x')) \in d_Y)$ (uniform continuity).

From (1) follows because of the completeness of Y: $\bigwedge x{\in}X \bigvee y(x){\in}Y \bigwedge d_Y{\in}D_Y \bigvee t(d_Y,x){\in}S \bigwedge t(d_Y,x) \leq t{\in}S$ $((f(t)(x), y(x)) \in d_Y)$ (uniform convergence). If Y is separated (i.e. $diag(Y{\times}Y) = \bigcap D_Y$), y(x) is unique and consequently a function $F: X \rightarrow Y$ is defined by $x \mapsto y(x)$. If Y is not separated, a set $Y_{[x]}$ of many $y(x)$ may exist and all $\widetilde{F} = (x, y(x))_{x \in X} \in \prod_{x \in X} Y_x$ are equivalent with respect to $\bigcap D_Y$. From (2) follows by the axioms of uniform structures and with (1): $\bigwedge d_Y \in D_Y \bigvee t(d_Y) \in S \bigwedge t(d_Y) \leq t \in S \bigvee \widetilde{d}_Y(d_Y) \in D_Y \bigwedge x,x' \in X(t)$ $(((f(t)(x'), f(t)(x)) \in \widetilde{d}_Y(d_Y) \wedge (f(t)(x), y(x)) \in \widetilde{d}_Y(d_Y)) \Rightarrow (f(t)(x'), y(x)) \in d_Y)$ and $\bigwedge x,x' \in X \bigwedge d_Y \in D_Y$ $\bigvee d_X(d_Y) \in D_X$ $(((x, x') \in d_X(d_Y)) \Rightarrow (y(x), y(x')) \in d_Y)$ (for separated Y this implies uniform continuity (stability) of F).

7.2 Adaptive systems

Iterations. $f: X \to X$, X a complete uniform space, parameter set P = logical time $S = \mathbf{N}_0$, $f_{[n]} = f^n$, $n \in \mathbf{N}_0$, $x_{[n+1]} := f^n(x_{[0]})$. Necessary for convergence is that the filter base $\{\{x_{[n]}, x_{[n+1]},...\} \mid n \in \mathbf{N}_0\}$ has a limit set X^*, $X^* = f(X^*)$. If $X^* = \{x^*\}$, then $x^* = f(x^*)$ is a fixpoint. Sufficient for convergence is the Cauchy criterion (f a contractive mapping).

Feed back controlled systems. S discrete or continuous, $p(t)$ an external parameter which also incorporates "noise", "disturbances" and which parameterizes var f, $x(t)$ the set state, $\widetilde{x}(t)$ the actual state, $d(x(t), \widetilde{x}(t))$ a defect distance, depending on the underlying topology, $q(t) = \varphi(t, x(t), d(x(t), \widetilde{x}(t)))$ the defect correction which parameterizes var x, $x(q(t)) \in$ dom $f(p(t))$, $\widetilde{x}(t') = f(p(t), x(q(t)))$. The structure is shown in diagram Fig. 27.

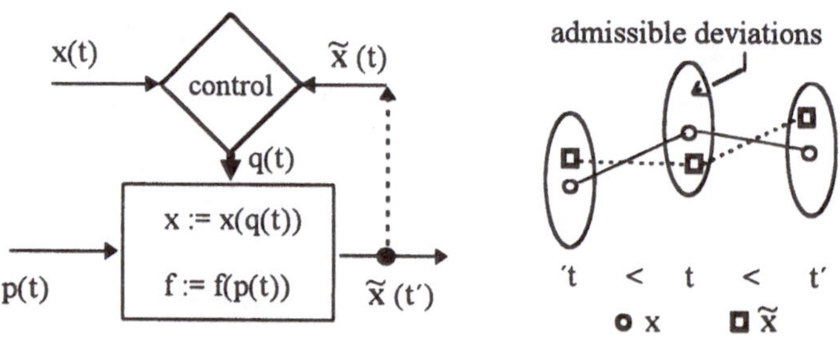

Fig.27

As a generalization, φ can also depend on past states and defects (in the continuous case on derivatives). A stability analysis is necessary.

Goal oriented systems. We consider a function variable $f(\text{var } p) : \{f(p) \mid p \in P \wedge f(p) : X \to Y\}$, var $x : \{x(q) \mid q \in Q \wedge x(q) \in X\}$, i.e. $R = P{\times}Q$, var $y := f(\text{var } p, \text{var } x) : Y$. A function $f^*: X \to Y$ is specified by its properties and assumed is $f^* \in$ dom var f. Wanted is a $p^* \in P$ such that $f^* = f(p^*)$. For an (approximate) solution one tries to construct a filter base $\{\{p_{[n]}, p_{[n+1]}, ...\} \mid n \in \mathbf{N}_0\}$ converging to p^*.

Example (artificial neural networks): For given finite L we consider for $l \in L$, $X_l \subseteq \prod_{i \in I_{[l]}} X_{l[i]}$, $Y_l \subseteq \prod_{j \in J_{[l]}} Y_{l[j]}$, var w_l : $W_l \subseteq \prod_{k \in K_{[l]}} W_{l[k]}$, function variables f_l (var w_l) : $X_l \rightarrow Y_l$, $\emptyset \neq C \subseteq \bigcup_{l \in L} J_l \times \bigcup_{l \in L} I_l$. l numbers "neurons"/"nodes", f_l(var w_l) is a "node function", parameters w_l are "weigths", C determines the "synaptic network" of "feed forward type". It is assumed that concatenation $\mathbf{K}((f_l(\text{var } w_l))_{l \in L}, E(C), C)$ with E(C) an equality relation yields a function variable f(var w) : X → Y, var w : $\mathscr{W} \subseteq \prod_{l \in L} W_l$. Classically, the structure $((f_l(\text{var } w_l))_{l \in L}, C)$ is subdivided into ordered connected "layers", $\bigwedge l,l' \in L \; (f_{[l]} = f_{[l']})$ and $f_{[l]}$ has a threshold behavior (see also the article of D.W. Pearson and G. Dray in this volume and the Example in **6.3** with Fig.17).

Anticipatory systems. As pointed out in **5.3**, **6.4**, **6.5**, a physical process $r_U = (r_t)_{t \in \alpha(U)}$, U a physical time, can have continuations var $\Delta r_{\text{var } V}$ into an unknown future time var V, U $<_{\wedge\wedge}$ var V. The domain of var $\Delta r_{\text{var } V}$ is in general itself a variable var dom var $\Delta r_{\text{var } V}$, depending on process history r_U, on not foreseen external events $(\text{var } e_t)_{t \in \alpha(\text{var } W)}$, var W $\sqsubseteq$ var V, and on the not yet known availability of physical resources needed for process continuation.

In case the object which controls var dom var $\Delta r_{\text{var } V}$ and var $\Delta r_{\text{var } V}$ (the "controller"), is "intelligent" and knows from memorized previous experience, or anticipates by extrapolation, or observes or senses some of the control parameters of var dom var $\Delta r_{\text{var } V}$ and var $\Delta r_{\text{var } V}$ already during time U, partial assignments to these variables can be made concurrently with (part of) process r_U. For example, possibly needed resources can be provided by the resource manager, input processes for the expected process continuation can be started. Though anticipated partial assignments to the domain and the variable can afterwards turn out to be redundant or to be wrong and then have to be corrected, anticipation can improve efficiency and can speed up the continuation of processes. A visualization is shown in Fig.28.

Anticipatory systems were first theoretically discussed in a book by R. Rosen 1985, although they were known and applied long before.

Examples:

(1) Numerical approximation processes: "predictor-corrector"-, "convergence acceleration"-, "Richardson extrapolation"-method. The mathematician is the controller, during one algorithmic step he can from intermediate results extrapolate future quantities for a better conditioned next algorithmic step.

(2) Micro-processor technology: "Cache", "look ahead", "snooping", "prefetch" techniques. The controller is the processor's control unit, which by implemented routines looks several instructions ahead, starts data transfers, decoding, pipelining of instruction phases concurrently to the instantly processed instruction. If the instruction causes a jump in the instruction sequence, anticipation failed.

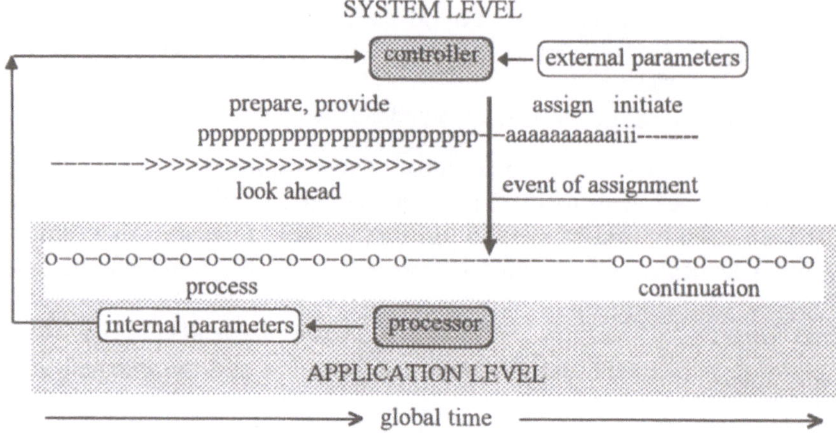

Fig.28

7.3 Topologized functional systems

In a functional physical system with variable objects we have essentially two kinds of functional processes:

(1) realization of functions and transports,

(2) realization of assignments to variables (val-functions, control operations).

In both cases the performing functional objects can be composite and hierarchically structured. As outlined in chapter **4**, neighborhoods to an object x in the domain X of the functional object $f: X \to Y$ can be introduced to express (generalized) distances or similarities between x and another object $x^* \in X$. This can be done by a valuated filter base $\mathbf{B}_x$ on pow X with $x \in \bigcap \mathbf{B}_X$. If we want to have the neighborhoods of distinct elements of X be comparable, the domain X of f has to be equipped with a uniform topological structure as described in **4.3**. If f is extended to a homomorphism $\tilde{f}$, mapping for any $x \in X$ the neighborhoods $\mathbf{B}_x$ of x into pow Y, $\tilde{f}(\mathbf{B}_x)$ is a filter base $\mathbf{B}_y$ of neighborhoods of $y = f(x)$. If Y carries a uniform topological structure in which y has neighborhoods given by a filter base $\mathbf{C}_y$, then continuity of f at x is defined by $\mathbf{B}_y$ finer than $\mathbf{C}_y$, i.e. to each $C \in \mathbf{C}_y$ exists a $B \in \mathbf{B}_x$ with $\tilde{f}(B) \subseteq C$. Instead of pow Y any complete lattice may be considered. These topological concepts are employed in cases (1) and (2) to express functional behavior for unprecise or uncertain ("fuzzy") data and control parameters.

Example (selection by properties as described in **2.4**, **6.3**): In **6.3** we considered the binary relation $S = ((y_j, w_{ji}), (x_i, v_{ji}))_{ji \in U}$ of valuated objects $y_j, x_i, (j,i) \in U \subseteq J \times I$, the relation $M = (w_{ji}, v_{ji})_{ji \in U}$ of the valuations $w_{[ji]} \in \mathbf{W}$, $v_{[ji]} \in \mathbf{V}$, and defined for a given "input pattern" $(x_i, \tilde{v}_i)_{i \in \tilde{I}}$, for $\tilde{I} \cap I_{[j]}, \tilde{v}_{[i]} \in \mathbf{V}$, an assignment $(x_i, \tilde{v}_i)_{i \in \tilde{I}} \mapsto ((y_j, w_{ji})_{i \in \tilde{I}_{[j]}})_{j \in J(\tilde{I})}$ by selecting those (j,i) for which $(\eta(\tilde{v}_{[i]}, v_{[ji]}) = "t")$, i.e. $\tilde{v}_{[i]}$

$= v_{[ji]}$ (perfect match). We introduced $\eta : \mathbf{V}\times\mathbf{V} \rightarrow \mathbf{B} = (\{"t", "f"\}, "f" < "t")$, $\mathbf{B}$ is a (trivial) ideal base. Let $\mathscr{V}$ be a filter base on $\mathbf{V}\times\mathbf{V}$ which defines a uniform structure on $\mathbf{V}$ (for definition see **7.1**), and which is in addition a complete lattice. Then $\mathscr{V}$ defines uniform generalized functional distances $d_\cap$, $d_\cup$ on $\mathbf{V}$ (see **4.1**). We assume, $(\mathbf{W}, +, *)$ as considered in **6.3** has neutral elements $\mathbf{o}$ for $+$, $\mathbf{e}$ for $*$, and a partial ordering $\leq$ which is compatible with $+$ and $*$. Let be $h:\mathbf{V}\times\mathbf{V} \rightarrow \mathbf{B}$ $=_{def} \{w \mid w \in \mathbf{W} \wedge \mathbf{o} \leq w \leq \mathbf{e}\}$ with set extension $\tilde{h}$ being an antimorphism of $\mathscr{V}$ with $\tilde{h}(\bigcap\mathscr{V}) = \mathbf{e}$, $\tilde{h}(\mathbf{V}\times\mathbf{V}) = \mathbf{o}$. We define $\eta(v, v') =_{def} \tilde{h}(d_\cap(v, v'))$ for all $(v, v') \in \mathbf{V}\times\mathbf{V}$. Then for $\tilde{I}_{[j]} = \tilde{I} \cap I_{[j]}$, $J(\tilde{I}) =_{def} \{j \mid j \in J \wedge \tilde{I}_{[j]} \neq \varnothing\}$, $\eta(\tilde{v}_{[i]}, v_{[ji]})$ expresses the similarity of $\tilde{v}_{[i]}$ with $v_{[ji]}$ (approximate match) and can be used as valuating factor to w_{ji}: $(x_i, \tilde{v}_i)_{i\in\tilde{I}} \rightarrow ((y_j, (\eta(\tilde{v}_{[i]}, v_{[ji]}) * w_{[ji]})_{ji})_{i\in\tilde{I}_{[j]}})_{j\in J(\tilde{I})}$.

Valuated filter bases $\mathbf{B}_x$ on pow X can e.g. be obtained by probability distributions. Let X be a probability space $(X, \mathscr{S}, P)$ with $\mathscr{S}$ a σ-field, P a probability measure on $\mathscr{S}$. Then $P:\mathscr{S} \rightarrow [0,1] \subset \mathbf{R}$, $X \in \mathscr{S}$, $P(X) = 1$, $P(\varnothing) = 0$. A filter base $\mathbf{B}_x = \{\mathscr{B}_{[k]} \mid \mathscr{B}_{[k]} \in \mathscr{S} \wedge k \in K\}$ can then be valuated by $P(\mathscr{B}_{[k]})$. If $h: X \rightarrow \mathbf{R}$, $\mathscr{R}$ the Borel field on pow $\mathbf{R}$, and $\overset{-1}{h}: \mathscr{R} \rightarrow \mathscr{S}$, then a filter base on $\mathscr{R}$ maps on a filter base on $\mathscr{S}$ which can be valuated by P (the general case without reference to probability was considered in **4.3**).

Example (2-dimensional normal distribution): Let be $X = \mathbf{R}\times\mathbf{R}$, $g = (2\pi\,\sigma_1\,\sigma_2)^{-1} \exp(-(x^2/\sigma_1{}^2 + y^2/\sigma_2{}^2)/2)$ the probability density of P. We set $r^2 = (x^2/\sigma_1{}^2 + y^2/\sigma_2{}^2)$, $z = g(x,y)$. Then $z = (2\pi\,\sigma_1\,\sigma_2)^{-1} \exp(-r^2/2)$. We have $(x_0, y_0) \mapsto z_0 = g(x_0, y_0)$, $\overset{-1}{g}(z_0) = \{(x, y) \mid (x^2/\sigma_1{}^2 + y^2/\sigma_2{}^2) = r_0{}^2\}$, $\overset{-1}{g}([z_0, (2\pi\,\sigma_1\,\sigma_2)^{-1}]) = E(r_0{}^2) =_{def} \{(x, y) \mid (x^2/\sigma_1{}^2 + y^2/\sigma_2{}^2) \leq r_0{}^2\}$. $P(\overset{-1}{g}([z_0, (2\pi\,\sigma_1\,\sigma_2)^{-1}])) = \int_{E(r_0{}^2)} g(x,y)\,dxdy = -\int_{r_0^2/2}^{0} (r^2/2) \exp(-r^2/2)\, d(r^2/2) = (1 - (1 + (r_0{}^2/2)) \exp(-r_0{}^2/2))$ is a valuation of $E(r_0{}^2)$. The (monotone) filter base $\{(E(r_0{}^2) \mid 0 \leq r_0{}^2 < \infty\}$ converges to $\{(0,0)\}$. n-dimensional normal distributions with general quadratic forms as well as simplifying approximations to these are widely applied in pattern recognition and classification.

References

Albrecht, R.F. (1980): Concept of a Multi-Processor Processing Unit. Computing 25, pp 1-16.

Albrecht, R.F. (1992): Design of an optional Control- and Dataflow Multiprocessor, Workshop on Parallel Processing, TU Clausthal, Informatik Report 92/1, pp 63-76.

Albrecht, R.F. (1994a) "Some Basic Concepts of Object Oriented Databases", System Science, vol.20, No.1, Wroclaw, pp 17-30.

Albrecht, R.F. (1994b): Modelling of Computer Architectures, Proc. of the First International Conference on Massively Parallel Computing Systems (MPCS), IEEE Computer Society Press, pp 434-442.

Albrecht, R.F. (1994c): Modelling of Discrete Systems, Proc. of the 1994 Human Interaction with Complex Systems Symposium, SPIE, Greensboro, N.C., pp 204-214.

Albrecht, R.F. (1995): On the Structure of Discrete Systems, L. N. in Computer Science, Pichler, F., Diáz, R.M., Albrecht, R. (Eds.), Computer Aided Systems Theory, Springer, Heidelberg-New York, pp 3-18.

Albrecht, R.F. (1996a): The Structure of Discrete Systems, Trends in Theoretical Informatics, Albrecht, R., Herre, H. (eds.), Austrian Computer Society, Vienna, pp 127-144.

Albrecht, R.F. (1996b): Hierarchical Data Flow Concepts, Proc. 2cd Int. Conf. on Massively Parallel Computer Systems, Ischia 96, IEEE Computer Society Press, Los Alamitos, Ca., pp 8-14.

Albrecht, R.F. (1996c) Remarks on logical modelling of manufacturing systems, Proc. of the Workshop "A new mathematical approach to Manufacturing Engineering", publ. by EICAS Automazione, Torino, pp 85-89.

Albrecht, R.F.(1997): Systems with Topological Structures. Proc. 1st Int. Conf. on "Computing Anticipatory Systems" (CASYS´97) Liège, Belgium, Aug.11-15, ed. D. Dubois.

Albrecht, R.F., Németh G. (1997): A Generic Model for Knowledge Bases. Proc. 9th Int. Conf. on Systems Research, Informatics and Cybernetics. Baden-Baden, Germany, Aug.18-23, ed. G.E. Lasker.

Allen, J. F. (1984): Towards a General Theory of Actions and Time. Artificial Intelligence 23, pp 123-154.

Bic, L. (1992): A Process-Oriented Model for Efficient Execution of Data Flow Programs, in Data Flow Computing, J.A.Sharp ed., Ablex Publ. Corp., Norwood, N.J., pp 332-347.

Bourbaki, N. (1951): Topologie Général, Act. Sc. In. 1142, Hermann & C^{de}, Paris, pp 40, 41.

Kulisch U. (1975): Formalization and Implementation of Floating Point Arithmetic, Computing 14, pp 323-348.

Kulisch U., W.L. Miranker (1981). Computer Arithmetic in Theory and Practice, Acad. Press, New York.

Kulisch U. (1996). Numerical Algorithms with Automatic Result Verification. Am. Math. Soc. Lect. in Appl. Math., 32.

Mesarovic M.D., Y. Takahara (1989): Abstract Systems Theory, L.N. in Control and Information Science, Springer, Heidelberg-New York.

Rosen R. (1985): Anticipatory Systems, Pergamon Press, New York.

Zadeh L.A (1965): Fuzzy Sets, Information and Control 8.

Acknowledgment. For useful discussions on the subject my thanks are due to G. Németh (Budapest), R. Moreno-Díaz (Las Palmas) and B. Quatember (Innsbruck), partners in an Austrian-Hungarian and an Austrian-Spanish cooperation project.

AN INTRODUCTION TO DISCRETE EVENT MODELING FORMALISMS

Fernando J. Barros and Bernard P. Zeigler

1. Introduction

Modeling formalisms offer several advantages over non formal methods in the representation of simulation models. These advantages include a compact and rigorous notation. Models specified in a modeling formalism are also independent of any simulation language.

Discrete event models play an important role in representing a large number of systems. The *Discrete Event System Specification* (DEVS) described in (Zeigler 76) and (Zeigler 84), is a compact notation to represent discrete event systems. DEVS models behave like independent units and thus support modular model building.

The DSDEVS formalism (Barros 95) provides the representation of networks of DEVS models. DSDEVS networks can have static or dynamic nature. Several DEVS models can also be coupled together while maintaining the same properties of simple models. This characteristic allows the construction in a hierarchical manner, a key factor for representing complex systems. The application of dynamic structure concepts to other type of systems, namely, discrete time and differential equation systems, is described in (Barros 97b).

This chapter is organized as follow: Section 2 presents the DEVS and the DSDEVS formalisms and a detailed version of DEVS models. Section 3 presents the DELTA simulation environment and describes in detail the implementation of basic models. Section 4 gives a detailed example of a static structure network. Section 5 describes structural inheritance and compares the DEVS approach with more classical state approaches. Section 6 provides a detailed example of a dynamic structure network.

2. Modeling Formalisms

We will introduce two modeling formalisms to represent discrete event systems: the Discrete Event System Specification (DEVS) created by (Zeigler 1976; Zeigler 1984) and the Dynamic Structure Discrete Event System Specification (DSDEVS) introduced by (Barros 1996; Barros 1997). The DEVS formalism will be used to represent basic systems and the DEDEVS formalism will be used to represent the network of DEVS basic models. The DSDEVS formalism can represent dynamic structure models and static structure networks as a special case.

2.1 The DEVS Basic Model

Basic models are defined in the DEVS formalism (Zeigler 1976), by the structure

$$M = \langle X, S, s_0, Y, \delta_{int}, \delta_{ext}, \lambda, \tau \rangle$$

where

X is the set of input values
S is the set of sequential states
s_0 is the initial partial state
Y is the set of output values
$\delta_{int}: S \rightarrow S$ is the internal transition function
$\delta_{ext}: Q \times X \rightarrow S$ is the external transition function, where
$Q = \{(s,e) \mid s \in S, 0 \leq e \leq \tau(s)\}$ is the total state set
e is the time elapsed since last transition
$q_0 = (s_0,0)$ is the initial state
$\lambda: S \rightarrow Y$ is the output function
$\tau: S \rightarrow \mathbf{R}_0^+$ is the time advance function

The external transition function describes how a model changes its state in response to external values. A model in state s, will change to state $\delta_{ext}(s,e,v)$, when it receives an input value v, after e time units elapsed since its last transition. The internal transition function describes how the model changes its state if no external value is received during the time interval specified by the time advance function. A model entering state s will change to state $\delta_{int}(s)$ if it does not receive any input during the interval $\tau(s)$. The values sent from a model to the output are given by the function λ.

EXAMPLE. We will show how to describe a simple model in the DEVS formalism. A basic delay is a model that holds an input item and releases it after a time interval. When this component holds an item it is said to be active. It is passive otherwise. Incoming items are ignored if the model is active. The basic delay is defined by

$$M_{BD} = \langle X, S, s_0, Y, \delta_{int}, \delta_{ext}, \lambda, \tau \rangle$$

where

$X = A = \{a_1, a_2, \ldots\}$ is a set of items
$S = \{\text{active}, \text{passive}\} \times \mathbf{R}_0^+ \times A^{\varnothing}$
$A^{\varnothing} = A \cup \{\varnothing\}$ is the event closure of A
$s_0 = (\text{passive}, \infty, \varnothing)$
$Y = A$
$\delta_{ext}((\text{passive}, \infty, \varnothing), e, v) = (\text{active}, T, v)$
$\delta_{ext}((\text{active}, \sigma, a), e, v) = (\text{active}, \sigma - e, a)$
$\delta_{int}(\text{active}, \sigma, a) = (\text{passive}, \infty, \varnothing)$
$\lambda(\text{active}, \sigma, a) = a$
$\tau(\text{active}, \sigma, a) = \sigma$

In Fig. 1 are represented the input, output and state trajectories of a basic delay when it receives a sequence of three inputs a_1, a_2, a_3.

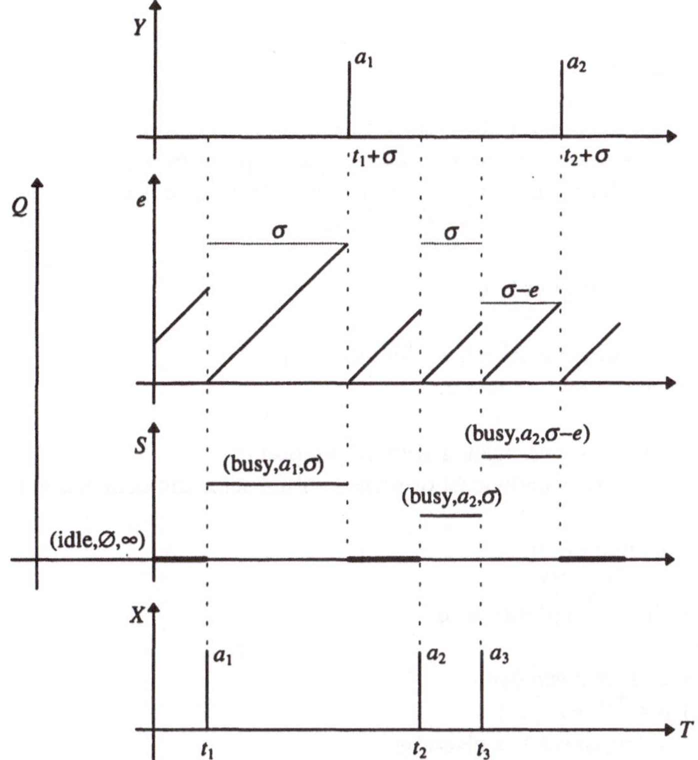

Fig. 1. Model trajectories.

Starting from the bottom, item a_1 arrives at time t_1 and it causes the phase to change from idle to busy. The model has a fixed delay of T and thus a_1 leaves the system at time $t_1 + \sigma$, or equivalently, when $e = \sigma$. The model next state is given by the internal transition $\delta_{int}(a_1,\sigma)$ = (passive,∞,∅). Item a_2 arrives at time t_2 and starts service. However, before a_2 has finished its service a new item a_3 arrives at time $t_3 < t_2 + \sigma$. Because the component has no buffer capabilities the new item must be ignored, thus the external function just changes σ to a new value $\sigma - e$, where e is the time a_2 already spent in service. Item a_2 finishes the service as if a_3 never arrived to the model.

A model is said to be in a *passive* state, s if $\tau(s) = \infty$, and is *active* otherwise. Internal and external transitions are commonly called *events*. An event can be defined as a change in model sequential state. External events change model sequential state in a consequence of an external cause. An internal event is triggered by the model itself without any external cause.

The DEVS formalism deals with arbitrary sets for inputs and outputs, and thus it can represent models in any specific domain. In a computer domain, input and output sets can be jobs, in modeling traffic these sets can represent cars and trucks, and in an ecology model we may need to represent soil and trees. The DEVS formalism is open to any interpretation of the domain and all kind of models can be represented within this framework.

2.2 Structured Input/Output

We have described the input and output sets of basic and network models using the unstructured sets X and Y. From an implementation point of view, it is convenient to structure these sets by the use of ports. A *structured interface* is defined by

$$IO = \langle {}^{X}P,\ \{{}^{X}_{p}V\}, {}^{Y}P\ , \{{}^{Y}_{q}V\} \rangle$$

where

${}^{X}P$ is the set of input port labels

for each $p \in {}^{X}P$

${}^{X}_{p}V$ is a value set associated with input port p

${}^{Y}P$ is the set of output port labels

for each $q \in {}^{Y}P$

${}^{Y}_{q}V$ is a value set associated with output port q

The sets ${}^{X}P$ and ${}^{Y}P$ are usually a set of names with a semantic connection to the model domain.

Given a structured interface

$IO = \langle {}^{X}P, \{{}^{X}_{p}V\}, {}^{Y}P, \{{}^{Y}_{q}V\} \rangle$

we can obtain the standard interface

$IF = \langle X,Y \rangle$

where the input set is given by

$X = \{(p,v) | \ p \in {}^{X}P,\ v \in {}^{X}_{p}V \ \}$

and the external output set Y is given by

$Y = \{(p,v) | \ p \in {}^{Y}P,\ v \in {}^{Y}_{p}V \ \}$

2.3 The DSDEVS Network Model

The DSDEVS formalism provides a specification for networks of DEVS models. In contrast to other simulation formalisms, the structure of a DSDEVS network can change over time. The DSDEVS *dynamic structure network* is defined by

$$DSDEVN_{\Delta} = \langle X_{\Delta}, Y_{\Delta}, \chi, M_{\chi} \rangle$$

where

Δ is the network name

X_{Δ} is the network input value set

Y_{Δ} is the network output value set

χ is the network executive

M_{χ} is the model of χ

The DSDEVS network is defined with a special component, the *network executive* χ. M_{χ}, the model of the executive, is a basic model and is defined by the DEVS structure

$$M_{\chi} = \langle X_{\chi}, S_{\chi}, s_{0,\chi}, Y_{\chi}, \delta_{int_{\chi}}, \delta_{ext_{\chi}}, \lambda_{\chi}, \tau_{\chi} \rangle$$

The information about the dynamic structure network is located in the state of the executive. Changes in the network-related state variables will be mapped onto changes in the network structure. A state $s_{\chi} \in S_{\chi}$ is defined by

$$s_\chi = (D^\chi, \{M_i^\chi\}, \{I_i^\chi\}, \{Z_{i,j}^\chi\}, \Xi^\chi, V^\chi)$$

where

D^χ is the set of components
M_i^χ is the model of component i, for all $i \in D^\chi$
I_i^χ is the set of influencees of i, for all $i \in D^\chi \cup \{\chi, \Delta\}$
$Z_{i,j}^\chi$ is the i-to-j translation function, for all $j \in I_i^\chi$
Ξ^χ is the select function
V^χ represent other state variables

The executive variables must obey to the following constraints:

$\chi \notin D^\chi$
$M_i^\chi = \langle X_i, S_i, s_{0,i}, Y_i, \delta_{int_i}, \delta_{ext_i}, \lambda_i, \tau_i \rangle$ is a basic DEVS model, for all $i \in D^\chi$
$i \notin I_i^\chi$, for all $i \in D^\chi \cup \{\chi, \Delta\}$
Ξ^χ: $(2^{D^\chi \cup \{\chi\}} - \{\}) \rightarrow D^\chi \cup \{\chi\}$, with $\Xi(D') \in D'$
$Z_{\Delta,j}^\chi$: $X_\Delta \rightarrow X_j$
$Z_{i,\Delta}^\chi$: $Y_i \rightarrow Y_\Delta$
$Z_{i,j}^\chi$: $Y_i \rightarrow X_j$
if $Z_{k,\chi}^\chi(y) \neq \varnothing$ then $Z_{k,j}^\chi(y) = \varnothing$
for $k \in D^\chi \cup \{\Delta\}$ and for all $j \in I_k^\chi - \{\chi\}$
$\varnothing$ represents the absence of value (the null value)

The last constraint states that if the executive receives an input then no other model can receive an input. This prevents structure from becoming ill-defined. This constraint and the select function have been removed by the newly defined DSDE formalism, (Barros 1997).

The 5-tuple $(D^\chi, \{M_i^\chi\}, \{I_i^\chi\}, \{Z_{i,j}^\chi\}, \Xi^\chi)$ defined in the executive state is referred to as the *network structure*. Any change in one of these variables is called a *change in structure*. A discussion of closure under coupling of the DSDEVS formalism, necessary to build models in a hierarchical and modular manner and the description of the DSDEVS abstract simulators, needed to interpret model implicit dynamic behavior, can be found in (Barros 1995; Barros 1996ac).

2.4 Structured Network Models

Output to input functions are not very amenable to a computer implementation. Usually one defines a set of *quasi-identity* functions commonly know by *links*. Let L^χ be the set of network links. We say that port (p_i, n_i) of component i is *linked* to port p_j of component j, if

$$((i,p_i),(j,p_j)) \in L^\chi$$

Given the set L^χ we can obtain $\{Z_{i,j}^\chi\}$, the set of output to input functions, by

$$Z^{\chi}_{i,j}(p_i,v) = \begin{cases} (p_j,v) & \text{if } ((i,p_i),(j,p_j)) \in L^{\chi} \\ \varnothing & \text{otherwise} \end{cases}$$

We indirectly define the select function by assigning priorities to each component. For each component $i \in D^{\chi}$ we assign a priority $R^{\chi}_i \in \mathrm{I}^{+}_{0}$. The select function is defined by

$$\Xi^{\chi}(D') = j, \text{ such that } R^{\chi}_j = \max\{R^{\chi}_i \mid i \in D'\},\ D' \subseteq D^{\chi}$$

With the help of the link set L^{χ} and the priority set $\{R^{\chi}_i\}$ we can define the network structure by

$$(D^{\chi},\{M^{\chi}_i\},L^{\chi},\{R^{\chi}_i\})$$

From this tuple we can retrieve the structure in the standard format

$$(D^{\chi},\{M^{\chi}_i\},\{I^{\chi}_i\},\{Z^{\chi}_{i,j}\},\Xi^{\chi})$$

where for each $i \in D^{\chi}$

$$I^{\chi}_i = \{j \mid \exists\, p_i,p_j\text{: } ((i,p_i),(j,p_j)) \in L^{\chi}\}$$

The set $\{Z^{\chi}_{i,j}\}$ and the select function Ξ^{χ} are obtained as defined before.

Connections must obey to some constraints. An output port q of component $i \neq \Delta$ can be connected to an input port p of component $j \neq \Delta$, if ${}^{Y}_{q}V_i \subseteq {}^{X}_{p}V_j$. An input port p of network Δ can be connected to an input port q of component $i \neq \Delta$, if ${}^{X}_{p}V_{\Delta} \subseteq {}^{X}_{q}V_i$. An output port p of component $i \neq \Delta$ can be connected to an output port q of network Δ, *if* ${}^{Y}_{p}V_i \subseteq {}^{Y}_{q}V_{\Delta}$.

3. The DELTA Simulation Environment

The DELTA modeling and simulation environment is an implementation of the DSDEVS formalism in the Smalltalk language (Barros 96b) that uses the structured concepts described in Sections 2.2 and 2.4. We will introduce this modeling environment by describing the DELTA implementation of the basic delay presented in Section 2.1. The block diagram of the basic delay is represented in Fig. 2, where the input and output ports in and out are represented by small rectangles.

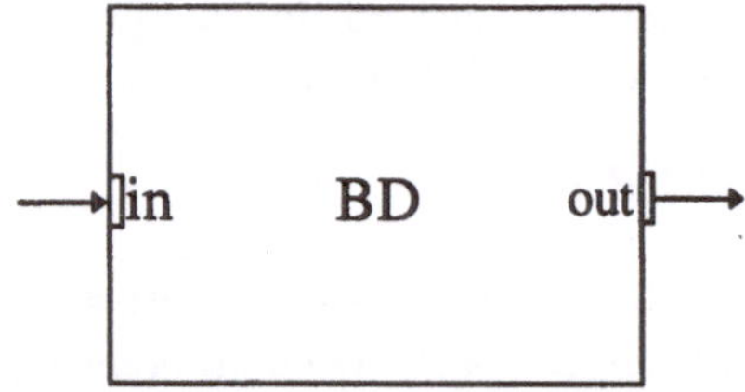

Fig. 2. Block diagram of the basic delay.

The model can be described using the structured input/output notation by

$$M_{BD} = \langle\{\text{in}\},\{A\},S,s_0,\{\text{out}\},\{A\},\delta_{int},\delta_{ext},\lambda,\tau\rangle$$

Model initial state is defined by the method initialize given by

```
initialize
   job := nil.                          "No job"
   self passivateIn: #idle.
```

where the method **passivateIn:** Π, passivates the model in phase Π, that is,

$$\text{passivateIn:}((\textit{phase},\textit{sigma},\textit{job}),\Pi) = (\Pi,\infty,\textit{job})$$

The external transition is defined by the method **external: elapsedTime port:** ***aPort*** **value:** ***aJob*** given by

```
external: elapsedTime port: aPort value: aJob
   phase = #idle ifTrue: [
      job := aJob.
      ^self holdIn: #busy sigma: 5
   ].
   phase = #busy ifTrue: [
      ^self continue: elapsedTime
   ].
```

where the method **holdIn:** Π **sigma:** t, holds the model in phase Π during time t,

$$\text{holdIn:sigma:}((\textit{phase},\textit{sigma},\textit{job}),\Pi,t) = (\Pi,t,\textit{job})$$

and the method **continue:** e, holds the model in the current phase without changing component previously scheduled time, i.e.,

$$\text{continue:}((\textit{phase},\textit{sigma},\textit{job}),e) = (\textit{phase},\textit{sigma}-e,\textit{job})$$

The internal transition is defined by the internal method

```
internal
   phase = #busy ifTrue: [^self passivateIn: #idle].
```

and the output function is defined by

```
output
   (phase = #busy) ifTrue: [^Port port: #out value: job].
```

where the message **Port:** p **value:** v, creates an output at port p with a value v.

A partial view of DELTA class hierarchy is represented in Fig. 3. Each basic model has a corresponding class in DELTA. Class **BasicModel** is the root class for all basic models. New basic models inherit the basic operations for time advance. For example, the **BDelay** class models basic delays and defines external, internal, start, and output functions. The variables ***phase*** and ***sigma*** are also defined for all basic models.

Network models are defined by the corresponding executive classes. Class **Executive** is the root class for all domain executives. This class provides all basic operations for structure management. These operations are of two types: creating an initial network structure; and, changing network structure at simulation run time. The class executive provides an empty structure by class method **structure**, that must be overridden by its subclasses. Methods to add and remove models and links are also provided.

Executive instance methods provide the basic operations to change model structure at run time. Its subclasses must define the methods for structure management. For example the class **Switch** defines an external method to change the structure. To create a network we must send the **new:** message to the respective executive class. This message creates all the models defined in network definition, and returns an instance of class Network.

In the subsequent sections we will show examples on how to build models in the DELTA environment.

```
BasicModel(Instance)  Variables            BasicModel(Class)
    timeAdvance          phase                 new:priority:
    holdIn:sigma:        sigma                 new:
    continue:
    pasivateIn:                              Executive(Class)
                                                 new:
  Executive(Instance)                            new: priority:
      link:port:to:port:                         structure
      addModel:                                  addModel:class:
      removeModel:                               removeModel:
      unLink:port:from:port:                     link:port:to:port:
                                                 replaceClass:by:
                                                 unLink:port:from:port:

      EF(Instance)                               EF(Instance)
                                                     structure

      Pipeline(Instance)                         Pipeline(Instance)
                                                     structure

          BSPipeline(Instance)                       BSPipeline(Class)
                                                         structure

      Switch(Instance)                           Switch(Class)
          external:port:value:                       structure

  BDelay(Instance)                           BDelay(Class)
      start
      external:port:value:
      internal:
      output:

Network(Instance)                            Network(Class)
```

Fig. 3. Partial DELTA class hierarchy.

4. Static Structure Models

A common combination of models is to connect them together in pipeline arrangement. This situation occurs when an item must pass through a sequence of operations. A car wash machine, a set of operations in a lab, or the sieve of Erastosthenes for finding prime numbers (Barros 97), can be described by this topology. In Fig. 4 is depicted a pipeline of 3 basic delays already described. The output of a delay is connected to the input of another. The network model receives items in port in and send them through port out. An incoming item passes through models M1, M2 and M3 before it can leave the pipeline. The pipeline model is formally defined by

$$DSDEVN_{\text{Pipeline}} = \langle X_{\text{Pipeline}}, Y_{\text{Pipeline}}, \chi, M_\chi \rangle$$

where

$X_{\text{Pipeline}} = \{\text{in}\} \times A$

$y_{\text{Pipeline}} = \{\text{out}\} \times A$

The model of the executive is given by

$$M\chi = \langle \{s_{0,\chi}\}, s_{0,\chi}, \tau_\chi \rangle$$

and the initial state $s_{0,\chi}$ is defined by

$$s_{0,\chi} = (D^\chi, \{M_i^\chi\}, L^\chi, \{R_i^\chi\}, (\text{passive}, \infty))$$

where

$D^\chi = \{\text{M1,M2,M3}\}$

$\{M_i^\chi\} = \{M_{\text{M1}}, M_{\text{M2}}\ M_{\text{M3}}\}$

$L^\chi = \{((\text{P,in}),(\text{M1,in})),$
$((\text{M1,out}),(\text{M2,in})),$
$((\text{M2,out}),(\text{M3,in})),$
$((\text{M3,out}),(\text{P,out}))\}$

$\{R_i^\chi\} = \{0,1,2,3\}$

$phase = \text{passive}$

$\text{sigma} = \infty$

This model has just one state and is always passive. Thus this network has a static structure. The time advance function is defined by

$\tau_\chi((D^\chi, \{M_i^\chi\}, L_1, \{R_i^\chi\}, (phase, sigma)) = sigma$

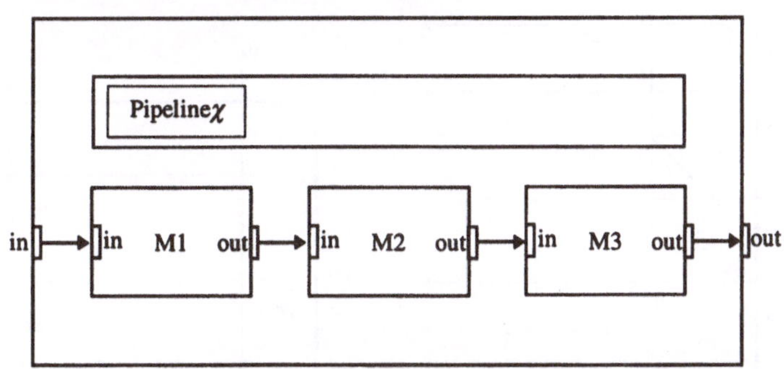

Fig. 4. Pipeline of delays.

The pipeline is implemented in DELTA by the structure method defined in class Pipeline. This network has three components of the class BDelay named M1, M2 and M3, where M1 is the first element and M3 the last element of the pipeline.

```
Pipeline::structure
  |s|
  s := super structure.                          "The empty structure"
  s addModel: #M1 class: BDelay.                 "Adds M1 of class BDelay"
  s addModel: #M2 class: BDelay.                 "Adds M2 of class BDelay"
  s addModel: #M3 class: BDelay.                 "Adds M3 of class BDelay"
  s link: #Network port: #in to: #M1 port: #in."Links network port #in to M1 port #in"
  s link: #M1 port: #out to: #M2 port: #in.      "Links M1 port #out to M2 port #in"
  s link: #M2 port: #out to: #M3 port: #in.      "Links M2 port #out to M3 port #in"
  "Links M3 port #out to network port #out"
  s link: #M3 port: #out to: #Network port: #out.
  ^s
```

Two basic operations were used to define this network model: adding models and create links. The method **addModel: *aName* class: *aClass***, adds a new model called *aName* of class *aClass*. When the simulation starts it creates instances of the classes that define the model. The method **link: *aName* port: *aPort* to: *bName* port: *bPort***, creates a link between port *aPort* of model *aName* and port *bPort* of model *bName*. The name #Network is a reference to the network model. To create a pipeline we send the message **new: *aName* priority: *aNumber*** to the class **Pipeline**. The corresponding method creates the pipeline executive and all models and connections, and it then returns an instance of class **Network**. During this process instances of the classes that define the network elements are created. A sample trajectory of the pipeline is represented in Fig. 5.

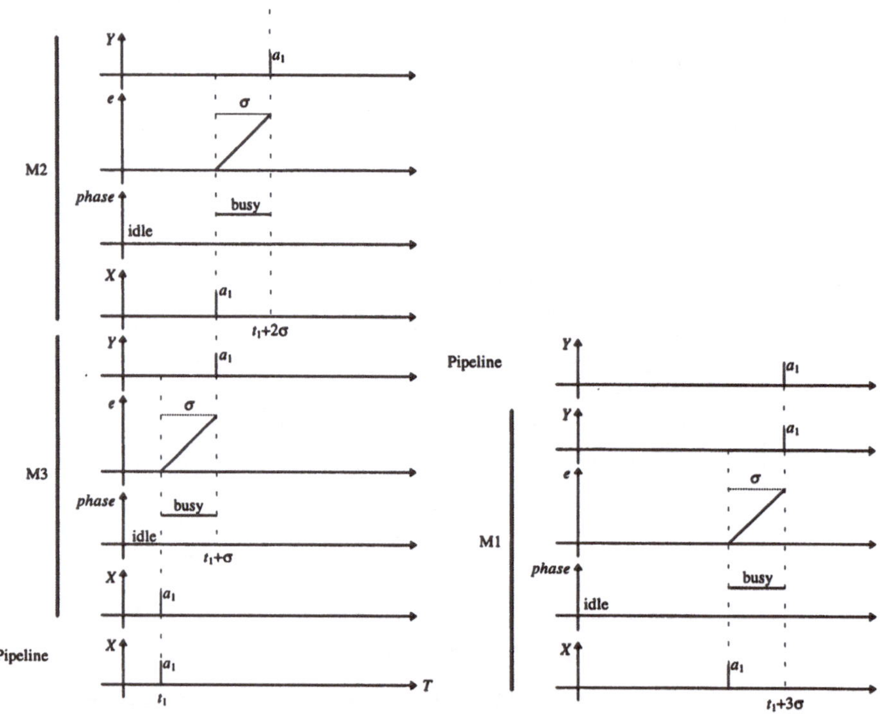

Fig. 5. Model trajectories.

For simplicity we assume that service time is the same for all the components in the pipeline. Item a_1 arrives at time t_1 to the network and it is sent to component M1. After σ time interval it leaves M1 and enters M2 that sends a_1 to M3 after a σ interval. When M3 releases a_1, this item is sent to the network that delivers it to the outside. This value can then be sent to another component if the pipeline is a part in a larger model.

The *Experimental Frame* (EF) concept has been introduced by (Zeigler 84), and provides a separation between models and experiments. The EF can be defined as the set of conditions under which the model will be operated. It is a very powerful concept since it permits to test the same model under different conditions of experimental frames, or to use the same EF to test different models. In workload applications, the overall goal of the EF is to generate items to a model and then compute some performance measures such as the time items spend inside the model.

A standard design for the EF, (Zeigler 1990), consists in a combination of a Generator (G), an Acceptor (A) and a Transducer (T) and is represented in Fig. 6. A generator creates new items following some probability distribution and releases them into the model. The Transducer keeps a record of item exit and return to the EF, and based on this results it computes several performance measure like cycle time and mean number of items in the model. The Transducer controls the length and the number of simulations to achieve an established confidence level.

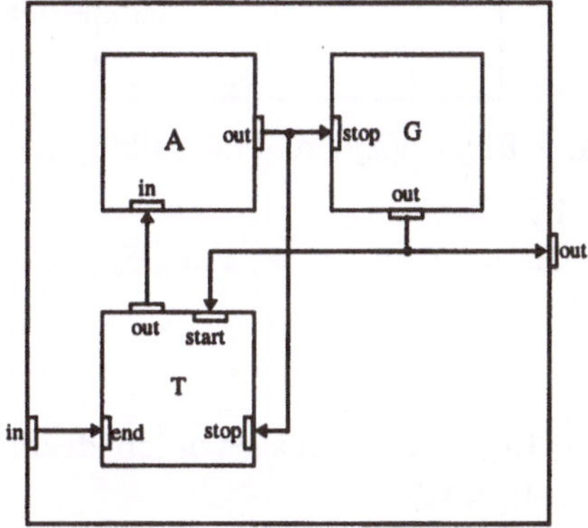

Fig. 6. Experimental frame.

To test the Pipeline we only need to connect it with the EF and run a simulation. This new model is represented in Fig. 7.

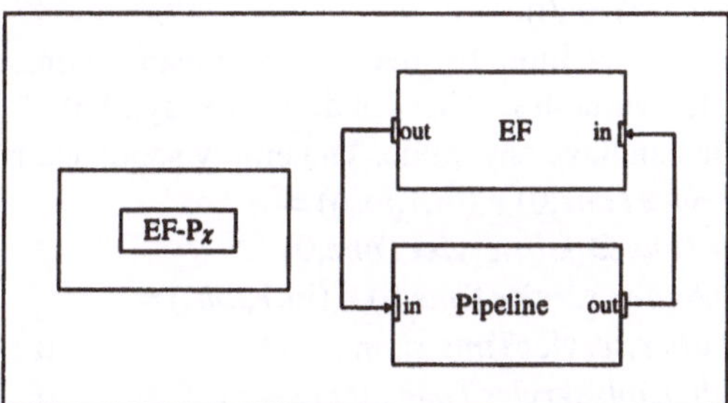

Fig. 7. Running a simulation.

5. Structural Inheritance

To show how structural inheritance works, we will first introduce a model with buffer capabilities and then we will show how a pipeline of buffered servers can be built from the pipeline of basic delays described in Section 2.1.

5.1 The Balking Server Model

The balking server represented in Figure, is a server with a limited buffer capacity. In this server, all incoming jobs that arrive when the buffer is full must balk. When a job finishes service, it is removed from the buffer and the server starts serving of another job in the buffer. If the buffer is empty the server becomes idle.

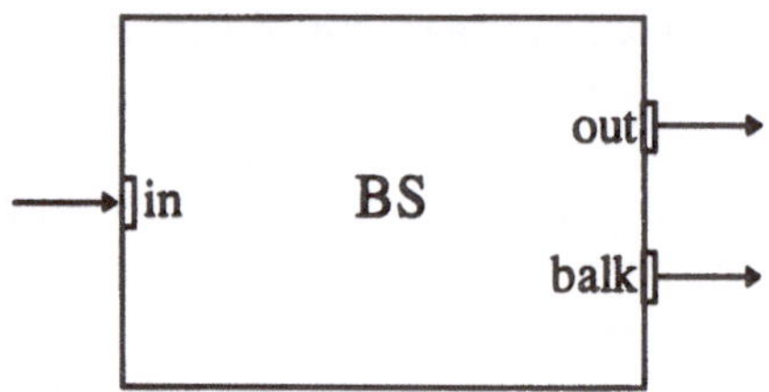

Fig. 8. Block diagram of the balking server.

The server is represented by

$$BS = ({}^{X}P, {}^{X}V, S, s_0, {}^{Y}P, {}^{Y}V, \delta_{ext}, \delta_{int}, \lambda, \tau)$$

where

${}^{X}P = \{\text{in}\}$

${}^{Y}P = \{\text{out,balk}\}$

${}^{X}_{\text{in}}V = {}^{Y}_{\text{out}}V = {}^{Y}_{\text{balk}}V = A = \{a_1, a_2, a_3, \ldots\}$, is a set of job identifiers

$S = \Phi \times \mathbf{R}_0^+ \times A^\varnothing \times A^\varnothing \times \mathbf{R}_0^+ \times \mathbf{R}_0^+$

$\Phi = \{\text{busy,idle,balk}\}$ is a set of model phases

A state $s \in S$ is given by

$s = (\mathit{phase}, \mathit{sigma}, \mathit{jobToReject}, \mathit{jobInService}, \mathit{serviceTime}, \mathit{remainingTime})$

The initial state s_0 is given by

$s_0 = (\text{idle}, \infty, \phi, \phi, \mathit{serviceTime}, 0)$

We define the external transition by declarative means, considering the different combinations of state variables. The underscore symbol "_" means that the corresponding variable can have any value. The empty sequence is represented by Λ.

$\delta_{ext}((\text{idle}, \infty, \varnothing, \Lambda, \mathit{serviceTime}, 0), e, (\text{in}, 1, \mathit{job})) =$

$(\text{busy}, \mathit{serviceTime}, \varnothing \mathit{job}, \mathit{serviceTime}, 0)$

$\delta_{ext}((\text{busy}, \mathit{sigma}, \varnothing, \mathit{buffer}, \mathit{serviceTime}, 0), e, (\text{in}, 1, \mathit{job})) =$

$(\text{balk}, 0, \mathit{job}, \mathit{buffer}, \mathit{serviceTime}, \mathit{sigma} - e)$ if $\text{size}(\mathit{buffer}) > 3$

$(\text{balk}, 0, \varnothing, \mathit{buffer}.\mathit{job}, \mathit{serviceTime}, \mathit{sigma} - e)$ if $\text{size}(\mathit{buffer}) <= 3$

The time advance function is defined by

$\tau(\mathit{phase}, \mathit{sigma}, \mathit{jobToReject}, \mathit{buffer}, \mathit{serviceTime}, \mathit{remainingTime}) = \mathit{sigma}$

The internal function is defined by

$\delta_{int}(\text{busy}, \_, \_, \mathit{head} \backslash \mathit{tail}, \mathit{serviceTime}, \_) =$

$(\text{idle}, \infty, \varnothing, \Lambda, \mathit{serviceTime}, 0)$ if $\text{size}(\mathit{tail}) = 0$

$(\text{busy}, \infty, \varnothing, \mathit{tail}, \mathit{serviceTime}, 0)$ if $\text{size}(\mathit{tail}) > 0$

$\delta_{int}(\text{balk}, \_, \_, \mathit{buffer}, \mathit{serviceTime}, \mathit{remainingTime}) =$

$(\text{busy}, \mathit{remainingTime}, \varnothing, \mathit{buffer}, \mathit{serviceTime}, \mathit{remainingTime})$

The output function is defined by

λ(busy,_,_,*head*|*tail*,_,_) = (out,1, *head*)
λ(balk,_,*jobToBalk*,_,_,_) = (balk,1,*jobToBalk*)

The method start defines model initial state and is given by

```
start
   jobToBalk := nil.                       "The null value"
   buffer := OrderedCollection new.        "Creates an empty buffer"
   remainingTime := 0.
   serviceTime := 5.
   self passivateIn: #idle.                "Model starts in phase idle"
```

The external function is given by

```
external: e port: p value: v
   (p = #in) & (phase = #idle) ifTrue: [
         buffer add: v.
         ^self holdIn: #busy sigma: serviceTime
   ].
   (p = #in) & (buffer size < 3) ifTrue: [
         buffer add: v.
         ^self holdIn: #busy sigma: (sigma - e)
   ].
   (p = #in) & (phase = #busy) ifTrue: [
         remaingTime = sigma - e.
         jobToBalk := v.
         ^self holdIn: #balk sigma: 0
   ].
```

The output function is defined by

```
output
   phase = #busy ifTrue: [^Port: #out value: (buffer first)].
   (phase = #balk) ifTrue: [^Port: #balk value: jobToBalk].
```

The internal function is defined by

```
internal
   jobToBalk := nil.
   phase = #busy ifTrue: [
         remainingTime := 0.
         buffer remove.              "Remove the first element of the buffer"
         (buffer size > 0) ifTrue: [
                  ^self holdIn: #busy sigma: serviceTime
         ]
         ifFalse: [^self passivateIn: #idle]
   ].
   phase = #balk ifTrue: [
          ^self holdIn: #busy sigma: remainingTime
   ].
```

The server moves among three different phases: *idle, busy* and *balk*, as represented by the *Phase Transition Diagram* (PTD) of Fig. 9. We remember that a model phase is just a state variable and is usually used as a compact designation for a set of states. For example a model in phase busy can have 1, 2 or 3 items in the buffer. The model starts in phase idle and jumps to phase busy after the arrival of a value. If an input causes a buffer overflow the model interrupts the current active job and changes to phase balk to reject the item. After this phase the model goes to phase busy. We note that with the help of the *e* variable (elapsed time) it is possible to resume the preempted job in service at the time it was interrupted. When the server finishes the last job it goes to the idle phase. We made no assumption about service time that can be any arbitrary distribution.

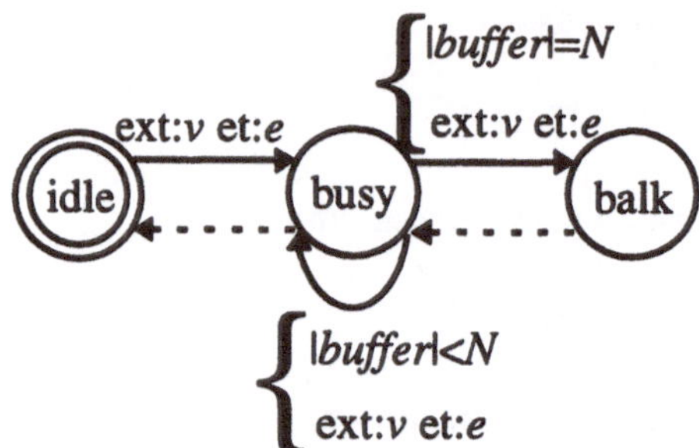

Fig. 9. Phase transition diagram of a balking server.

5.2 Power of DEVS Representation

In Fig. 10 is represented the *State Transition Diagram* (STD) of the same model. All distributions are assumed to be exponential and so there is no need to represent the elapsed time since last transition. The input rate is λ and the service rate is μ. The STD diagram provides a detailed representation of the systems where there is a different state for each different number of jobs in the buffer. Thus if buffer size is limited to N the corresponding STD will have $N + 1$ states.

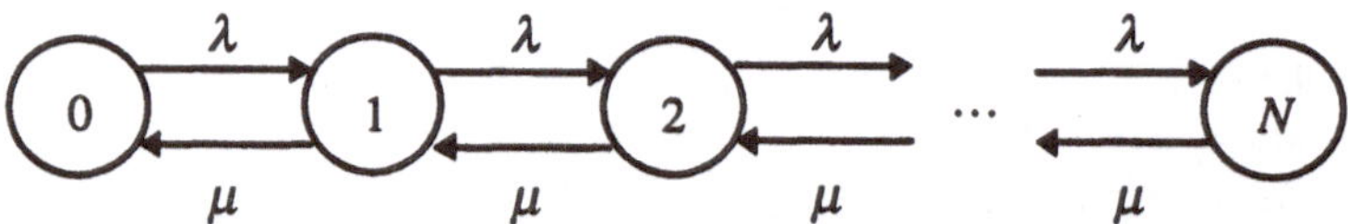

Fig. 10. State transition diagram for a balking server.

We will see why the STD **does not** provide a good notation for complex models. If we connect n basic models with $N + 1$ states in a pipeline fashion we will obtain a state transition diagram with $(N + 1)^n$ states. However, the modular approach taken by the DEVS formalism we only deal n independent models each one with its set of phases. While constructing a state transition diagram for the first case is virtually impossible for large models, building DEVS models in a hierarchical manner will be possible for almost any system. We note that DEVS models are descriptive by nature and no analytical tool has been created yet to provide answers to the same type of questions that STD can solve. Simulation models trade analytical power for power of representation. STD however have a major limitation: they can only deal with exponential distributions (Kleinrock 76; Cassandras and Strickland 89). Because this distribution has the memoryless property, STD do not represent the time elapsed since the last transition. Thus STD cannot deal with arbitrary distributions.

5.3 Model Reuse

An important feature of modeling environment is the ability that it offers for reusing existing models. In DELTA models can be reused by structural inheritance, (Barros 96). A new network model is built upon an existing one and only differences need to be coded. In Section 4 we have described a pipeline of three basic delays and we will show how a pipeline of balking servers can be built using structural inheritance. This network is depicted in Fig. 11 and can be defined in DELTA by

```
BSPipeline::structure
  |s|
  s := super structure.                    "Obtains the parent structure"
  s replaceClass: BDelay by: BServer.      "Replace class BDelay by class Bserver"
  s link: #M1 port: #balk to: #Network port: #balk.    "Create new links"
  s link: #M2 port: #balk to: #Network port: #balk.
  s link: #M3 port: #balk to: #Network port: #balk.
  ^s
```

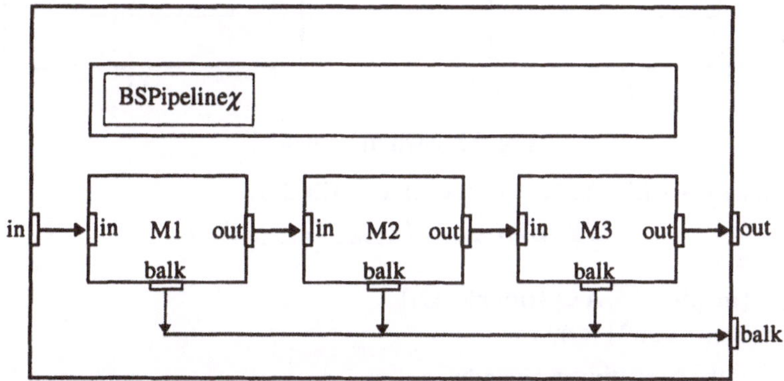

Fig. 11. Creating a new pipeline by structural inheritance.

The call replaceClass: BDelay by: BServer, replaces all the occurrences of models of class BDelay by models of class BServer. After this call we have only BServers connected in the same way as BDelay's. The remaining lines only add the 3 new links from the balk port of the new servers to the balk port of the network. In Fig. 11 these new links are represented in gray. The pipeline of balking server has reused most of the definition of the pipeline of basic delays. Model reuse provides not only savings in development time, but also reduces the effort for debugging new models because they are generally built upon tested models.

6. Dynamic Structure Models

The DSDEVS formalism can be used to represent networks that change their structure at simulation run time. We will describe a dynamic structure switch network, represented in Fig. 12. This network controls the inputs to basic delays and it has two configurations. In the first configuration, represented in Fig. 12a, incoming items arriving at port in:1 and in:2 are sent to BD1 and BD2, respectively. In the second one, incoming items arriving at port in:1 and in:2 are sent to BD2 and BD1, respectively. The change between the two configurations is initiated by values arriving at port in of the Switch executive.

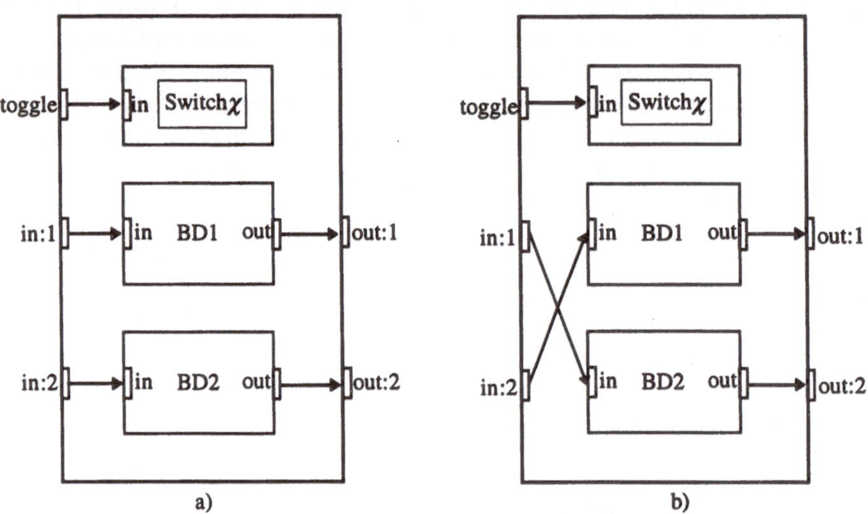

Fig. 12. Switch network.

The Switch network is a DSDEVS network described by

$$DSDEVN_{\text{Switch}} = \langle X_{\text{Switch}}, Y_{\text{Switch}}, \mathcal{X}, M_\chi \rangle$$

where

$X_{\text{Switch}} = \{\text{in:1}, \text{in:2}\} \times A \cup \{(\text{toggle}, \varnothing)\}$

$Y_{\text{Switch}} = \{\text{out:1}, \text{out:2}\} \times A$

The model of the executive is given by

$$M_\chi = \langle X_\chi, S_\chi, s_{0,\chi}, \delta_{ext\chi}, \tau_\chi \rangle$$

where

$X_\chi = \{\text{in}\} \times \varnothing$

A state $s_\chi \in S_\chi$ is defined by

$$s_\chi = (D^\chi, \{M_i^\chi\}, L^\chi, \{R_i^\chi\}, (phase, sigma, C))$$

where

$D^\chi = \{\text{BD1}, \text{BD2}\}$

$\{M_i^\chi\} = \{M_{\text{BD1}}, M_{\text{BD2}}\}$

$L^\chi \in \{L_1, L_2\}$

$L_1 = \{((\text{Switch, toggle}), (\mathcal{X}, \text{in})),$
$((\text{Switch}, \text{in:1}), (\text{BD1}, \text{in})),$
$((\text{Switch}, \text{in:2}), (\text{BD2}, \text{in})),$

((BD1,out),(Switch,out:1)),
((BD2,out),(Switch,out:2))}

L_2 = {((Switch,togle),(c,in)),
((Switch,in:1),(BD2,in)),
((Switch,in:2),(BD1,in)),
((BD1,out),(Switch,out:1)),
((BD2,out),(Switch,out:2))}

$\{R_i^\chi\} = \{3,2,1\}$
phase = passive
sigma = ∞
$C \in$ {up,down}

Variable C just indicates which configuration is currently being used. The remaining components of the executive are defined by

$$s_{0,\chi} = (D^\chi,\{M_i^\chi\},L_1,\{R_i^\chi\},(\text{passive},\infty,\text{up}))$$
$$\tau_\chi((D^\chi,\{M_i^\chi\},L_1,\{R_i^\chi\},(phase,sigma,\text{up})) = sigma$$
$$\delta_{ext\chi}((D^\chi,\{M_i^\chi\},L_1,\{R_i^\chi\},\text{up}),e,(\text{toggle},\varnothing)) = (D^\chi,\{M_i^\chi\},L_2,\{R_i^\chi\},\text{down})$$
$$\delta_{ext\chi}((D^\chi,\{M_i^\chi\},L_2,\{R_i^\chi\},\text{down}),e,(\text{toggle},\varnothing)) = (D^\chi,\{M_i^\chi\},L_1,\{R_i^\chi\},\text{up})$$

The ability to change structure is strikingly clear in this example. Note that when the toggle input is received, the external transition function of the executive changes the coupling slot of the structure from L_1 to L_2 or conversely. Such global change in structure is rather difficult to describe and verify for correctness if done without a dynamic structure facility.

In the DELTA environment each executive has a class method called structure that defines an initial network structure. Models are added by the method addModel: *aName* class: *aClass*. When simulation starts it creates *aClass* instance named *aName*. Links are defined by the method link: *aModel* port: *aPort* to: *bModel* port: *bPort*. Two special aliases are defined: #Network refers to the network and #Executive refers to the network executive.

```
Switch::structure              "Class method that defines the initial structure"
   |s|
   s := super ses.
   s addModel: #BD1 class: BDelay.
   s addModel: #BD2 class: BDelay.
   s link: #Network port: #toggle to: #Executive port: #in.
   s link: #Network port: #in1 to: #BD1 port: #in.
   s link: #Network port: #in2 to: #BD2 port: #in.
   s link: #BD1 port: #out to: #Network port: #out1.
   s link: #BD2 port: #out to: #Network port: #out2.
   ^s
```

The switch starts in the up configuration as defined by

```
start                          "Initial state"
   c := #up.
   self passivateIn: #passive.
```

For convenience of implementation and to follow standard rules for classes, the initial network structure is not set in the model start method but in the class method structure. The external method changes the network between two configurations. First it removes the links from network ports in1 and in2 ,then it creates the appropriate new links.

```
external: elapsedTime port: p value: v
   self unLink: #Network port: #in1 from: #BD1 port: #in.
   self unLink: #Network port: #in2 from: #BD2 port: #in.
   c = #up ifTrue: [
      self link: #Network port: #in1 to: #BD2 port: #in.
      self link: #Network port: #in2 to: #BD1 port: #in.
      c := #down.
   ]
   ifFalse: [
      self link: #Network port: #in1 to: #BD1 port: #in.
      self link: #Network port: #in2 to: #BD2 port: #in.
      c := #up.
   ].
   self continue: elapsedTime.
```

A typical input and output trajectory of this model is represented in Fig. 13.

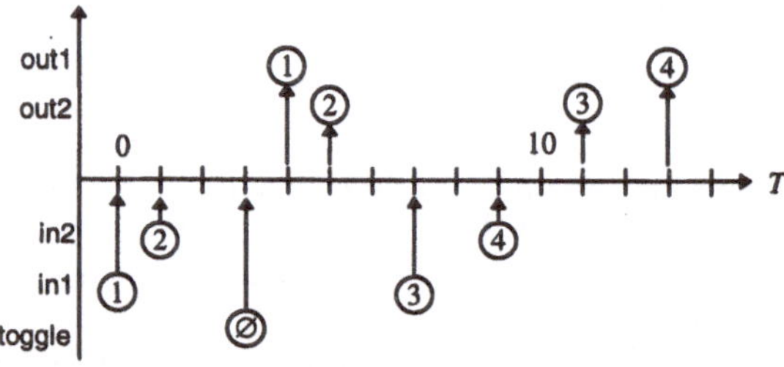

Fig. 13. Input and output trajectories of the Switch network.

Job1 arrives at time 0 at port in1 and job2 arrives at port in2 at time 1. A toggle input arrives at time 3 but does not affect job1 and job2 because the switch changes only the input connectors. Thus job1 and job2 are sent to ports out1 and out2 respectively. Jobs arriving next are affected by the changed input wires: job3 arrived at in1 leaves by port out2, and job2 that arrived at port in2 leaves the network, at time 13, by port out2.

This example shows only the ability to change links in run time, however any type of structural change can be achieved in the DELTA environment.

7. Conclusions

We have described two simulation formalisms able to represent discrete event systems. The DEVS formalism can represent basic models. The DSDEVS formalism can represent static and dynamic structure networks of basic DEVS models. These formalisms support the construction of modular and hierarchical simulation models, permitting the representation of complex systems. The expressive power of these formalisms was discussed relative to analytic methods that deal only with state transition representations and make assumptions about stochastic behavior in order to obtain solutions. The DELTA modeling and simulation environment is an implementation of the DSDEVS formalism. Several examples of model construction were given. Structural inheritance, an important factor for design reuse, was described. A variety of other simulation environments based on the DEVS formalism have been developed and are described in (Zeigler, Praehofer and Kim, 1998). Further information is also available at web site: www-ais.ece.arizona.edu.

References

Barros, F. J. (1995): Dynamic Structure Discrete Event System Specification: A New Modeling and Simulation Formalism for Dynamic Structure Systems. Proceedings of the 1995 Winter Simulation Conference, Arlington(VA), pp. 781-785.

Barros, F. J. (1996a): Dynamic Structure Discrete Event System Specification Formalism. Transactions of the Society for Computer Simulation, 1: 35-46.

Barros, F. J. (1996b): Dynamic Structure Discrete Event System Specification: Structural Inheritance in the DELTA Environment. Proceedings of the Sixth Annual Conference on AI, Simulation and Planning in High Autonomy Systems, San Diego(CA), pp. 141-147.

Barros, F. J. (1996c): Modeling and Simulation of Dynamic Structure Discrete Event Systems: A General Systems Theory Approach. Ph.D. Dissertation, Department of Informatics Engineering, University of Coimbra.

Barros, F. J. (1997a): Dynamic Structure Modeling and Simulation of the Erasthostenes Sieve for Prime Numbers. 30th Annual Simulation Symposium, Atlanta(GA), pp. 184-189.

Barros, F. J. (1997b): Modeling Formalisms for Dynamic Structure Systems. ACM Transactions on Modeling and Computer Simulation, 4. (Accepted for publication).

Cassandras, C.G. and Strickland, S.G. (1989): Sample Path Properties of Timed Discrete Event Systems, Proceedings of the IEEE, 1: 59-71.

Chow, A. C. (1996): Parallel DEVS: A Parallel, Hierarchical, Modular Modeling Formalism and its Distributed Simulator. Transactions of The Society for Computer Simulation, 2: 55-67.

Kleinrock, L. (1976): Queueing Systems, Wiley, New York.

Zeigler, B. P. (1976): Theory of Modelling and Simulation. Wiley, New York.

Zeigler, B. P. (1984): Multifaceted Modelling and Discrete Event Simulation. Academic Press, London.

Zeigler, B. P. (1990): Object-Oriented Simulation with Hierarchical, Modular Models: Intelligent Agents and Endomorphic Systems. Academic Press, Boston.

Zeigler, B. P., H., Praehofer, and T.G., Kim. (1998): Theory of Modelling and Simulation (Revised Edition). Academic Press, Boston (in preparation).

Design of Microsystems: Systems-theoretical Aspects

Franz Pichler

1 Introduction

Advances in Microelectronic technology, especially sensor- and actuator technology, have resulted in the development of integrated circuits which, in addition to microelectronics and transducers, contain mechanical and optical components. Consequently it has become possible to develop complex systems by coupling these new components with conventional microelectronic components and systems. The construction of these systems necessitates changes in the design and manufacturing process. This paper concentrates on the changes necessary, especially in the design process, to meet the challenges of this new field of microsystems (i.e., micro-electromechanical systems, micromechatronics).

In this context, we are primarily interested in the kinds of models which can be used to develop the building blocks and couplings for systems. Here, formal mathematical constructions for model-building and model-transformations deserve our interest. With this goal in mind we will consider topics of applied mathematics, specifically those from mathematical systems theory, which can be useful in the design of microsystems. As usual, we can expect that the introduction of the newly developed microsystems-technology will require the development of new theories in applied mathematics and mathematical systems theory. Although existing methods can be adapted to meet the modeling requirements of microsystems design, it will be necessary, as experience of the past with microelectronics confirms, to develop new formal modeling concepts and related methods. However, we can expect to find a foundation in the rich source of existing theory in pure mathematics. To discover which existing parts of mathematics are appropriate to the given task, requires a certain practical familiarity in applied mathematics and systems theory; a fact which is often neglected by pure mathematicians who often consider the question "trivial".

2 The Microsystem-Modeling Process

Modeling of a desired microsystem begins with defining the "requirements" which have to be fulfilled according to the clients point of view. We can distinguish between two types of requirements: "functional" and "nonfunctional". Functional requirements are generally defined in engineering as the properties of a system which guarantee a desired function. In the case of a microsystem (MS) they define the properties which are expected to contribute to a proper embedding of it into the overall system (i.e., the environment of the MS). Very often they define an input/output relation (I/O relation) in the form of a "black box" which has to be realized in the final MS. Non functional requirements, on the other hand, define conditions which have to be met by the microsystem. Usually they concern domain specific properties (e.g., properties which express economic or physical features). Nonfunctional requirements are often expressed formally by restricting the validity of specific variables to certain domains. Contrary to functional requirements, the evaluation of nonfunctional requirements often requires detailed knowledge of the domain-specific model.

Requirements, as expressed by the client, are very often formulated in a natural language. In order to incorporate these requirements into Computer Aided Design tools (CAD tools) it is necessary to translate them into a formal form written in some given programming language. We call the first formal model for the microsystem MS in which it's requirements are described in a programming language R, MSR. The next step in the design of MS is the most difficult one. It is the construction of a feasible microsystem architectural model (MSA) for the given MSR requirement model, which fulfills all of it's requirements. The step from MSR to MSA is usually decomposed into several steps, where each individual step consists of a refinement of the architecture. Starting from MSR we develop different architectural models: MSA_1,...MSA_{n-1}, MSA_n = MSA, where MSA_{i+1} is a refinement of MSA_i (i=1,2,...,n-1). These models form a hierarchical model of height n. While for all the architecture models MSA_1 to MSA_n the fulfillment of the given functional requirements is mandatory, only when reaching MSA_n = MSA it is assured that in addition the nonfunctional requirements are met. MSA consists of components and the couplings (i.e., interfaces) between them and known instance-models (e.g., models which allow the evaluation of the nonfunctional requirements and which assure physical reliability).

Considering the current state of the art in microsystems design, we must confess that our picture of stepwise top-down design of the architecture is too ideal. In many practical cases of design, modeling the architecture above the physical level is not possible. The reason for this is partly given by the fact that the design of complex microsystems in a systematic way is rather rare. Such systems are often still invented by a creative designer using his intuition. Another reason is that models of components, and subsystems of coupled components, are not available in a "stable sense" (i.e., with fixed material properties). It is still necessary to design "in the material" and not "on the material" to use an expression coined by Hugo de Man. In most cases of microsystems design, the properties of the material have to be considered as unknown, and will only become fixed during the design activity.

For these reasons, constructing MSA from MSR is difficult, since the fulfillment of the requirements only occurs when models which specify the material properties are reached. However, given the rate of progress in microsystems development, we can expect in the future that more and more microsystems components and building blocks will exist; allowing for "requirement driven", top down design as we have sketched above. With the specification of MSA we have finished the design phase and can set forth on the next phase, implementation, in which we convert the "blue print" of MSA into an engineering project.

3 Formal Microsystem Architectural Models

According to the goal of this paper, we will now investigate possible formal modeling means for the design of MSA. Principally, since microsystems are generally very rich with respect to their physical structures, their mathematical models potentially can make use of any kind of mathematical structure. However, since we are modeling within a systems-theoretical framework, specific mathematical theories are available. The question of which mathematical models are suitable for the construction of MSA starting from MSR is equivalent to the question of which kind of homomorphic constructions can be developed to model the components and couplings of the sequence MSA_1, MSA_2, ... of architectural models which leads to MSA.

Differential equation systems are the preferred mathematical tool for creating "highly structured models" of the components of MSA. In microelectronics, systems of ordinary differential equations are applicable for modeling many kinds of electronics, while partial differential equation systems are needed for modeling micromechanics and microoptics. These methods are also useful for certain instance models, where nonfunctional requirements which concern properties in space and time have to be evaluated. Regarding the type of models to be used for MSR, we might assume that their formal nature will be specified by graphs, relations, functions, etc. In conclusion we can assume that the model-types for MSA_1,...MSA_{n-1} will be some kind of homomorphic images of differential equation systems which relate to each other and which allow certain nonfunctional requirements to be evaluated. For MSA_1 it is required that it can be imbedded mathematically into MSR such that the fulfillment of the functional requirements can be proven.

In the case of microelectronics, where the art of modeling the design process has reached a mature stage, the use of homomorphic images in this way is well known. For digital microelectronics we mention as examples of classes of model types:

- Boolean functions,
- binary sequential switching circuits,
- finite state machines,
- extended finite state machines,
- petri nets.

For analog microelectronics, the following examples of model types are used as the building blocks of information transmission systems:

- 2-port circuits,
- multi-port circuits,
- transfer functions,
- frequency filters and equalizers,
- models for modulation devices,
- A/D transducers.

Similarly we might assume that such examples can also be found for the field of micromechanics and microoptics. In micromechanics we can consider primitive machines like the inclined plane, crank, wedge, and screw and derived machine elements such as transmissions, gears or ball-bearing as candidates for defining such homomorphic images.

At the current stage of research and development in microsystems, we are not able to completely describe the kinds of formal models in terms of mathematical systems theory which will be required. We have to first await further practical results in microsystems design. As soon as certain building blocks of microsystems become stable, in the sense that they can be considered as reusable parts, it will make sense to formalize them and prepare a related theoretical framework, so that they can be identified as homomorphic images of more structurally refined formal models. The theory of automata is an example of where such an empirical approach in the development of mathematical systems theory has taken place before. It's development started in the early 1950's when digital electronic circuits with memory first became practical and a functional description of the behavior of such circuits was needed. Another example is given by the development of the use of complex numbers in electrical engineering by Charles Proteus Steinmetz in connection with the introduction of alternating current in homes and in industry in the 1880's. Today both are fields of mathematical systems theory, namely the theory of finite automata and the frequency theory of electrical networks, and are important means of formal modeling.

However, for formally modeling microsystems, we should consider in particular a topic of mathematical systems theory which has already received careful mathematical treatment. It is the field of "dynamical systems" as introduced many years ago in the famous monograph of Birkhoff. Dynamical systems can be considered as homomorphic images of systems of solutions of differential equations. The existing theory consists of a wide range of topics drawn from the mathematical fields of algebra, topology, differential geometry and functional analysis. At this point we want to express the hope that applied mathematicians and mathematical systems theorists will in the future pay attention to the field of dynamical systems in building a formal framework for modeling microsystems at the different architectural levels used in the top down design process.

In the following sections we will try to outline the general approach for such a research orientation.

4 Systems-Morphisms

A fundamental concept to relate formal models from different architectural levels is the concept of systems-morphisms. A systems-morphism maps a formal model of a certain type to a more abstract model usually of the same type. By definition it is required that this mapping preserve certain structural properties of the model. As a simple example consider the mapping of a finite automaton $M=(A,B,Q,\delta,\lambda)$ onto its quotient $M/\sim$ by the map $h=(h_1,h_2,h_3)$ where $M/\sim=h(M)=(A,B,h_1(Q),h_2(\delta),h_3(\lambda))$. The preservation of the structure of M by h is assured by commutativity as shown in the following diagram.

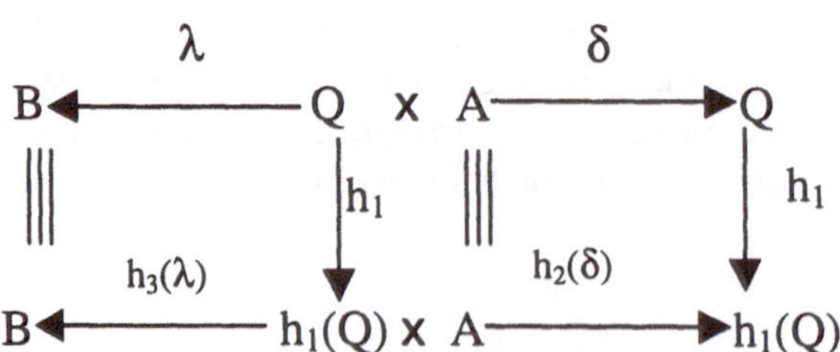

Similarly we might define a systems morphism $h=(h_1,h_2)$ for a function $f:X\rightarrow Y$ using the following commutative diagram.

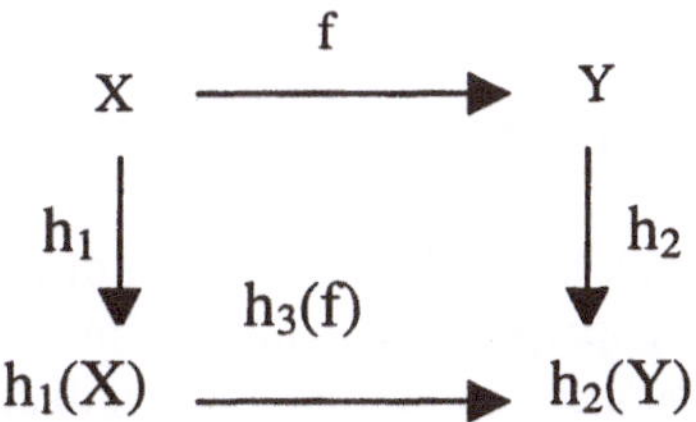

It is known that any equivalence relation ~ of X allows the construction of a morphism of this kind. In the case of the "natural morphism" of h(n) for f, the equivalence relation n on X is given by $xn\overline{x}:\Leftrightarrow f(x)=f(\overline{x})$ for all $\overline{x},x\in X$. In this case we have $h_1(X)=X/n$, $h_2(Y)=Y$ and $h_3(f)$ is given by the map $h_3(f):X/n\rightarrow Y$ which is given by $h_3(f)([x]_n):=f(x)$ where $x\in[x]_n$. It is possible to construct additional examples of this kind which show how to use systems-morphisms to reach a coarser structural model of a given formal model. The construction of this mapping is based on the knowledge of certain equivalence relations. As an example, in the case of finite automata, the knowledge of a so called congruence relation defined on the state set Q is required.

It has already been pointed out that the most refined models used in microsystems, which are used for modeling the functional architecture or the nonfunctional properties are given by ordinary or partial differential equations systems. Generally we can assume that the so called "differential equations of physics" will be sufficient to do the job. The solutions of these types of models form a large variety of dynamical systems. On the other hand we know that for a top down design of microsystems we have to start from a MSR and we have to be able to construct formal models at the different levels of architecture and to relate them by morphisms. In consequence we see that in microsystems design, the knowledge of the morphisms associated with the dynamics of "differential equations of physics" is essential. Michael Arbib, an eminent promoter of mathematical systems theory and cybernetics, appropriately calls such morphisms dynamorphisms.

Therefore, from a purely mathematical point of view, the task is to study the different kinds of possible dynamorphisms of certain dynamical systems that appear as solutions of differential equations describing the physical phenomena in microsystems. Although such a program would be desirable for it's theoretical results, at the current stage of microsystems engineering it is still of little practical value. The problem is, as we pointed out earlier, that currently only a few reusable components for microsystems are known. This means that there are not enough practical examples of successful microsystem design currently available in order to elaborate a relevant theory of dynamorphisms. The theoreticians will have to wait until practitioners develop examples of these kind of components for the higher levels of microsystems modeling.

It should be mentioned that the development of the formal models used in the field of microelectronic design required the same steps. The development of the formal methods for logic synthesis and high level synthesis, including the development of appropriate design languages such as VHDL, could only take place after several years of practical experience in the design of complex integrated microelectronic circuits "by hand" (i.e., by the skillful and innovative work of a designer).

5 Inductive and Deductive Decomposition

The task of refining an architectural model MSA_i of level i into a model MSA_{i+1} of level i+1 requires knowing appropriate decomposition methods. We distinguish here between two classes of decomposition methods: inductive methods and deductive methods. We first discuss the principal approach of both methods and then concentrate on deductive decomposition methods, which from the standpoint of mathematical systems theory are more interesting.

Inductive decomposition methods, as we call them here, are based on the following heuristic procedure; for a given model M of a component of the architecture MSA_i of level i we select "by experience" from the known candidates of components of the refined architecture MSA_{i+1} k models $M_1,M_2,...,M_k$ together with an appropriate coupling concept K such that the network $K(M_1,M_2,...,M_k)$ of coupled models fulfills the requirements R(M) of M. The selection of $M_1,M_2,...,M_k$ and of K depends mainly on the skill and experience of the designer and is therefor considered an "art". Concerning requirements of $K(M_1,M_2,...,M_k)$ we have, however, to consider the following fact; the components $M_1,M_2,...,M_k$, and the coupling model K, generally create additional requirements $R_a(M_1)$, $R_a(M_2)$,..., $R_a(M_k)$, $R_a(K)$ which have to be fulfilled by the microsystems MS. It is left to the skill of the designer to take this fact into account when finding the best suited decomposition $K(M_1,...M_k)$ of M.

The second class of decomposition methods, inductive decomposition methods, are based on a computing method which derives $K(M_1, M_2, ..., M_k)$ from the component model M. The construction of the models M_1, M_2, ..., M_k is done using the appropriate morphisms h_1, h_2, ..., h_k which map M, which is part of the architectural model MSA_i, into the models $M_1 = h_2(M)$, ..., $M_k = h_k(M)$ contained in the architectural model MSA_{i+1}. In addition, an associated mathematical computing method k must be known, which derives for given M, M_1, M_2, ..., M_k the coupling model $K = k(M,M_1,M_2,...,M_k)$. With respect to the discussion of the requirements of the decomposition $K(M_1, M_2, ..., M_k)$ of M which is derived by a deductive decomposition method, the same arguments which we gave in the case of inductive decomposition remain valid.

6 Deductive Decomposition of Dynamical Systems

Deductive decomposition methods of the kind which we discussed above do exist for specific classes of dynamical systems. This is the case if a dynamical system Dyn(M) allows the construction of the morphisms (i.e., dynamorphisms in the sense of Arbib) $h_1,h_2,...,h_k$ which enable the computation of a decomposition $K(M_1,M_2,...,M_k)$ as we discussed above. It is known that with any dynamical-invariant equivalence relation π of the state space of Dyn(M) we can associate a dynamorphism h_π by the map h_π:

$Dyn(M) \rightarrow Dyn(M)/\pi$ of Dyn(M) into the quotient-dynamical system $Dyn(M)/\pi$. In that way the question concerning the finding of the suitable dynamorphisms $h_1, h_2, \ldots, h_k$ is reduced to the task of the construction of proper dynamical-invariant equivalence-relations $\pi_1, \pi_2, \ldots, \pi_k$ on the state space of Dyn(M). The following examples show that the determination of $\pi_1, \pi_2, \ldots, \pi_k$ is most effectively done by means of a "generative concept" Gen(M) of Dyn(M), (e.g., a model which determines the (global) state transition function of Dyn(M) only locally such that Dyn(M) can be derived from Gen(M) by "integration"). For the following four specific classes of dynamical systems we sketch the associated deductive decomposition method based on dynamical invariant equivalence relations.

Decomposition of Finite State Machines (Hartmanis-Stearns (1966))

For each Finite State Machine FSM, the set of all possible dynamical-invariant equivalence relations is given by the lattice L(FSM) consisting of the set of all congruence relations of FSM. The computation of L(FSM) is generally a computationally hard problem. However, for most of the practical cases in microelectronics and microsystems the determination of the appropriate congruence relations $\pi_1, \pi_2, \ldots, \pi_k$ can be done and an associated decomposition $K(FSM_1, FSM_2, \ldots, FSM_k)$ with $FSM_i = FSM/_{\pi i}$ $(i=1,2,\ldots,k)$ can be computed (Müller-Wipperfürth (1994a,b)).

Decomposition of Linear Differential systems LDS=(A,B,C) (Pichler (1976), Zunde-Pichler (1982)

Each A-invariant subspace U of the state space $\mathbb{R}^n$ of a linear time-invariant differential system LDS=(A,B,C) defines an associated dynamical-invariant equivalence relation π on $\mathbb{R}^n$ by $q\pi q :\Leftrightarrow q-q \in U$ for $q,q \in \mathbb{R}^n$. For each such π we have an associated linear system LDS/π which uses the quotient space $\mathbb{R}^n/U$ as a state space. The function h which maps LDS to LDS/π is a dynamorphism. By selection of a set of appropriate A-invariant subspaces U_1, U_2,...U_k it is possible to reach a related decomposition $K((LDS/_{\pi_1}, \ldots, LDS/_{\pi_k})$ of LDS.

Decomposition of ordinary nonlinear differential systems (Krener (1975a,b))

For ordinary nonlinear differential systems with corresponding differential equation system of the form $x' = f(x,u,t)$; $x(0) = x_o$ it is possible to use the associated Lie algebra to compute decompositions.

Decomposition of general dynamical systems with input and output (Mesarovic-Takahara (1975), Salovaara (1967), Pichler (1975,1983))

It has been shown how to construct for a general causal input/output function f, an associated dynamical system Dyn(f). As soon as the function f allows the computation of a dynamical invariant congruence relation on the state space of Dyn(f) it is possible to represent f by a decomposition of Dyn(f).

The examples indicate that for specific classes of dynamical systems there do exist associated deductive decomposition methods with possible applications to the design of the microsystems architecture model. It is obvious that additional mathematical work will be necessary to extend and adapt such methods so that they are of practical value to microsystems engineering. Future development in microsystems design and in the associated microsystems technologies will provide the proper motivation to continue such theoretical research.

7 Conclusion

This paper dealt with conceptual and theoretical issues related to microsystems design. Microsystems, as it is well known, are highly integrated circuits which are based on different technologies, especially: microelectronics, micromechanics, microacoustics, microhydraulics, micro-vacuum tubes, and micro optics. Using these different technologies it is possible to implement different kinds of "machines" and to integrate them by coupling them into a complex system on a single silicon chip. For modeling a microsystem we need, depending on the different realization technologies for its components and their couplings, a variety of different modeling concepts and tools. The complexity of the modeling process provides new challenges during systems design. The paradigms which exist for modeling and tool making (i.e., those used in microelectronics, mechanical engineering or control engineering) need to be revised and adapted for the new requirements (DeMan (1990)). In addition to the development of CAD tools for the engineering support of the designer, formal mathematical methods and related tools have to be investigated and elaborated. This related research constitutes for the near future an important task in applied mathematics and mathematical systems theory.

Microsystems, as it was pointed out in this paper, are rapidly developing. Currently the development of different microtechnologies and material science oriented research are the main focus. With regard to building blocks for components, only "intelligent" sensors and "intelligent" actuators have reached some level of stability in design. The example of microelectronics design and related tool development suggests that the final development of systems theoretical approaches in microsystems design has to wait until a certain degree of maturity in design practice has been reached. A benchmark for this could be the existence of a full catalogue of reusable building blocks and components and related models resulting from the successful design of microsystems. However, we suggest to start with fundamental studies in this field today in order to be better prepared for future cooperation with design practitioners.

References

DeMan, H. (1990): Microsystems: A Challenge for CAD Development. In: Microsystems Technologies 90, H. Reichl ed., Springer Verlag Berlin, pp. 3-8

Hartmanis, J., Stearns, R.E. (1966): Algebraic Structure Theory of Sequential Machines. Prentice Hall, Inc. Englewood Cliffs, N.J.

Krener, A.J. (1975a): Local approximation of control systems. Journal of Differential Equations 19: 125-133

Krener, A.J. (1975b): A decomposition theory for differential systems. In: Proceedings of the IFAC 6th World Congress, Boston, August 24-30, 1975. Part 1, Theory, paper 36.5

Mesarovic, M.D., Takahara, Y. (1975): General Systems Theory: Mathematical Foundations. Academic Press, New York

Müller-Wipperfürth, T. (1994a): Finite State Machine Structuring with CAST.FSM* applied to VLSI controller Design. PhD thesis, Johannes Kepler University, Linz, Austria

Müller-Wipperfürth, T. (1994b): On the Integration of CAST.FSM into the VLSI Design Process. In: T.I. Oeren and G.J. Klir, editors, Proceedings CAST'94, Ottawa, Springer Verlag Berlin, pp. 363-372 (Lecture Notes in Computer Science, volume 1105)

Pichler, F. (1975): Mathematische Systemtheorie: Dynamische Konstruktionen. Walter de Gruyter Berlin New York

Pichler, F. (1976): On the decomposition of linear dynamical systems. In: Proceedings of the 3rd European Meeting on Cybernetics and Systems Research, pp. 445-454 (Hemisphere Publishing Corporation Washington)

Pichler, F. (1983): Dynamical Systems Theory. In: Cybernetics: Theory and Applications (ed. R. Trappl), Springer Verlag Berlin, pp. 43-56

Salovaara, S. (1967): On Set Theoretical Foundations of System Theory. Acta Polytechnica Scandinavia, Mathematics and Computing Machinery Series No. 15, Helsinki

Zunde, P., Pichler, F. (1982): Complete characterization of State space decompositions of linear dynamical systems. In: F.R. Pichler and R. Trappl, editors, Progress in Cybernetics and Systems Research, volume VI, Hemisphere Publishing Corporation, Washington, pp. 380-388

A formal representation of DSS generator

Yasuhiko Takahara
Xiaohong Chen

1 Introduction

It is usual that an MIS (management information system) is categorized by the three levels, TPS (transaction processing system), MIS in a narrow sense and DSS (decision support system). A TPS of the first level constitutes the infra-structure of an MIS, which handles transaction process of a business, and maintains a corporate DB (data base). This DB supplies basic information for the MIS.

An MIS of the second level represents the problem solving activity of the MIS. The target of this level is mostly a well structured problem which is modeled and solved by help of the management science.

A DSS of the third level is also concerned with problem solving of a corporation. It uses a model, furthermore, to attack a problem. But the main target of a DSS is not a well structured but a semi-structured. Although the second level MIS is supposed to work by itself on generation of a solution (an optimal solution), a DSS works with a user interactively supporting his problem solving activity and indicates an alternative, which can be rationalized by the model used, to help the user to find a satisfactory solution. Since the problem is supposed not well structured, the concept of optimal solution is not applicable to a DSS.

If the central function of an office or management of a corporation is decision making, a DSS is a computer based information system which directly addresses the function. Its significance has been well recognized but its practical implementation is still at an early stage.

A DSS is designed for each specific problem. There is no DSS which is universally applicable to any problem. A DSS specifically generated for a given problem is sometimes called specific DSS. A DSS generator which is the topic of this chapter is not a specific DSS but a system which behaves as a specific DSS when a target problem is embedded into it. A DSS generator may be considered as a universal DSS in the weak sense.

Fig.1 shows a functional scheme of a DSS generator. A traditional paradigm of a DSS generator says that it consists of three components, DiaS (dialog system), MMS (model management system) and DMS (data management system) (Sprague 1983).

Since a specific DSS works with a user interactively, a DSS generator should have a good user interface which is called a dialog system. A user sends a command and receives a response through the DiaS.

Since a specific DSS explores a problem based on a model, the DSS generator has an MB (model base) and its MMS. The MB supplies necessary models and solvers (LP is a typical) for a given specific problem.

Since a model cannot work without data, a DSS generator is also associated with a DB and its DMS.

If the target problem of a DSS is complicated and ill-structured, it is natural to construct the corresponding model by the so-called model integration approach, where the problem is first decomposed into subproblems, the subproblems are then represented by submodels and a final model is built by integration of the submodels (Dolk, et al., 1993). This operation should be done easily, spontaneously and instantly. The model space of Fig.1 is devised for that purpose. It is a platform for model construction and model manipulation.

These four components must be well organized to generate a useful specific DSS. Then, a coordination scheme is naturally required for them. The scheme is realized by OS in Fig.1.

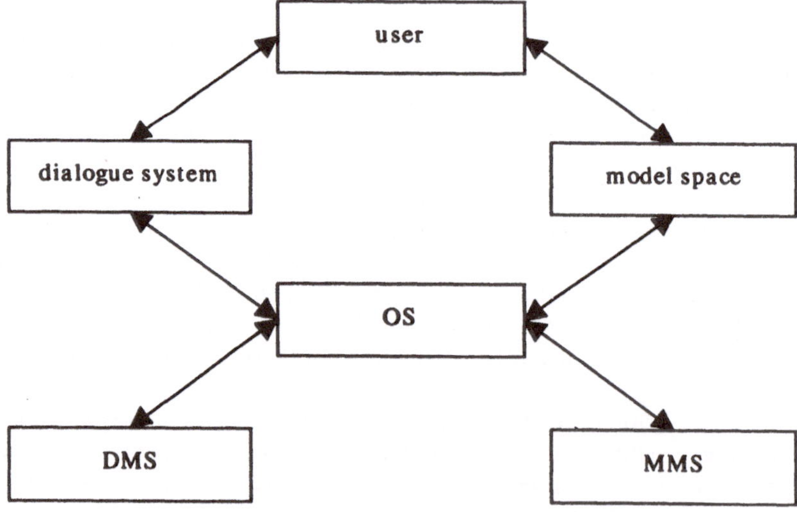

Fig.1. Functional structure of DSS generator

The idea of model integration, model platform or OS of DSS are well recognized but their implementations are not well understood. This chapter will investigate a DSS generator of Fig.1 on the system theory (Mesarovic, et al.,

1989) and present a formal description of it as a theoretical basis of a DSS generator. The description is derived as an extention of the Turing machine. The head which will be called automaton block covers DiaS and OS while the tape corresponds to the model platform. Actually, the authors' group has developed a DSS generator based on the consideration of this chapter (Takahara, et al. 1996).

Formalization is usually presented in the standard logico-mathematical terms but since they are at a too low level to describe a DSS generator, this chapter will adopt Prolog descriptions in addition to the logico-mathematical terms.

2 A universal model of DSS generator

The most abstract understanding of a DSS generator is that it is an object which is transformed into a specific DSS when a model (including data) is embedded into it. The generated specific DSS is an event driven system whose input is a command from a user and whose output is a response to the user. An event driven system can be formalized as an automaton (Mesarovic, et al. 1989).

Let

$$A = \text{ set of user's commands(input alphabet)},$$

$$B = \text{ set of DSS's responses (output alphabet)}.$$

Suppose the input output behavior of the specific DSS is given by the function:

$$DSS : A^* \to B^*$$

where A^* and B^* are the free monoids on A and B, respectively. Then the specific DSS can be realized by the following automaton (which is not necessarily finite):

$$DSS^* =< A, B, C, \delta, \mu >$$

where

$$C = \text{ set of states},$$

$$\delta : C \times A \to C,$$

$$\mu : C \times A \to B.$$

C can be equal to A^* and then

$$\delta(c, a) = ca \text{ (concatenation of c and a)},$$

$$\mu(c, a) = DSS(ca).$$

We need a theoretical model which generates the above automaton when a problem model is embedded. The structure of Fig.2 will be used for the purpose. The model is an extension of the Turing machine for a DSS generator.

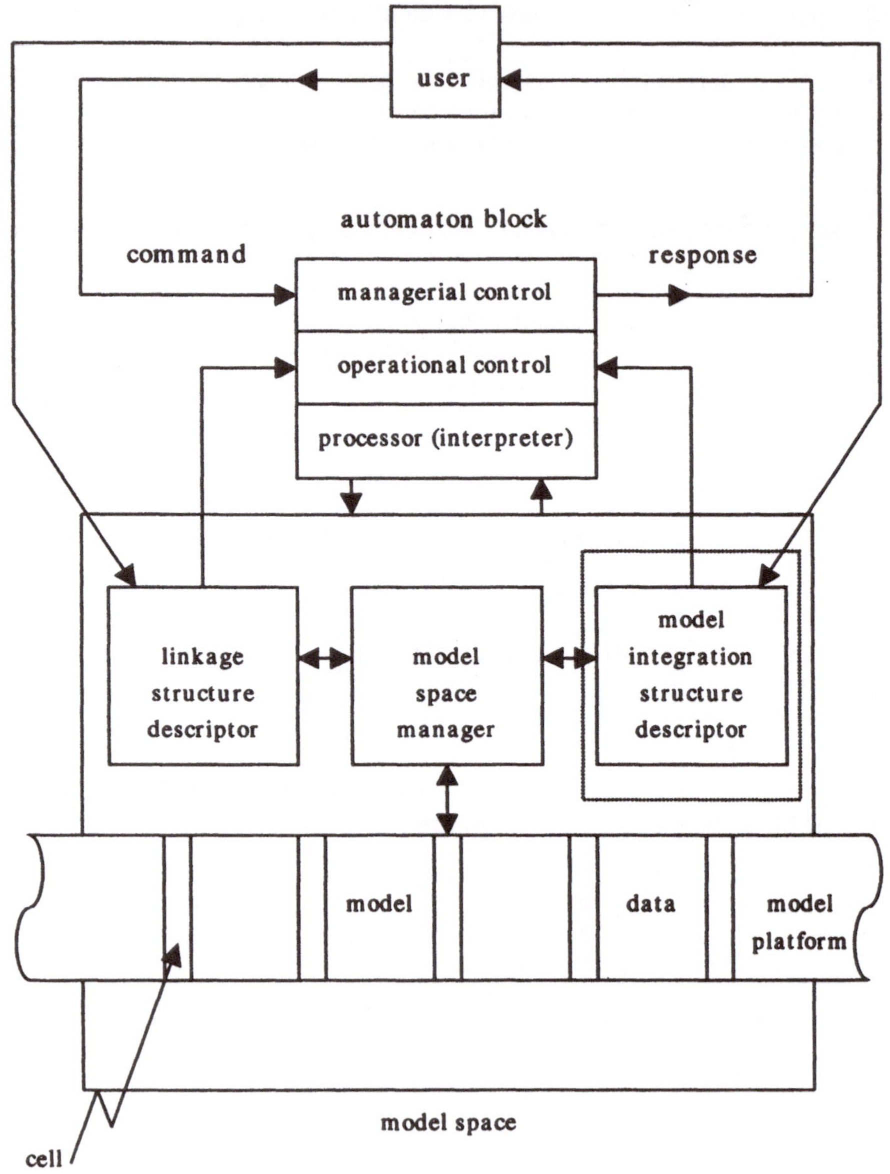

Fig.2. Universal machine structure of DSS generation

The model consists of two parts, automaton block and model space, which correspond to the head and the tape of the Turing machine, respectively. The above function 'DSS' is coded on the model space. The automaton block interprets the code to generate the behavior of the corresponding automaton.

The model space covers the model space and some functions of MMS and DMS of Fig.1. The automaton block covers the OS and the DiaS of Fig.1. Fig.2 is, hence, a model to address DiaS, OS and the model space where MMS and DMS are hidden as lower level functions.

The model space consists of three components, model platform, linkage structure descriptor and model space manager (In our real inplementation there is the fourth component, integration structure descriptor. In this chapter, however, the integration concept is omitted for the sake of simplicity).

The tape of the Turing machine, strictly speaking, corresponds to the model platform. Embedding of a problem model is realized by describing it on the model platform. The model platform consists of cells. As the subsequent formalization shows, submodels and data can be normalized as lists of atomic statements or predicates (Prolog rules). Each atomic statement is stored in a cell of the platform.

Submodels on the model space are assumed input-output systems. Their connection is, then, realized by linkages between inputs and outputs of them. The linkage structure descriptor holds the information, which output is connected to which input. When a linkage structure is established for a group of submodels, we say the group is organized.

The model space manager interprets the linkage structure descriptor and provides the organized behavior of the group when the automaton block processes the model platform. It is important to notice that submodels are defined on the model platform independently but the model space manager produces a virtual image of the organized structure for the automaton block using the linkage description as if there were a real organized model. The linkage description is supplied by a user.

The automaton block consists of an interpreter (processor) of the atomic statements and two management components, operational control and management control, and processes the model space according to the user's request and yields a response to the request. The multilayer concept, process (interpreter), operational control and management control is clearly based on the hierarchical formulation of a complex system of the systems theory. These are conceptual explanation of Fig.2.

3 Organized model representation in GST (general systems theory)

Model connection of input output models which is the basis of Fig.2 has been studied in GST (Mesarovic, et al. 1989, Wymore 1993). The formulation of the universal model is based on the result.

Suppose there exists a class of input-output models $P = \{P_i =< X_i, Y_i, P_i > |i \in J\}$ where

$J = \{1, .., n\}$ is an index set,

$X_i =$ input set of the i-th model,

$Y_i =$ output set of the i-th model,

$P_i : X_i \rightarrow Y_i$.

A desired model is realized by linking the family $P = \{P_i | i \in J\}$.

Suppose, after a link operation, X_i has the following structure:

$$X_i = M_i \times W_i$$

where $m_i \in M_i$ represents a decision variable (and plus a parameter) and $w_i \in W_i$ a linked (interaction) variable.

Let

$$X = X_1 \times ... \times X_n \text{ and}$$

$$Y = Y_1 \times ... \times Y_n.$$

Then, the interaction W_i is represented by a mapping K_i as:

$$K_i : X \times Y \rightarrow W_i.$$

Let

$$K = \{K_i | i \in J\}.$$

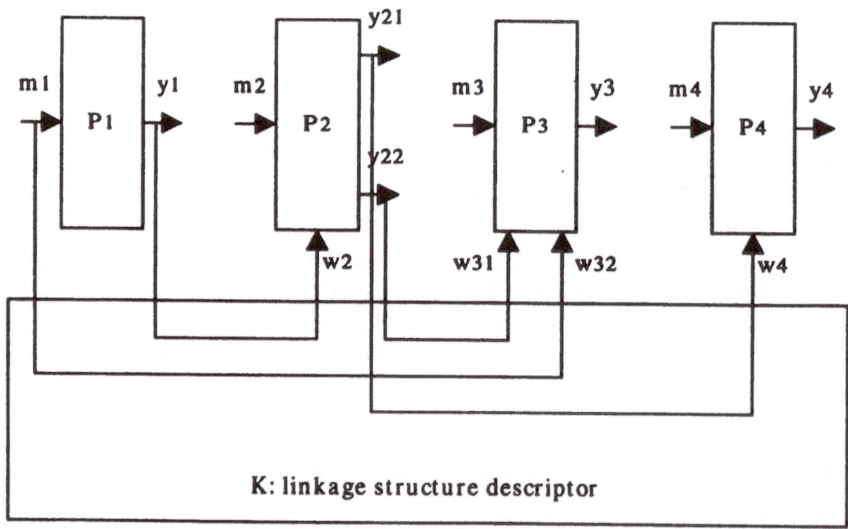

Fig.3. Example of organized model

Let us illustrate the above formulation by a simple example of Fig.3 where variables are vectors. The organized model consists of four submodels, P_1, P_2, P_3 and P_4. For instance, P_3 is given by: $X_3 = M_3 \times W_3, P_3 : (M_3 \times W_3) \rightarrow Y_3, (w_{31}, w_{32}) \in W_3$ and $y_3 \in Y_3$. Although K_i is defined formally on $X \times Y$, the real contents are given by:

$$K_1 : Empty,$$

$$K_2 : w_2 = y_1,$$

$$K_3 : w_{31} = y_{22}, w_{32} = m_1,$$

$$K_4 : w_4 = y_{21}.$$

Realization of the model space of the universal model is based on the above formulation. The model platform and the linkage structure descriptor correspond to P and K ,respectively. $< P, K >$ is a formulation of an organized model in GST.

4 Model space

Now we will investigate a computer realization of $< P, K >$ introducing detailed structures.

4.1 Model and model platform

Suppose a model is described by a functional language or let a general form of a user's model m be given by the following:

$$m = \{y_i = f_i(x_{i1}, ..., x_{in}) | i \in I\}$$

where y_i and x_{is} are variables (or variable names), f_i is an any function and I is an index set. Let N(m) be the set of the variable names of m.

Let a binary relation $\leq \subset N(m) \times N(m)$ be:

$$y_i = f_i(x_{i1}, ..., x_{in}) \rightarrow x_{ij} \leq y_i (\mathrm{j=1,...,n}).$$

Then $\leq$ represents a causal relation among variables. Let $\leq^*$ be the transitive closure of $\leq$. We assume $\leq^*$ can be expanded as a tree, tree(m), or loopless structure in the usual way.

Let

$$Leaf(m) = \text{the set of leaf variables of the tree(m)},$$

$$Root(m) = \text{the set of root variables of the tree(m)}.$$

Then, the set of input variable names of the model m $X^n(m)$ is given by Leaf(m) and the other are output variable names $Y^n(m)$. That is,

$$X^n(m) = Leaf(m) \text{ and}$$

$$Y^n(m) = N(m) - X^n(m).$$

Root(m) will be used for formulation of model execution (Refer to Appendix 1).

Let

$$X^n(m) = \{x_1^n, ..., x_p^n\} \text{ and}$$

$$Y^n(m) = \{y_1^n, ..., y_q^n\}.$$

Let

$$X_i(m) = \text{the set of values of the variable } x_i^n \text{ and}$$

$$Y_i(m) = \text{the set of values of the variable } y_i^n.$$

Let

$$X(m) = X_1(m) \times ... \times X_p(m) \text{ and}$$

$$Y(m) = Y_1(m) \times ... \times Y_q(m).$$

Then the input output relation of the model m is represented by the function

$$P(m) : X(m) \rightarrow Y(m).$$

It should be noticed that processing the model m really means computing P(m) for a given $x \in X(m)$.

Conventionally the symbol x_{ij} is used to represent the name of the variable as well as the value of it. Unless there is no confusion, we will follow the convention.

$y_i = f_i(x_{i1}, ..., x_{in})$ is the atomic statement to be stored in a cell of the model platform. Or as shown below, the functional form is easily transformed into a Prolog rule and then the generated predicate is stored in a cell.

For each $y_i = f_i(x_{i1}, ..., x_{in})$ let its corresponding Prolog rule be:

$$R_i : y_i(Y_i) : -x_{i1}(X_1), ..., x_{in}(X_n), f_i'(X_1, ..., X_n, Y_i),$$

$$assert(y_i(Y_i) : -!).$$

where Y_i and X_i are Prolog variables and

$$f_i'(X_1, ..., X_n, Y) \leftrightarrow Y = f_i(X_1, ..., X_n).$$

Notice that Y_i and X_j represent values of y_i and x_{ij}, respectively.

Let

$$Pr(m) = \{R_i | i \in I\}.$$

Then the following Prolog program

$$Pr(m)$$

$$? - y_i(Y).$$

Produces values for the variables $\{y_j | y_j \leq^* y_i\}$ and the values are equal to these derived from computation of m (Takahara, et al. 1987).

4.2 Linkage structure descriptor

Suppose a problem model is described on n submodels $\{m_1, ..., m_n\}$ on the model platform. Let us use notations X_i, Y_i and P_i instead of $X(m_i), Y(m_i)$ and $P(m_i)$, respectively.

Since an element of X_i is in general a vector, let $(x_{i1}, ..., x_{ip}) \in X_i$ be a typical element of it (p certainly depends on i). Let the name of the k-th variable of X_i be denoted by x_{ik}^n.

Similarly, the output Y_i is assumed to consist of q variables and the name of the j-th variable of Y_i will be denoted by y_{ij}^n.

Let

$$X_i^n = \{x_{ik}^n | k = 1, ..., p\},$$

$$Y_i^n = \{y_{ij}^n | j = 1, ..., q\}.$$

Since X_i is decomposed into M_i and W_i when a linkage structure is specified (refer to Section 3), X_i^n is also partitioned into two classes, M_i^n and W_i^n, respectively.

Let

$$N_i = X_i^n \cup Y_i^n = \text{set of variable names of } P_i,$$

$$N = \cup N_i,$$

$$V_i = X_i \cup Y_i = \text{set of values of the varaiables of } P_i,$$

$$V = \cup Vi = \text{super set of the values.}$$

Let the variable v_j of the model P_i be denoted by $[P_i, v_j]$ where $v_j \in N_i$.

Let

$$varP = \{[P_i, v_j] | v_j \in N_i, i = 1, ..., n\}.$$

Let

$$F = varP \cup \{[]\}.$$

Then, for each P_i let us define a mapping $IORep(P_i)$ as:

$$IORep(P_i) : N_i \to (F \times V)$$

such that

1. If $v^n \in M_i^n \cup Y_i^n$ then

$$IORep(P_i)(v^n) = ([], v)$$

where v is the current value of the variable v^n.

2. If $v^n \in W_i^n$ then

$$IORep(P_i)(v^n) = ([P_j, u^n], u) \text{ for some } P_j, u^n \text{ and } u$$

where $u^n \in N_j$ and u is the value of the variable u^n of P_j or evaluation of $[P_j, u^n]$.

$IORep = \{IORep(P_i)|i = 1, ..., n\}$ is the linkage structure descriptor of Fig.2 or an implementation of K on a computer. It contains additional information, the current value distribution of the organized model. Initial data are assumed stored at the value parts of IOReps.

Let us consider an example. Suppose $IORep(P)(v^n) = ([P', u^n], u)$. Then it means that the variable v^n of the model P is linked to a variable u^n of the model P' where u is the value of u^n or the evaluation of $[P', u^n]$.

4.3 Model space management

There are three basic tasks for the model space manager, update of IORep, support of loading of P_i onto the model platform and support for creation of the linkage structure descriptor. Since the last two tasks are related with a real implementation of MMS and DMS, we will consider only the first. In other words, the following discussions starts from the assumption that submodels are loaded on the platform and a link structure is specified by the user.

The model space manager updates $IORep(P_i)$ or updates values of the linked variables of $IORep(P_i)$ when the automaton block starts processing P_i. Since computation of P_i is, then, done based on the current values of the linked variables, the behavior of the link structure is automatically realized.

This task is represented by a function $updateIORep : P \times IORep \rightarrow IORep$ such that for each i and for each j

$$updateIORep(P_i)(IORep(P_j)) = IORep'(P_j)$$

where if $i <> j$ then $IORep'(P_j) = IORep(P_j)$ and if $i = j$ and if $IORep(P_j)(v^n) = ([P_k, u^n], u)$ then $IORep'(P_j)(v^n) = ([P_k, u^n], u')$ where u' is the current value of $[P_k, u^n]$. P_i is the target submodel to be updated.

5 Automaton block

As mentioned in Section 2, the automaton block consists of two parts, processor and management. The management part is composed of two subsystems, operational control and management control. The operational control is concerned with operational works, for instance, model execution. The operational control is an extention of an MMS. The management control interfaces with the user and coordinates activities of the operational control. This covers the DiaS and the OS of Fig.1.

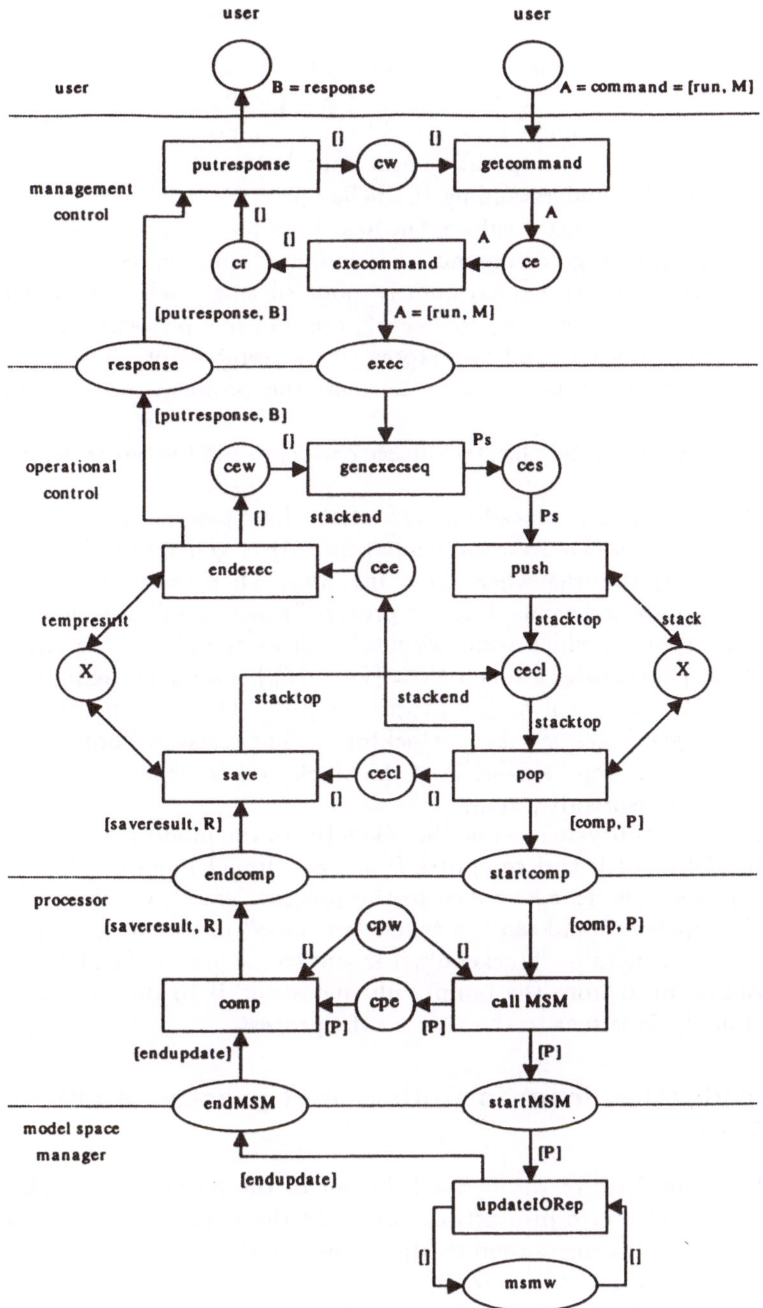

Fig.4. Petri net formulation of DSS generator process

5.1 Petri net formulation of process of automaton block

Fig.4 represents the process of the automaton block as a colored Petri network (Hee 1994) where the user's commnad is typically specified as model execution and the problem model is assumed to be a composite of submodels. Since the three levels work as a cooperative process, a Petri network representation is the most suitable for understanding the behavior.

The management control subsystem has three places, cw, ce, cr and three processes (transitions), getcommand, execcommand, putresponse.

The operational control subsystem is modeled as a push down automaton which has five places, cew, ces, cec1, cec2, cee and five processes, genexecseq, push, pop, save, endexec, and two stores, stack, tempresult. The push down automaton construction is necessary because the problem model consists of submodels.

The processor subsystem has two places cpw, cpe, and two processes callMSM, comp.

Initially, one token is placed on each of the four places, cw, cew, cpw and msmw. Suppose a token A=command=[run, M] is generated to the process "getcommand" by the user where $M = m_1...m_n$. Then the process "execcommand" is fired by A and sends A to the process "genexecseq" as well as [] to the process "putresponse" which waits for a token from the place "response".

"genexecseq" generates a token $Ps = [P_1, ..., P_n]$, a sequence of input output functions for M. Ps is sent to the process "push". The process saves Ps into the stack and generates a token "stacktop". The process "pop", accepting "stacktop", gets the top element P of the stack and transmits [comp, P] as a token to the processor subsystem.

The processor subsystem sends the token [P] to the model space manager to update the IORep of P and computes P using IORep(P) and yields a result R and sends [saveresult, R] as a token to the process "save", which saves R into the store "tempresult" and sends a token "stacktop" to the place cec1.

When "pop" generates "stackend", it is sent to the process "endexec", which gets a total result B from the tempresult and sends B to the process "putresponse". Finally, B is sent to the user by the process.

5.2 Production form formulation of process of automaton block

The Petri net form of the above behavior of the automaton block can be transformed directly into production forms. In the following production form the places and tokens correspond to states and inputs, respectively.

Let $C = \{cw, ce, cr\}, Ce = \{cew, ces, cec1, cec2, cee\}$, $Cp = \{cpw, cpe\}$ and $Cm = \{msmw\}$. Then since each subsystem has one token at any time the global state set is given by $C \times Ce \times Cp \times Cm$. In order to make the production form readable, we use the following notation:

$$< input >:< state > \rightarrow < action >< output >< nextstate > .$$

The meaning of the notation should be clear.
Fig.4 can be represented by the production form as follows:

1. Management control subsystem

$$command : [cw, cew, cpw, msmw] \rightarrow$$
$$[getcommand(A)]A[ce, cew, cpw, msmw]$$

Suppose $A = [run, M]$.

$$[run, M] : [ce, cew, cpw, msmw] \rightarrow$$
$$[intializestore(stack, tempresult)][genexecseq, M]$$

$[cr, cew, cpw, msmw]$

$$[response, B] : [cr, cew, cpw, msmw] \rightarrow$$
$$[putresponse(B)]command[cw, cew, cpw, msmw]$$

2. Operational control subsystem

$$[genexecseq, M] : [cr, cew, cpw, msmw] \rightarrow$$
$$[execseq(M, Ps)]Ps[cr, ces, cpw, msmw]$$
$$Ps : [cr, ces, cpw, msmw] \rightarrow$$
$$[push(Ps)]stacktop[cr, cec1, cpw, msmw]$$
$$stacktop : [cr, cec1, cpw, msmw] \rightarrow$$
$$[pop(stackend)]stackend[cr, cee, cpw, msmw]$$
$$stacktop : [cr, cec1, cpw, msmw] \rightarrow$$
$$[pop(P)][comp, P][cr, cec2, cpw, msmw]$$
$$[saveresult, R] : [cr, cec2, cpw, msmw] \rightarrow$$
$$[saveresult(R)]stacktop[cr, cec1, cpw, msmw]$$
$$stackend : [cr, cee, cpw, msmw] \rightarrow$$
$$[getresult(B)][putresponse, B][cr, cew, cpw, msmw]$$

3. Processor subsystem

$$[comp, P] : [cr, cec2, cpw, msmw] \rightarrow$$
$$[callMSM(P)][update, P][cr, cec2, cpw, msmw]$$
$$[update, P] : [cr, cec2, cpw, msmw] \rightarrow$$
$$[comp(P, R)][saveresult, R][cr, cecr2, cpw, msmw]$$

A capital letter indicates a variable or any value is substituted for it.

6 Realization in Prolog

The production form representation can be directly transformed into a Prolog representation. If we have a Prolog processor, then we can implement the above scheme, that is, a DSS generator with a model integration scheme. The above production form corresponds to the following Prolog rule:

$$product(< state >, < input >, < output >, < nextstate >) : - \qquad < action > .$$

For instance, the behavior of the processor subsystem is represented by:

$$product([cr, ccc2, cpw, msmw], [P], [saveresult, R],$$

$[cr, cec2, cpe, msmw]) : -callMSM(P)$.

If we add the following rule, which specifies the basic cyclic behavor of a DSS, to the generated Prolog rules, we can have a complete DSS generator in Prolog.

$$dss(C, A) : -product(C, A, B, NC), !, dss(NC, B).$$

Appendix 1 shows a complete list of the DSS generator program in Prolog. As mentioned in Section 4.1, if Root(P)=v, unification of v(X) implies evaluation of all the variables of P. This is used for comp(P, R) in Appendix 1.

A Implementation of the formulation in Prolog

A.1 DSS generator in Prolog

```
/* two submodels test1 and test2 are on model space */
/* test1:yY1=xX1+5;test2:yY2=xX2*2 */
/* link relation K:xX1=yY2;xX2=yY1 */
/* organized model=< [test1, test2], K > */
```

A.2 management control subsystem

```
/* basic cycle of DSS */
/* C=state;A=input */
   dss(C,A):-
/* execution and state transition */
/* B=output;NC=next state */
   product(C,A,B,NC),!,
/* start next cycle */
   dss(NC,B);
/* process of management control */
/* wait for user input A */
   product([cw,cew,cpw,msmw],command,A,[ce,cew,cpw,msmw]):-!,
```

```
    getcommand(A);
/* command is model run */
    product([ce,cew,cpw,msmw],[run,M],[genexecseq,M],[cr,cew,cpw,msmw]):-!,
/* initialize stack and tempresult */
    initializestore(stack,tempresult);
/* model run is over;yield response to user */
    product([cr,cew,cpw,msmw],[putresponse,B],command,[cw,cew,cpw,msmw]):
    -!,putresponse(B);
```

A.3 operational control subsytem

```
/* get execution sequence Ps for M */
    product([cr,cew,cpw,msmw],[genexecseq,M],Ps,[cr,ces,cpw,msmw]):-!,
    execseq(M,Ps);
/* push Ps into stack */
    product([cr,ces,cpw,msmw],Ps,stacktop,[cr,cec1,cpw,msmw]):-!,push(Ps);
/* stack is empty */
    product([cr,cec1,cpw,msmw],stacktop,stackend,[cr,cee,cpw,msmw]):-
    pop(stackend),!;
/* pop the top content P */
    product([cr,cec1,cpw,msmw],stacktop,[comp,P],[cr,cec2,cpw,msmw]):-!,
    pop(P);
/* save computation result for P and */
/* try next contents of stack */
    product([cr,cec2,cpw,msmw],[saveresult,R],stacktop,[cr,cec1,cpw,msmw]):-!,
    saveresult(R);
/* computation of M is over */
    product([cr,cee,cpw,msmw],stackend,[putresponse,B],[cr,cew,cpw,msmw]):-!,
    getresult(B);
```

A.4 processor subsystem; interpreter of model platform

```
/* ask model space manger to update IORep(P) */
    product([cr,cec2,cpw,msmw],[comp,P],[endupdate,P],[cr,cec2,cpw,msmw]):-!,
    callMSM(P);
/* compute P,save result in IORep and yield result R */
    product([cr,cec2,cpw,msmw],[endupdate,P],[saveresult,R],
    [cr,cec2,cpw,msmw]):-!, comp(P,R);
/* initialize stack */
    initializestore(stack,tempresult):-
    assign(stack,[stackend]),assign(tempresult,[]);
/** action of management control **/
    getcommand(A):-
    xwriteln(0,"input alphabet(command) please!!"),
/* read command from the dialog window */
    xread(0,A1),parse(A1,A);
```

```
    putresponse(A):-
/* display response on dialog window */
    xwriteln(0,"RESULT=",A);
    getresult(A):-
    tempresult(A);
/** actions of operational control **/
/* stack manipulation */
/* save submodel list into stack */
    push(Ps):-
    stack(L0),
    append(Ps,L0,L),
    assign(stack,L);
    pop(P):-
/* get top element */
    stack([P|Ps]),!,
/* save the others */
    assign(stack,Ps);
/* save computation result */
    saveresult(R):-tempresult(R1),
    append(R1,[R],R2),
    assign(tempresult,R2);
/** actions of processor **/
    comp(P,Result):-
/* get top node variable of P */
    root(P,[Y]),
    univ(Pred,[Y,Result]),
/* evaluate the top node variable */
/* evaluation of top node implies evaluation of all child nodes */
    Pred,
/* save result into IORep(P) */
    updateOutIORep(P),
/* clear temporary data on the model platform */
    inputlist(P,InL),
    retract(InL),
    outputlist(P,OutL),
    retract(OutL);
```

A.5 model space subsytem

```
/** model platform **/
/* test1 and test2 are stored in two cells */
    yY1(Y1):-xX1(X1),Y1:=X1+5,assign(yY1,Y1);
    yY2(Y1):-xX2(X1),Y1:=X1*2,assign(yY2,Y1);
/** linkage structure descriptor **/
/* initial values: xX1=0;yY1=0;xX2=[1,2,3];yY2=[4,5,6] */
```

```
/* they are stored in ioreps */
/* xX1 of test1 is linked to yY2 of test2:xX1=yY2 */
   iorep([test1,xX1],[[test2,yY2],0]);
   iorep([test1,yY1],[[],0]);
/* xX2 of test2 is linked to yY1 of test1:xX2=yY1 */
   iorep([test2,xX2],[[test1,yY1],[1,2,3]]);
   iorep([test2,yY2],[[],[4,5,6]]);
/** integration structure descriptor **/
/* atomic process */
   execseq(test1,[test1]):-!;
/* atomic process */
   execseq(test2,[test2]):-!;
/* total=composite mode of test1 and test2 */
   execseq(total,[test1,test2]):-!;
/** model spcae manager **/
/* following predicates are representations of Leaf(P) and Root(P) of Sec.4 */
   inputlist(test1,[xX1]):-!;
   outputlist(test1,[yY1]):-!;
   inputlist(test2,[xX2]):-!;
   outputlist(test2,[yY2]):-!;
   root(test1,[yY1]):-!;
   root(test2,[yY2]):-!;
/* process of model space manager */
   callMSM(P):-
   updateIORep(P),
   geninpred(P);
/* update value of linked varaible */
   updateIORep(P):-
/* get variable N of P and its linkage information */
   iorep([P,N],[[P2,N2],V]),
/* get current value of link variable */
   iorep([P2,N2],[L,V2]),
/* assign current value to variable N */
   assign([iorep,[P,N]],[[P2,N2],V2]),fail;
   updateIORep(P);
/* save computation result into IORep */
   updateOutIORep(P):-
/* get output variable list of P */
   outputlist(P,L),
   assign(reg0,L),
   repeat,
   reg0([Y|Ys]),
   univ(Pred,[Y,V]),
/* get computed value of variable Y */
   Pred,
/* assign computed value into IORep */
```

```
    assign([iorep,[P,Y]],[[],V]),
    assign(reg0,Ys),
    Ys=[],!,
    retract([reg0]);
/* generate input predicate and place it on the model platform */
    geninpred(P):-
    inputlist(P,[]),!;
    geninpred(P):-
/* get input variable list */
    inputlist(P,L),
    assign(reg0,L),
    repeat,
/* X=input variable */
    reg0([X|Xs]),
/* get value of Xi */
    iorep([P,X],[Lin,V]),
/* assign input variable predicate X(V) on the model platform */
    assign(X,V),
    assign(reg0,Xs),
    Xs=[],!,
    retract([reg0]);
/* [cw,cew,cpw,msmw]=initial state */
    ?-dss([cw,cew,cpw,msmw],command);
```

References

[1] D. R. Dolk, et al., "Model Integration and Theory of Models", *Decision Support Systems 9*, 1993.

[2] K. M. Van. Hee, *Information Systems Engineering*, Cambridge UP, 1994.

[3] M. D. Mesarovic and Y. Takahara, *Abstract Systems Theory*, Lecture Notes in Control and Information Science, Springer, 1989.

[4] R. H. Sprague, *Building Effective Decision Support System*, Crolier Computer Sciemce Libarary, 1983.

[5] Y. Takahara, J. Iijima and N. Shiba, "Model Management System", *Systems Science*, 1987.

[6] Y. Takahara, N. Shiba and H. Tanaka, "An implementation of unified programming on actDSS", *Decision Support Systems 18*, 1996.

[7] A. W. Wymore, *Model-Based Systems Engineering*, CRC Press, 1993

Structure and Functions of Operating Systems

Horst D. Wettstein

1 Introduction

The operation of a modern computer cannot be imagined without the employment of an operating system. Even in small real-time computers dedicated to a single task some general parts independent of the specific application can be identified. These auxiliary parts assist in bridging the gap between the mostly rather uncomfortable hardware level and the needs of higher levels induced by the application environments. Furthermore, in many systems there exists the necessity for an additional task. Modern computers perform very fast, much faster than is needed for most applications especially for those with interactive controlling. In order to efficiently use the computer's power we have to organize the concurrent execution of many more or less independent courses of events. Hence, a number of objectives arises for which solutions have to be offered by the operating system. Some of the major topics can be identified as follows:

- Accepting sets of self-contained units capable of concurrent execution
- Organizing a multiplexed usage of processors
- Supplying facilities for interaction between execution units
- Organizing the competition for system resources
- Organizing access to remote recources (this hints at distributed systems)

Most of these topics are not specific to operating systems though very typical. Larger application systems, for example, also comprise facilities for logical resource management or management of concurrent execution tracks. Thus, we are tempted to rather speak of systems architecture in general compared to operating systems architecture as a special instantiation.

The subsequent sections will discuss the above mentioned main topics of a modern systems architecture. There is no desire, however, to dig into details. Rather, we like to expedite the general concepts and ideas to such a degree that they may be transferred to design processes for arbitrary systems.

2 General system structure

A large system in general and especially an operating system deals with many activities at the same time. For example, during an interaction with a user via display a file may be printed and an electronic mail may be received and prepared for being viewed. Some of these functions are entirely separate from each other, some others do depend on a certain amount of information exchange. The greatest challenge, however, in managing and controlling such a set of tasks is the fact, that in most cases this set is not static, i.e. not known in advance. New tasks appear as a result of human interaction, others disappear after having fulfilled their specific function. One of the major concerns of an operating system is to deal with a number of more or less independent functional units.

2.1 Functional units

Large systems can only be mastered if they are broken down into sets of perhaps hundreds of self-contained tasks, independent of each other or, at most, loosely coupled. The notion for such an organization is *concurrency*, which in this context does not necessarily mean simultaneous execution (see, however, section 5). It only hints at the fact that the tasks as functional units coexist at the same time competing for and sharing needed resources.

It is not hard to imagine that this notion of a unit of function carries on as the unit of distribution, of protection, of replication, of verification, and, in a broader context, as the unit of purchase. Functional units statically consist of a *code segment* and one or more *data segments*. The former represents a description of a sequence of state transitions applied to the information present in the data segments. In order to realize such a sequence the program has to be executed by a physical *processor*. We call such an execution an *activity*. Organizationally, a functional unit may be prepared to act as a carrier of an activity. A unit with such an attribute is called a *process*. Each unit within a system may possess that capability. In this case, all units would indeed execute concurrently. However, there may exist units with a relatively short execution time, for example, which makes it unnecessary to associate with them the ability for an own execution track, albeit the feature of functional independence is given. Those units may be invoked by other units, that is dynamically incorporated in the course of an existing activity. In contrast to a process we call this form a *procedure* (as a unit of work on system level). For the sake of simplicity in this paper we concentrate mainly on processes. They cause the greater impact on the design of an operating system and indeed establish the foundation for satisfactory performance of our computer systems. Only occasionally we will contrast processes to procedures.

Having introduced eventually a large number of processes as functional units of work we must be aware that these units cannot exist solely by themselves, that is by their code and data segments. At least they have to be identified such that they can be used by other processes possibly in the form of general services. Also, their demand on common resources has to be known to resource managers for scheduling purposes. Thus, for the operating system a process p is given as a set of attributes

$$p ::= \langle a_1, a_2, \dots, a_n \rangle$$

some of them attached at creation time and remaining constant other ones changing their value over time with relevance only within certain phases of a process's life. More specifically, the attributes may be partitioned into a group describing the functional characteristics and another group relevant to the activity part. The former are very likely similar for processes and procedures. Hence, one may state the formula *process ::= procedure + activity*.

2.2 Infrastructure

The attributes of a process are physically or logically collected into a *process description block* (also called task control block). Such a block represents a process to the operating system as a member of a community. The values of the attributes may be used, for instance, for strategic decisions and may be set as a consequence of such decisions. Decision units may themselves be realized as processes. Hence, in general, the attributes of processes may be read and written by other processes, which means that the operating system has to supply mechanisms to maintain process descriptions as well as operations to access these descriptions in a uniform manner. This kind of process management constitutes a system part logically separated from the processes and existing outside or between processes. This leads us to introduce what is called an *infrastructure*. Process management represents the first major constituent of such an infrastructure.

A second part evolves from the necessity for processes to exchange information. An operating system has to supply facilities for process *interaction*. As in daily life, such interaction may occur in manifold variants. In some cases plain signals fulfill the needs for synchronization, in other cases interaction consists of the exchange of messages or the processing of common data. Interaction by its bare nature cannot find its place within functional units, at least not the elementary variants. Thus, we again identify parts outside the processes, parts that show even more typically the characteristics of an infrastructure.

We have thus worked out the necessity for two main strutural regions of an operating system: the region of functional units and the region of infrastructure. Fig. 1 depicts this first structural view. When ordering the two main parts of a system we consider the infrastructure as the fundamental base of the system. No activity is possible without such a base. The processes are seen as users existing and acting above the infrastructure. By calling operations of the infrastructure processes are able to create new processes, change attributes of other processes, or send information to their companions. The infrastructure part is often called the *kernel* or the *nucleus* of the system. Sometimes, the system designer aims at a relatively small kernel, a so-called *micro kernel*. In other cases it might be useful to offer a rich spectrum of infrastructural concepts in order to assist functional solutions for ease and efficiency.

All events and activities in a system can proceed within the framework of this two-region architectural paradigm.

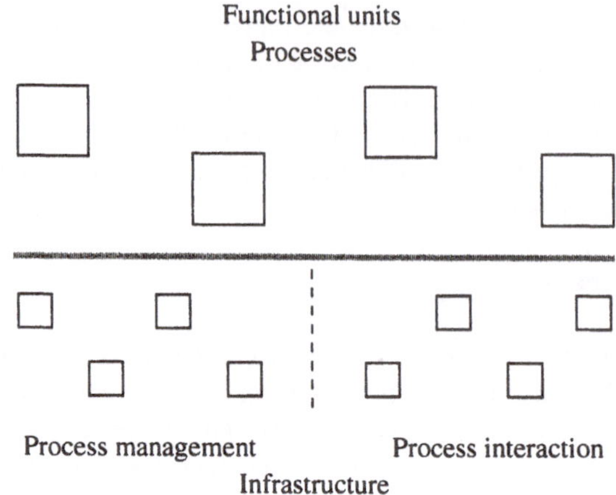

Fig. 1. Two-region structure of an operating system

2.3 Layers and partitions

At first glance, we consider a process as performing part of an application. However, many system oriented tasks such as resource management or interfacing to the human user can also be furnished as processes. Thus, a system comprises classes of processes, where each process of a class fulfills a certain kind of work. It is possible to recognize seven classes according to Fig. 2. The lower four classes generally contain resource services. In a system we deal with *real resources*, that is with peripheral devices and central memory, and with *logical resources*, that is with convenient views organized upon real resources. For both kinds we separate into functions for *handling* the resources and for the *management* of competition.

The upper two classes constitute controllers. A greater amount of work is usually divided into several steps which can be executed sequentially or concurrently. For each step, for instance, an execution environment has to be installed or, depending on the result of a step, a continuation decision has to be made. This *work flow management* appears again as a self-contained task. The highest level of controlling comprises all the *interaction mechanisms* with the environment of the computer, that is the human user in one case or a technical facility in another one.

The typical application is thus encapsulated between the controlling parts and the server parts. We consider an application as a service to a controller and at the same time as a client to the resource managers.

Induced by their kind of work the classes form a linearly ordered set which immediately suggests a layered architecture numbered from bottom to top (cf. Fig. 2).

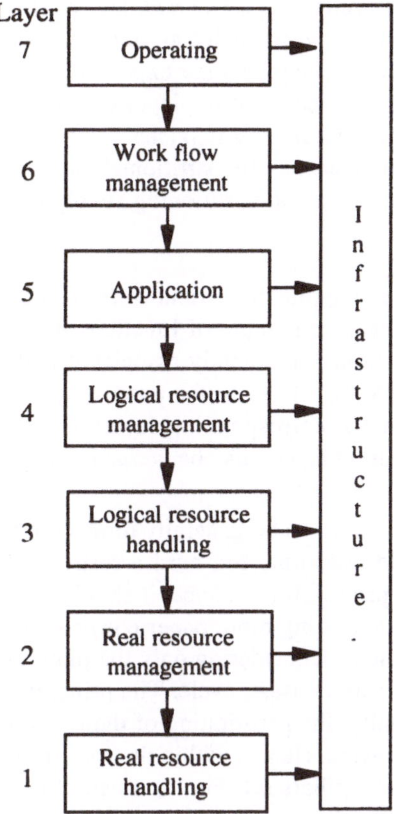

Fig. 2. The seven layers of an operating system

Among the seven layers a *usage relation*

layer i "uses" layer j

exists with i > j. The greater sign indicates that a process on a certain layer may skip lower layers, that is directly use functions further down in the line. For example, an application may access a physical device without any mapping of views, which is frequently the case in real-time systems. We speak of a *weak layering*. Mathematically we would express this feature with the term *transitivity*. The usage relation thus behaves as an order relation. The transitivity also applies to the usage of the infrastructure. Any functional unit from any layer is allowed to directly call upon the kernel. If we like we can imagine a two-dimensional structure as is already indicated in Fig. 2.

Above it has been stated that an architectural layer is equivalent to a class of specific work units. In certain cases some of these classes may be further divided into subgroups corresponding to finer type attributes. The most typical case is the grouping of the resource layers according to specific real devices (disks, tapes, sensors, actuators). Speaking in architectural terms such a grouping corresponds to a *partitioning* of a layer. For similar reasons it may become expedient to structure a layer into sublayers. A prevailing example occurs in the layer of logical resource handling when several features are transformed step by step (see section 6.4). Taken together, we can state that any layer or sublayer may be partitioned and any partition or subpartition may be layered. Thus, layering and partitioning supply tracks for an orderly growth of a system.

On the other hand, in many application frameworks systems are rather simple with respect to complexity as well as number of functions. For example, in dedicated systems real resources are often permanently associated with application processes so that no competition occurs. This leads to the possibility that certain work groups are empty which means that the corresponding layers are missing. Similarly, simple functions of two neighbouring layers may be merged to appear as a single layer.

The usage relation so far seems to be generally directed from top to bottom (cf. Fig. 2). This, however, would constitute too hard a restriction. From a process on the lowest layer operating a printer, for instance, it should be possible to invoke a message preparing process (requesting more paper e.g.) located on the application layer (system oriented application). Considering only the plain layers such an upward call appears to produce a dangerous usage cycle. The principle of partitioning, however, solves this problem. Usually, the partitioning of the real resource layer continues up into the logical resource layers. Hence, within the four resource layers corresponding partitions constitute a set of pillars (cf. Fig. 3). Each of these pillars supplies a complete comfortable service package to specific applications, for instance a display service to a message forming process (s.a.). Thus, it can be allowed for a process on a lower layer of one pillar to call on a process of a higher layer but in a different pillar. We can imagine that the pillars are linearly linked together thus dynamically keeping the usage relation acyclic.

The infrastructure region can similarly be divided into four layers (cf. Fig. 4). The topmost *kernel functions* define what is called the *system interface*, that is the set of operations known to the functional units. The next layer deals with *process switching*, which is the topic of the following section. Above we have seen that the kernel actually handles the representations of functional units and of interaction objects. This handling is primarily a matter of maintaining appropriate *data structures* in combination with efficient algorithms. A good deal of the performance quality of a system depends on the quality of these data structures. The best ones are just good enough for a system kernel. (It is a proper recommendation to implemenr at least some of them in hardware.) This holds even more stringent in dynamic systems when space for kernel objects has to be allocated and released within the framework of a kernel *space management*.

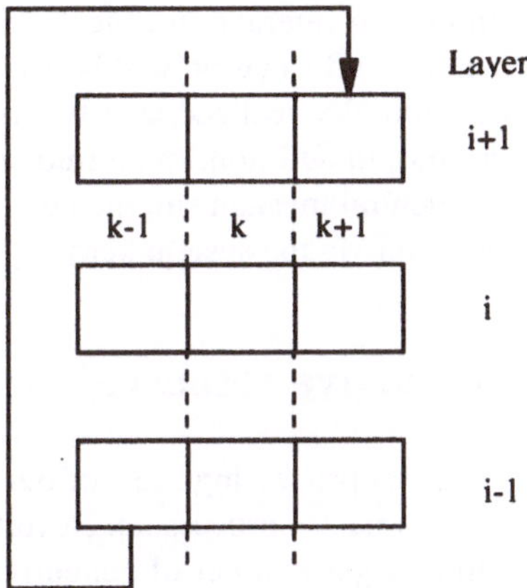

Fig. 3. Calling a higher layer

As in the region of functional units one or the other of the kernel layers may be missing (not the topmost). Partitions are almost natural if we think of the two main areas of kernel operations introduced above or of different data structures such as arrays, sequences, or trees.

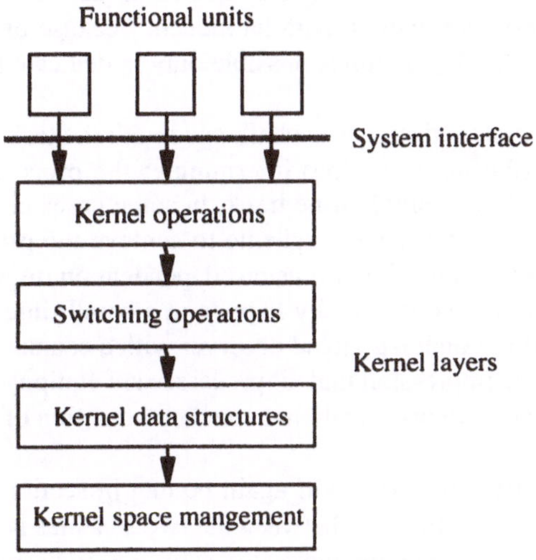

Fig. 4. The four layers of a system kernel

So far we have deduced the contents of a system kernel by a natural argument. Management of and interaction among functional units are per se designated for an infrastructure region. Kernel operations are typically short with respect to time and space. Sometimes, however, even an infrastructure mechanism may algorithmically grow and become more complex. Then, it may suggest itself to migrate such a mechanism to the region of functional units. An interaction object, for example, which has to analyse and mediate the transmitted information will better be endowed with its own activity. Similarly, if an object initially realized as a functional unit turns out to be temporally and spatially short and, in addition, to be frequently used, thus revealing the characteristics of an infrastructural mechanism, then the system designer need not hesitate to supply such an operation via the system kernel.

3 The processor as an active resource

In spite of the concept of concurrent processing most of our computers contain a single processor only, occasionally some 2-4 processors. Nevertheless, we want to structure our software systems into a larger number of concurrent units because systems designed in that way reveal much greater qualities with respect to stability, security, and modifiability. Hence, the necessity evolves to organize a scheme to execute many processes by only a small number of central processors.

3.1 Virtual processor

It would be very easy to organize a multi-process system if we had a processor available for every single process. We could attach a process to a processor permanently and have all processes execute in parallel indeed. Because of the general shortage of processors such an ideal situation is possible only in dedicated situations.

In most cases the system designer is challenged with the task of organizing a scheme that dynamically (during execution) intermingles the processing of all processes in execution mostly along a single time track, in some cases along k tracks if there are k processors available. For such a scheme to achieve the processors have to switch between processes at appropriate occasions dependent on time or function (cf. Fig. 5). Since those switching events usually happen at a small time scale (every some tens or hundreds of milliseconds), a global observer with a coarser grained time sensitivity may indeed have the impression that all processes are equipped with their own driving force. We use for such an organizational concept the notion of *virtual processor*.

The notion of a virtual resource will again be met in section 6.3 in connection with passive resources. The difference between the two schemes is the fact that a processor as an active resource attracts the work it is supposed to carry out, whereas a passive resource is assigned to a requester by some other unit. This difference naturally influences the make-up of a resource manager.

A resource manager for a processor doesn't really exist as an explicit unit (this may be somewhat different in parallel computers). Rather, the processor management appears as kind of an open routine within the execution track of the processes. The relevant functions are called *process dispatcher*.

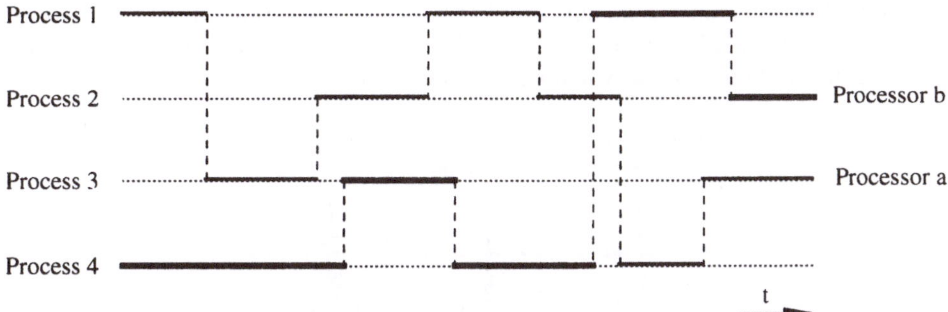

Fig. 5. Piecewise intermingling of four processes with two processors

3.2 Process switching

When designing a process dispatcher, one has to pay attention to three important details:

- When should a process switch take place?
- Should a process switch happen explicitly or automatically?
- Which policy should be applied to select a process to be switched to?

Besides that, the dispatching function has to be secure, efficient and independent of the individual processes.

The most important design principle is that all switching events are restricted to places within kernel operations. Whenever a process calls upon an infrastructure function a transfer to another process may take place. The primary reason is that within many kernel functions appropriate occasions for process switching can be identified, for instance, when the transport of a data portion from one environment to another has not been finished yet. The decision has two additional implications. Firstly, if a process should explicitly have the possibility to yield a switching by itself it should do so by invoking a kernel function supplied specifically for that purpose. Secondly, since switching situations frequently occur as a reaction to peripheral *interrupts*, these interrupts also have to invoke a kernel operation in the same way as if such an operation were explicitly placed in the body of a process's code segment. Thus, the mechanism of an interrupt is structurally mapped onto a regular kernel call. Explicit and automatic calling of kernel functions are made alike.

Appropriate situations for process switching are the following:

- End of a peripheral operation (a driver can launch a next operation)
- End of a time slice (a next process should receive its time slice)
- Creation of a new process (the new process may have higher priority)
- Synchronization event in an interaction operation (a process may have to wait)
- Explicit yield operation (especially in real-time applications)

The first four situations can be classified as automatic switching induced by system relevant conditions. Fig. 6 shows which kind of operation types are necessary in order to delegate all of the process switching events to the system kernel.

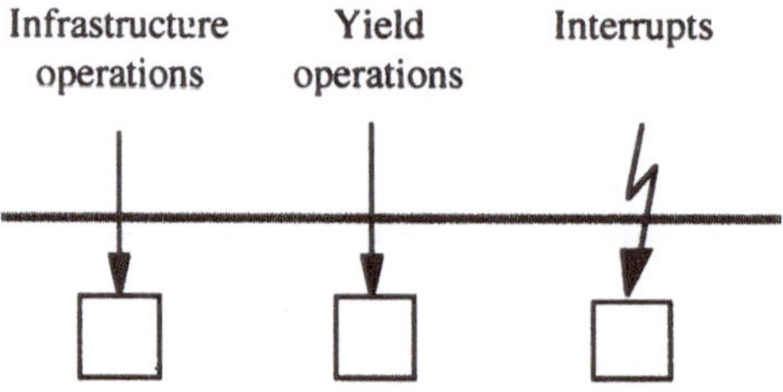

Fig. 6. Classes of kernel operations

The principal mechanism of process switching is depicted in Fig. 7. If a process invokes, for instance, a communication operation (see section 4) a certain condition may be satisfied (e.g. message arrived) or not satisfied. If it is, the flow of control returns to the calling process for continuation (path *a* in the picture). If the condition is not yet satisfied (message not yet arrived), the process cannot proceed. This constitutes a typical situation for the execution track to transfer to another process in order not to waste any processor time (path *b*). The switching function is the same no matter which kernel operation induces it. Therefore, it can be extracted from the kernel operations and placed on a next lower level within the kernel. One can imagine that the (virtual) execution tracks of all processes touch each other within the switching part such that it is easily possible for the real processor to cross over from one track to another. At some time later some other process switches back to the one that has just been suspended. At that moment control returns to the condition mentioned above for rechecking (path *c*). This procedure may be repeated until eventually the condition has been satisfied (variations may be exploited). From the view of a temporarily suspended process nothing has happened but a somewhat longer execution time within the switching box. Of course, new processes must be properly initialized in order to smoothly participate in that switching scheme.

Immediately before a cross over a process to be continued has to be selected. In simple cases the processes to be executed constitute a fixed sequence. In other cases a *selection policy* may consider individual priorities, accumulated progress, amount of waiting time, or any other behavioural characteristics in order to determine a continuation decision. Selection policies may become very sophisticated when including parameters varying over time. The highest level of strategical impact is being reached when processes in execution will be *preempted* in favour of newly arrived processes with higher priority. The selection function may be timely relieved, if the set of all processes is partitioned into subsets of runnable, of waiting (for satisfaction of conditions), and possibly of running processes. Only in that case, one identifies and maintains so-called *process activity states*.

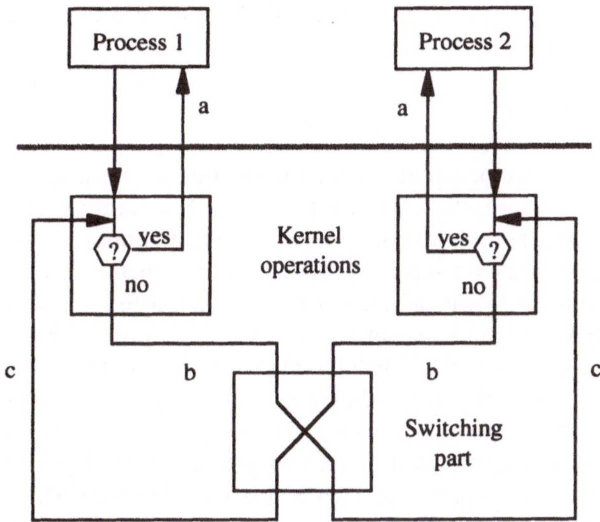

Fig. 7. Mechanism of process switching

Since part of a process's *execution state* is present in processor registers when the process is in execution, the switching part finally contains an exchange of these registers and similarly of address space information or, more generally, of an execution environment. This leads to the widely used synonymous notion of *context switching*. When a process is not running its processor execution state is kept in the process description block, this being an example of an activity attribute.

4 Interaction

If the entire work a system has to fulfill has been structured into specialized functional units, then it appears as a logical consequence that these units occasionally have to interact in order to reach a common target. Interaction in this context means exchange of information. Similar to every day life interaction within an information system is extremely manifold. In order to cope with broadly varying requirements of both the applications and the system itself, the operating system has to supply a rich spectrum of interaction mechanisms. Generally speaking, we consider at least the elementary forms of interaction as part of the system kernel, not excluding, however, the possibility of realizing more complex and more intelligent variants of interaction also within the framework of functional units.

In general, we distinguish three forms of interaction namely *coordination*, *cooperation*, and *communication*. The first one performs within the notion of time, that is with the fact, that several concurrent activities occasionally have to be synchronized with each other. The other two forms deal with the flow of data and can be associated with the dimension of space.

4.1 Coordination

When two processes are proceeding concurrently but are working on some common task, it may be necessary, that this task has to reach a certain state, established by one of the processes, before the other one is allowed to continue with its part of the goal. The two processes are temporally related to each other, they have to be synchronized. Considering Fig. 8 we can formulate the relationship by postulating that part A of process 1 has to be fully executed before part B of process 2 is allowed to commence execution ($A <_{exec} B$). When the event of termination of part A happens in the context of process 1 this fact is being signalled to process 2 which is expecting this event to occur by eventually waiting in front of its part B. This signalling is established by both processes calling on the system kernel. The kernel offers a "signal object" SO with the operations Signal(SO) and Wait(SO). The semantics of SO ensures the intended relation. When the signal operation is called before the corresponding wait operation, only this fact is deposited in the SO; process 1 is allowed to proceed. When process 2 calls the wait operation at some time later, the fact of the fulfilled prerequisite is sensed and process 2 is also allowed to continue resetting the signal. In the opposite case, when process 2 is the first one to execute the wait operation its execution is halted until the corresponding signal operation, literally speaking, delivers the signal. Only at that time the kernel finishes the wait operation and returns to process 2.

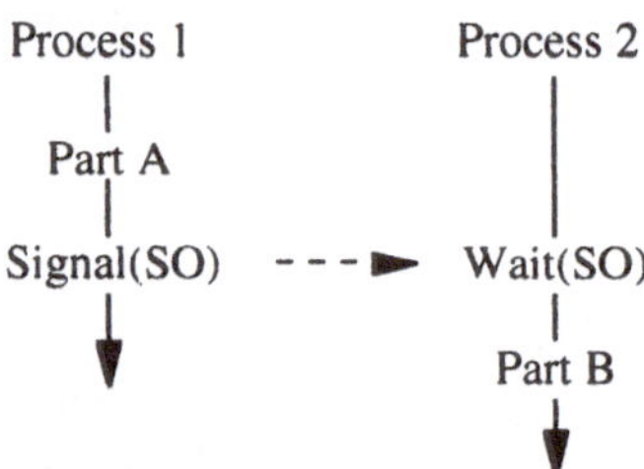

Fig. 8. The simplest form of a temporal dependency

The signal object SO establishes the simplest interaction mechanism we can think of. It can be realized with a data structure consisting of only two components, a binary quantity S with the possible values "signal present" and "signal not present", and an identifier quantity P with the values "process P waiting" and "no process waiting" (only necessary when activity states are applied). Conceptually, at an object SO any two processes can be in contact at a certain time, the pair varying over time. Many variations for special purposes can exist, such as, for instance, efficient ones for dedicated process pairs. Many copies of each kind can coexist. In dynamic systems, as the need arises, signal objects may be created and deleted, this being the case also for all of the following interaction objects.

The signal object mentioned so far is an elementary or primitive object. Two processes only are involved one as a sender and the other one as a receiver of a signal. In more complex situations also more processes can participate in a single interaction relation. Hence, signals and waiting processes may be queued within the object. Several signals from different sources may have to be accumulated before an effect on one or several target processes is allowed to take place. In such a case, for the design of the coordination object the *principle of grouping* is being applied. A group of processes on the sender side and/or the receiver side takes the role of a single process. If, for instance, on both sides the group members are connected in a conjunctive fashion, we speak of an *and-to-and-coordination* (cf. Fig. 9). Further variations may differ in the way how the groups are formed, statically at creation time or dynamically at execution time. All of these variations still display an elementary character and are, therefore, candidates of being realized within the system kernel. Finally, all signal objects have to be properly initialized. "no signals pending" and "no processes waiting" appear as appropriate values.

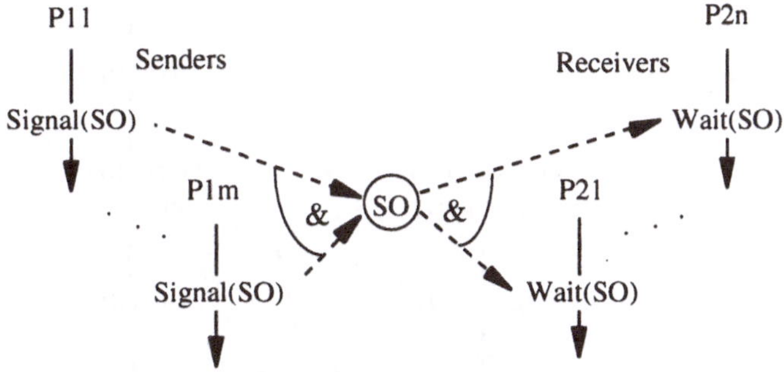

Fig. 9. An and-to-and-coordination situation

4.2 Cooperation

Normally a process comprises a piece of program and an area of storage, eventually isolated from all other processes. That means, that this storage can only be accessed from the program of the process itself. No information flow from one process to another one is possible. However, if it can be achieved that an additional amount of storage can be attached to more than one process simultaneously in such a way that all of these processes can refer to that piece of space, then information exchange becomes feasible. With respect to such a common data deposit we call this kind of interaction a *cooperation*. It is a symmetric mechanism because each participating process can apply the same and, principally, any available operation. For example, the common element may be used as a counter incremented and decremented arbitrarily by two processes or, if convenient, advanced by one process and reduced by the other one. There are many appearances of the cooperation paradigm. Global variables, common sections, or shared files are frequently used notions. It is also a very efficient mechanism due to the fact that powerful operations can be applied directly to the common data.

However, when used with processes cooperation may cause inconsistencies. Suppose one activity has partially changed a data structure and a second activity reads the data structure at such an instant. It very likely senses an incorrect value. Even worse, if both processes are about to change the data structure simultaneously, it will probably be destroyed. In order to avoid such a malfunction, the activities have to mutually exclude each other from operations on such a shared data structure. That is, if one process has proceeded beyond the begin of a so called *critical section* a second one may not cross the same point. This is a classical application of synchronization introduced above. Fig. 10 shows how the signalling concept can be applied as a protection scheme to data under simultaneous access. In simple cases when the data operations are indivisible by themselves an uncoordinated cooperation may also occur.

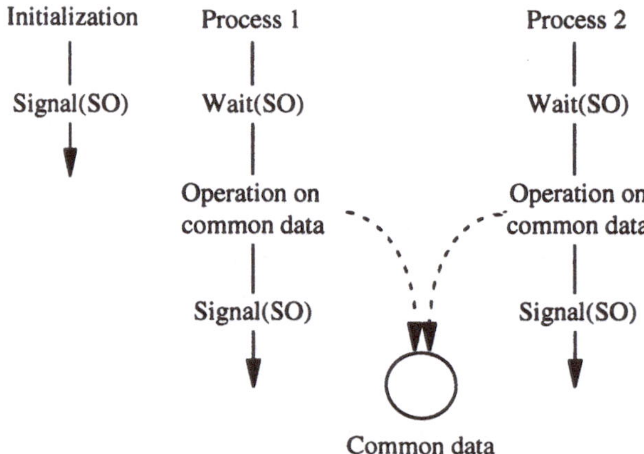

Fig. 10. Protection of common data from simultaneous access

Usually, one adapts the notion to the situation. Instead of speaking of waiting and signalling one uses the names Lock(LO) for entering and Unlock(LO) for leaving a critical section. LO stands for "lock object". The lock may take the values "locked" and "not locked" with the latter value as an initialization. Locks are kernel objects, too. Another frequently used notion is (binary) *semaphore*. Again, many variants of a lock object can be designed. Examples are counting locks which grant entrance to a predefined number of processes and keep additional ones outside, or read/write locks, which behave as an empty operation for processes only to read the data structure and exclude all other combinations of readers and writers from each other. Recently, also locks which grant anonymous fractions of a limited capacity have come into use.

4.3 Communication

The last interaction form to be described is the widespread *communication*. Actually, communication is a transportation paradigm. A piece of date (a message) from the data area of one process is transported (in most cases copied) to the data area of another process. As with cooperation we can distinguish communication with and without coordination. Uncoordinated communication is the proper means between units within the context of a single activity. The most frequent application is the transfer of parameters from a calling to a called procedure unit and the return of results. When programming in a high level language this kind of communication is usually hidden by the runtime system.

More explicit is communication between processes (cf. Fig. 11). One uses the operations "Send(CO, source_buffer)" and "Receive(CO, target_buffer)". CO means "communication object". Interprocess communication, too, may consist of copying the message from the sender to the receiver only. In other cases the processes need to be synchronized with certain events, the most important one being the arrival of the message in the target buffer. Usually, the receiver waits for the message to be available. However, other synchronization conditions are possible and occasionally appropriate to achieve useful effects. We list the possible behaviours:

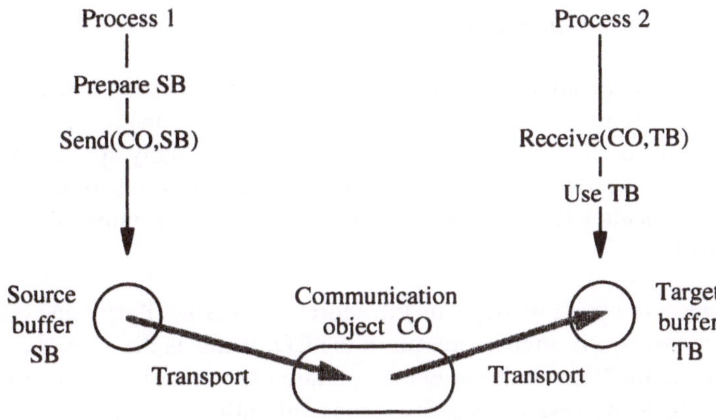

Fig. 11. The simplest communication situation

On the sender side:

- The send operation checks whether the receiver (actually target buffer) is known at the CO. If so, the message is transmitted, otherwise no action is performed (*trying send*).

- The send operation deposits the message in the CO and forwards it to the target buffer if its address is already known. In any case the sending operation terminates (*asynchronous send*).

- The send operation deposits the message and, additionally, a diversion address in the CO. If the message arrives in the target buffer, either immediately or at a later time under control of the receive operation, the sender process is diverted to the supplied address. In any case, the send operation terminates (*diverting send*).

- The send operation deposits the message in the CO and eventually waits until the target buffer becomes known from the receiving side. If it is already known, the action is the same as in the asynchronous case (*synchronous send*).

On the receiving side:

- The receive operation checks whether a message is available in the CO. If so, the message is accepted, otherwise the operation behaves as a null operation (*trying receive*).

- The receive operation deposits the address of a target buffer. If a message is already available it is accepted immediately. In both cases the receive operation terminates (*asynchronous receive*).

- The receive operation deposits the address of a target buffer and, additionally, a diversion address. If the message is not yet available, the receiver process gets the chance to continue. However, if the message finally appears in the target buffer the receiving process is diverted probably to a reaction routine. If the message is already available the diversion takes place immediately (*diverting receive*).

- The receive operation deposits the address of a target buffer and checks whether a message is already present in the CO. If this is the case, the message is copied to the target buffer and the operation terminates, the process proceeds. In the opposite case the activity is delayed until a corresponding send operation delivers a message (*synchronous receive*).

The reader may notice that the descriptions are ordered according to increasing coupling of a process to the partner process. The loosier couplings combine pairs with one trying operation (a pair with both operations trying never succeeds in transferring a message). The tightest coupling takes place when both operations behave synchronously. Then the communicating processes wait for each other and continue simultaneously (also called *rendez-vous*). Combinations establish 15 versions of coupling. The most frequent one comprises an asynchronous send together with a synchronous receive. Two such pairs, one for calling a process and one for returning results to the caller, establish the familiar *client-server-pattern*. An asynchronous receive may be useful if the target process must not be delayed but wants to be prepared for an eventually incoming new message. If this expectation fails probably an older message can be used instead. The diverting sends resp. receives can be used to realize so-called *exceptions* (software interrupts). If something unexpected happens, the diversion can be triggered, otherwise the activity proceeds normally.

Communication objects may be designed in even more versatile variants compared to coordination or cooperation. The following parameters can be regarded:

- Length of messages: constant or variable
- Coupling: various degrees between loose and tight (s.a.)
- Capacity : exactly one, up to a constant number of, or an unlimited number of messages or processes
- Grouping (messages or processes): construction of messages from several sources, copying messages to several targets
- Overflow behaviour (when capacity remains constant): overwriting, rejection, or waiting.
- Grouping (communication objects): logical buses or mailboxes.
- Association form: exit port, channel, or entry port.

The grouping together of several communication objects facilitates among other features connections changing over time (e.g. varying subsets of receivers). Transportation objects normally exist independent of processes (channels). However, for reasons of creation or administration they may be associated with the senders (exit ports) or with the receivers (entry ports). With ports only m:1 or 1:n-relations are possible. The highest flexibility offer channels (m:n-relationships). When the amount of data to be transmitted grows larger and messages may pile up the CO can be realized as a peripheral file. In that case, because of longer waiting times, it becomes more appropriate to realize the communication operations as processes.

5 Performance

In spite of the fact that computers are steadily becoming faster situations may exist in which performance with respect to a certain requirement is still too weak. Suppose a certain task making use of several peripheral processors in addition to the central processor. Its execution time may be too long even if it is having the entire machine at its disposal exclusively. In other cases a bottleneck may have developed somewhere in a subsystem, that is a load, measured in number of orders, has grown too high to be processed in due time. Hence, the need for performance improvement arises.

In many cases such an improvement can be achieved by organizing more concurrency in addition to that already established by processing different processes simultaneously. That concurrency is to specifically improve the performance measures *response time* (total execution time of a task) or *throughput* (number of orders fulfilled in a time unit). Such an organization will be successful if the entire work of a task can be divided into smaller pieces distributed onto several central (if available) and peripheral processors in such a way that as many of them as possible are executing the various pieces of the same task. Improving performance thus means making use of existing technical parallelity.

With respect to the above mentioned performance measures we distinguish between *internal* and *external concurrency*. The first paradigm aims at the shortening of the response time of a single task by temporally overlaying internal execution pieces of it. The second one improves throughput by processing several orders of the same kind at the same time.

Any kind of such concurrency can by achieved (1) by casting pieces of work into the execution context of processes and (2) by applying proper coordination or communication patterns. Thus, the desired specific concurrency presents itself as an application of interaction. For the sake of simplicity in the sequel we will show these patterns using the more elementary coordination operations.

5.1 Internal concurrency

Internal concurrency comes in two variants. The first one is indicated in Fig. 12. Starting from a sequential execution of a process we assume a sequence of sections S_1, ..., S_{p-1} data independent of each other and, in addition, also independent of a section A and a section B enclosing the S_i's (cf. Fig. 13(a)). In such a case the sections S_i ($1 \leq i \leq p-1$) can be cut out and organized as server processes. These can be triggered simultaneously from a (fork) point in front of section A. At some (join) point after section B the results must be collected. The resulting time diagram is depicted in Fig. 13(b). The scheme establishes a *p-fold internal concurrency*. The reader may notice that signal objects with an and-grouping on one side have been applied.

Sometimes, a certain subprocess (server) has to be engaged many times consecutively indicated by placing the triggering signal operation in a loop of the main process (client) (cf. Fig. 14). This establishes as a fundamental pattern a twofold concurrency of the above mentioned kind applied many times in sequence. However, if the execution times of the overlaid parts vary relatively to each other, an additional effect can be achieved, when the server process, as a supplier for example, can run ahead of the client process for some cycles, thus stocking up results from which the client can serve itself when the individual speeds change to the opposite. A similar behaviour occurs when the server acts in a consumer role.

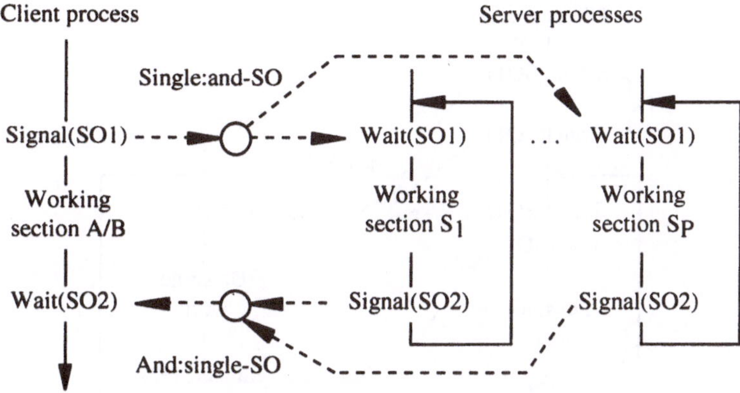

Fig. 12. Multiple overlay structure as a form of internal concurrency

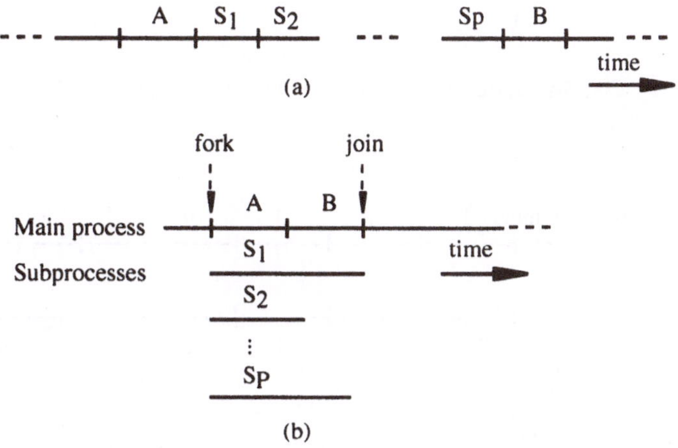

Fig. 13. Effect of internal concurrency (a) before and (b) after overlaying

For starting up such a scheme the server process has to be initialized with a couple of orders outside the main loop in the client. (For equalizing orders and deliveries the same amount of reception operations has to be placed at the trailing part in the client.) If up to k orders or deliveries can be queued we speak of a *k-fold buffering* scheme. The feature is established by applying signal objects with a k:1-capacity together with an appropriate placing of coordination operations according to Fig. 14. Fig. 15 shows a time diagram with a 4-fold buffering.

Of course, the two variants can be combined for a multiply overlaid buffering pattern.

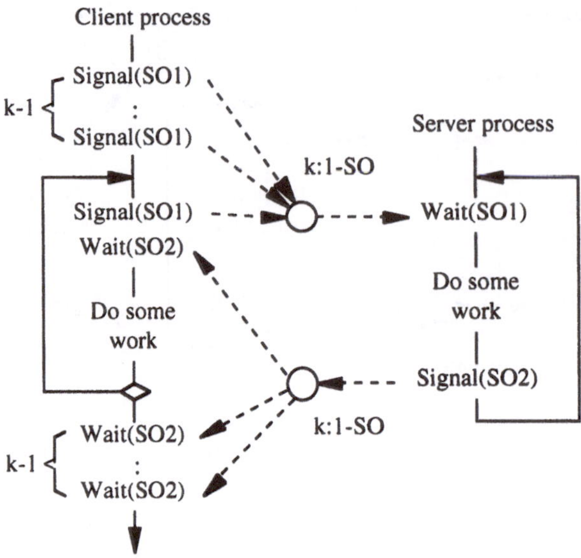

Fig. 14. Buffering as a form of internal concurrency

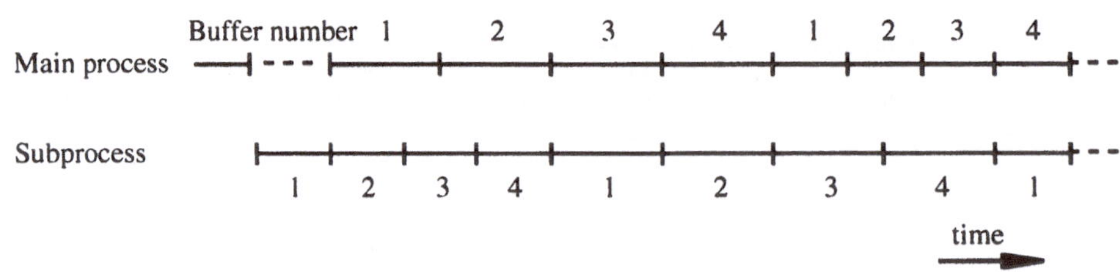

Fig. 15. Smoothing out speed changes by buffering

5.2 External concurrency

With respect to external concurrency we distinguish three well known variants: *reproduction*, *pipelining*, and *multiplexing*. Reproduction is the easiest kind to realize. We just install a server process several times and attach their order reception operations to an m:p-signal object (cf. Fig. 16). As long as there are enough orders waiting in the signal object a process finishing a previous order will pick up a next order and proceed execution immediately. Hence, as many orders as there are reproduced processes may be kept in execution.

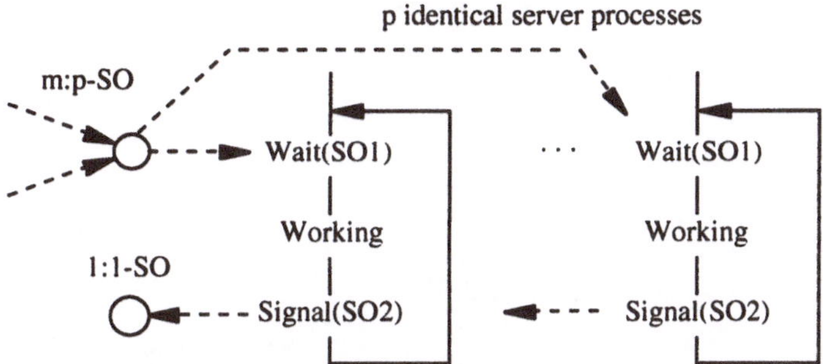

Fig. 16. Reproduction as a form of external concurrency

To gain a pipeline we take a process and cut it in several, say p sections S_1, ..., S_p. Each section is then completed to become a full process and section S_i is linked with section S_{i+1} via a signal object. The resulting structure can be viewed in Fig. 17. After an initialization phase all members of the pipeline may be working on different orders. Some partially fulfilled orders may have piled up in the intermediary signal objects. Hence, an arbitrary number of orders may have been taken into processing, restricted only by a limited capacity of the involved interaction objects.

If a server process contains a sequence of places where subserver processes are called and deliveries from there are awaited with work phases in between, then the wait operations can be moved together to form a structure depicted in Fig. 18. Original orders and results from subservers are accepted at the combined wait operation. If incoming signals (resp. messages) are identified according to their source, the process can branch to the corresponding processing parts. The structure intermingles new orders with steps of previous orders, thus multiplexing its time onto arbitrarily many orders. In other words we can say that the waiting times within the processing of one order are used to process other orders. Here the same principle is applied to orders within a process as it is the case with processes around a processor.

Usually, a certain bottleneck can be resolved with any of the three schemes. However, in most cases when considering the application context a certain form offers itself more appropriate. All kinds of internal or external concurrencies can be applied hierarchically. For example, a step in a pipeline may be reproduced, whereas another one may take advantage of a buffering scheme.

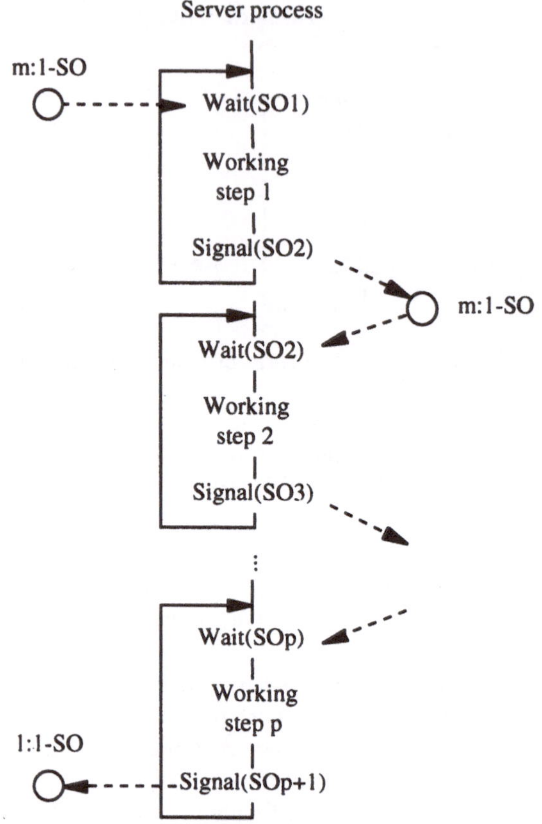

Fig. 17. Pipelining as a form of external concurrency

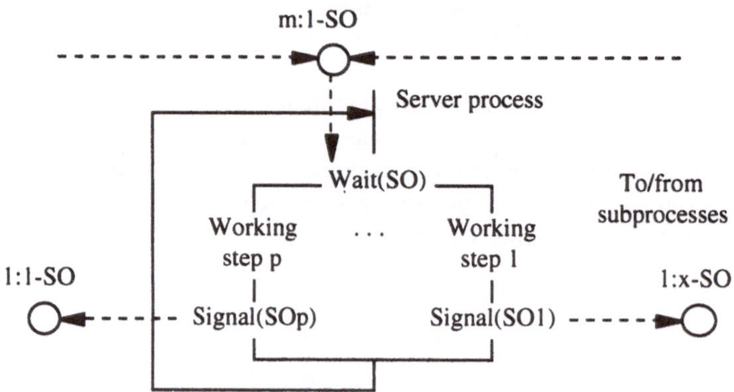

Fig. 18. Multiplexing as a form of external concurrency ($x \in \{1,c,n\}$)

6 Resource Management

One of the main and most versatile tasks of an operating system is the management of resources. If we have hundreds of processes running concurrently and each one, for instance, needs a clock for controlling time dependent activities, it is usually not possible to supply each one with a physical time unit. Rather we must establish some organizational means, such that processes can at least have an impression to possess sufficient resources to reach their aims. In this section such organizational concepts will be developed.

First of all we distinguish between *single-unit* and *multiple-unit* resources. The former are allocable only as a whole, the latter in smaller, sometimes definable pieces.

With respect to usage we recognize the *handling* of a resource, that is running and controlling a resource according to its physical, geometrical, and technological features and the *administration* of a resource in such a way that multiple usage does not produce errors, destruction, or inconsistencies. In most situations we can separate these components very distinctly. Only in some distinguished organizational modes it becomes necessary to interlink the two functions. In this section we will discuss the administrational aspects only.

Furthermore, we classify into *real* and *logical* resources. The former comprise physically visible units (printers, displays etc.), communication links, technical components in real-time processing (sensors, actuators, telephone wires), and many others. The latter class consists of all objects which are defined by mapping a real resources to views more convenient for applications. Such mappings can offer more suitable information organization, better access time, greater number of units, or any other useful feature. With respect to operating mode, we speak of *direct*, *preemptive*, and *virtual* organization. Primarily real resources, partially however also logical resources can be operated in any of these modes. Finally, a resource manager can be realized as a *kernel operation* (again in simple cases) or as a *process*, a *process team*, or a *procedure unit*. Combinations of these construction options span an exploiting design space from which specific resource manager may be chosen. The number of different variants is really not a small one.

6.1 Real resources

The simplest resource manager deals with a *single-unit resource* which can be used at a time only by a single process exclusively. The manager has to take care of the following details:

- Indicating whether the resource is free or occupied
- Granting the resource to the next requesting process if it is currently free
- Delaying additional processes presenting demands
- When the resource is returned by a process selecting one of the waiting processes, if any, giving it the possibility to continue

These functions are incorporated into two resource management operations. Fig. 19 depicts a simple flow diagram of them. Actually, these operations are very similar to the lock operations explained in section 4.2. Indeed, resource management is a widespread form of cooperation. Therefore, in simple cases a resource manager can be realized within the system kernel just as a cooperation object is. Only when management becomes more complicated and time consuming for algorithmic or policy reasons, resource managers can themselves be realized as processes.

A more complex resource to be managed is one that exists of an ordered set of allocable units (any kind of memory or pools of single-unit resources). We call it a *multiple-unit resource*. A standard request is a demand for a subset of consecutively indexed units. The manager has to keep track of which units are occupied and which are free (realized for instance by an occupancy vector). Furthermore, two additional tasks have to be dealt with:

- At allocation time, selecting a subset from the currently free parts under an appropriate strategy. Strategy goals could be (a) find a subset most rapidly or (b) keep *fragmentation* of the resource (amount of pieces too small to be used) as low as possible.

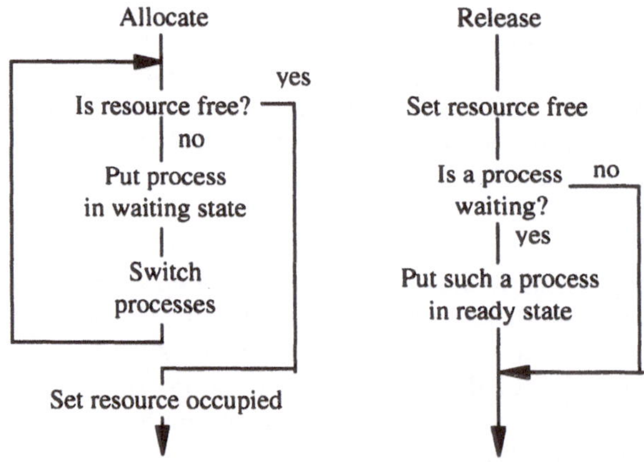

Fig. 19. Resource management as kernel operations

- At release time, selecting a subset from the waiting requesters under an appropriate strategy. Strategy goals could be (a) serve the requesters fairly or (b) keep *utilization* of the resource (relative busy time) as high as possible.

Many algorithms exist, partially rather sophisticated, to solve especially the first problem. In the literature they are specifically known as *memory management* algorithms.

6.2 Preemptive usage

So far a resource ora piece of it was considered assigned to a process all the time from an allocate to a release event, no matter whether other processes have posted demands on that resource, even if such processes are equipped with higher priorities. This kind of organization is called *nonpreemptive*.

Now, if a system is short on a certain resource and there are many processes needing the resource, its usage may be organized *preemptively*. Preemption may be time or demand driven. Under time controlling the resource is switched among the processes in regular time intervals. More common, however, is switching when a next process posts an allocate operation and this next process possesses a higher priority than the current owner with respect to the resource. The preemption is applied transitively. When a resource is finally released by a process it may be reallocated to a preempted user or to a new requester depending on their position in a priority queue.

Preemption, however, is an expensive mechanism. Usually, the resource behaves as an information carrier that contains part of the execution state of a process. If a current owner is preempted this state must either be saved to an auxiliary resource or recomputed when the resource is regained. Which mechanism is chosen probably depends on the amount of information deposited on the primary resource. If an additional depot is used preemption makes sense only if that secondary resource is available more amply.

6.3 Virtual resources

The notion of a preemptively organized resource usage may be further developed. If a number of processes is running concurrently and if all of them make use of a certain resource in an equally urgent manner, one can argue that the resource should be granted to a process at the moment it needs it. That is, any posted demand causes a preemption of the resource if it is not already attached to the requester and if it is currently not free. Naturally, such an organization is functioning only if the demand rate is not excessively high. If it works properly, each process can gain the impression of being furnished with the resource exclusively. Thus, a view of more resources than actually exist in reality is established, a view of a *virtual resource*. The only difference a process may recognize is an occacionally increased access time. Thus, the concept of a virtual resource is a means to cover the shortage of the resource by a proper organizational scheme.

The greatest effect of a virtual resource is gained when it is applied to a multiple-unit resource. Usually, the processes have a maximum demand for a certain number of units, however, only smaller amounts are actually needed most of the time. These subsets are varying over time and by size. This fact can be used to satisfy an increasing demand of one process with the decrement in the needs of another one. Hence, more processes can be supplied with so-called *working-sets* than it would be the case when considering all of the maximum amounts.

Of course, when such a preemption is applied in a multiple-unit environment, it is not immediately clear, which unit will be preempted. A selection policy must be applied which should minimize the number of unit exchanges. RNU (recently not used) or NFU (not frequently used) may be explorated. Furthermore, a mapping is needed from virtual unit indices into real unit indices. Usually, a process maintains a comprehensive virtual index space for its access operations. Opposed to that, from time to time different real units may be underlaid this virtual space (cf. Fig. 20).

Lastly, it may be necessary to install a controlling scheme in order not to overburden the concept. If too many processes participate, the number of unit exchanges may become to high with the possibility for the system to thrash (doing nothing but exchanging units).

The most famous example of a virtual resource is the virtual main memory, nowadays a feature of almost any computer system. However, the concept may be applied to any other resource as well. Interesting applications may turn up, for instance, with virtual diskettes or even virtual printers.

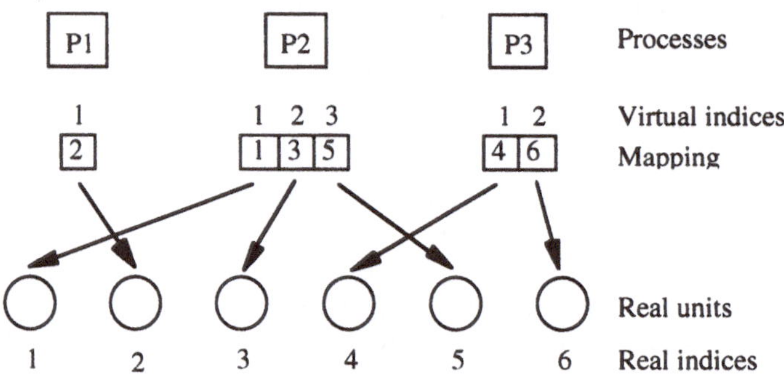

Fig. 20. Mapping of virtual to real resource units

One may notice, that within the framework of a virtual resource, a process always uses the real resource directly, that is in a way given by the physical, technical and geometrical features. This is not the case with the concept of logical resource which will be discussed next.

6.4 Logical resources

Because of its physical, technical, and geometrical restrictions the direct use of a real resource is mostly rather inconvenient. This primarily influences the system architect when dealing with device drivers. However, also an application programmer may be concerned if, for instance, he would have to write his data onto a disk reflecting the number of tracks, the capacity of a track, the sequential nature of access, the positioning of the read/write head and the like.

In order to avoid such incommodities one encloses the real resources with transformation units which create a view of a resource more adapted to the needs of the applications. Nevertheless, the demanding processes use such views in the same way they use real resources, that is via read and write operations. We call such a view a *logical resource*, because it represents an ideal resource existing as an organizational means only. Possible features to be transformed are among others:

- Size of transportation portions
- Number of units
- Power of functions
- Data set organization
- Combination of resources
- Access speed

If several features are to be transformed around the same resource, this could nonetheless be established within a single functional unit. If otherwise the single transformation steps become too complex a chained transformation can be organized, thus giving reason for sublayering (see section 2.3).

Fig. 21 shows a first kind of generating a logical resource. A using unit accesses a real driver via an additional functional unit the sole task of which is to form the better view. We call it a *logical driver*. Sometimes the goal can be reached with the assistance of another system service only. With this scheme, being sufficient in many cases, the original resource is the final target of an access operation albeit via an extra transformation step. The access speed is the one of the real resource. In contrast, if we want to gain the impression of a higher access speed we have to insert a faster resource into the access path in such a way that, for example, outgoing data is temporarily deposited on that intermediate container. Only with a certain time offset will the data be shipped to the real resource under control of still another functional unit working independently and with its own rhythm. Fig. 22 shows this second kind of logical resource creation.

Fig. 23 depicts as an example the creation of the logical resource "traffic light". It is formed by the combination of three simple lamps being operated with the primitive functions "switch on" and "switch off". The traffic light driver offers more powerful functions for turning to a specific colour (which means switching one lamp off and another one on) or to a flashing or non flashing mode (with parameters frequency and light/dark ratio). Within an application for traffic control it will be much more convenient to deal with traffic lights instead of dealing with the original lamps.

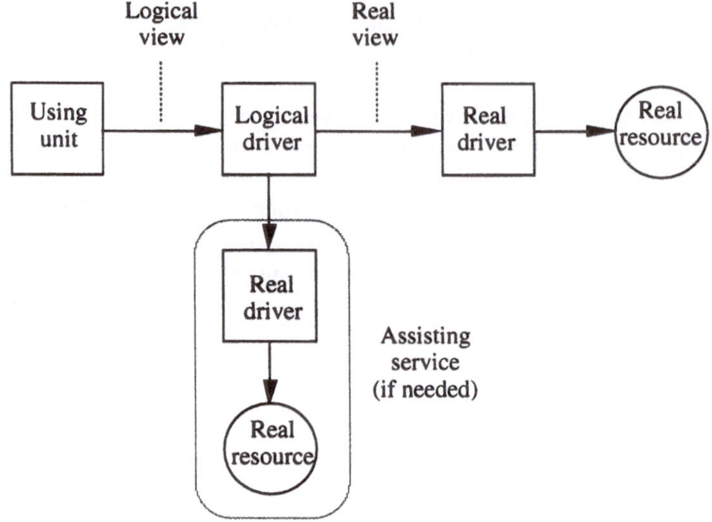

Fig. 21. First kind of forming a logical resource

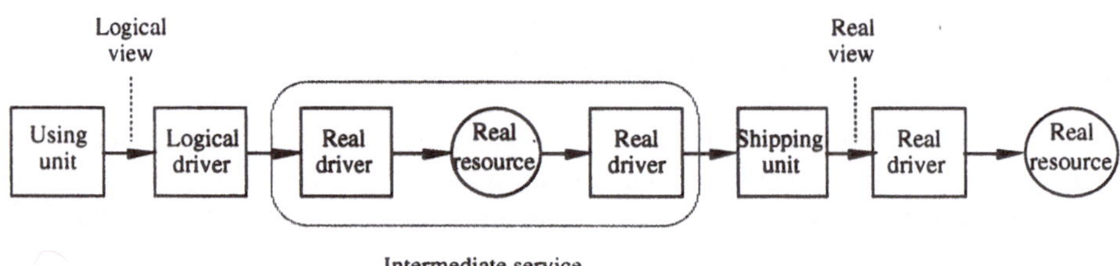

Fig. 22. Second kind of forming a logical resource

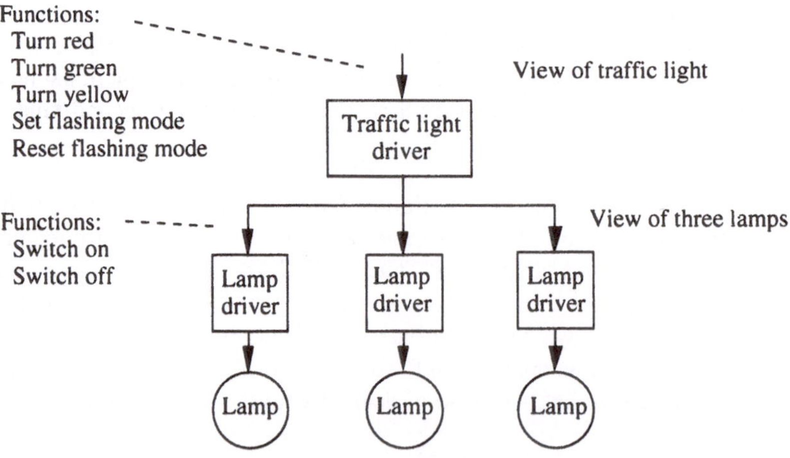

Fig. 23. Generating a logical resource "traffic light"

7 Controlling

In the privious section we have presented the various aspects of resource management. The handling of real resources will be slightly dealt with in the section to follow. With that the four resource layers (cf. Fig. 2) will have been covered. In the current section we want to say something about the controlling layers.

7.1 Work flow management

The principle task of a work flow controller has already been mentioned in section 2.3. A greater application is usually divided into a number of single steps, each step very likely realized as a process or as a group of processes. Each process solves a certain part of the entire task. In order to reach the ultimate solution all processes have to deliver their contributions. In general, the processes can be executed concurrently, however, in many cases some of the processes need as an input the results of other processes. Those have to execute serially. The work flow controller has to organize an appropriate execution pattern. In any case, if we suppose a dynamic environment each process has to be created, its program has to be loaded, needed resources have to be allocated, the parameters have to be supplied, and finally the process must be prodded and its result has to be awaited. If the execution behaves as expected, the result can be transferred to following processes. Otherwise, if an error has occurred the next steps will probably be skipped and an error indication to the user has to be created. The resources formerly assigned to the process have to be freed. In a distributed system this initial preparation of a work step and its final clear up will be supplemented by the search for a proper execution location and the transfer of programs and data to and from such a location. The more locations are involved in an application the more complex the work flow manager will develop.

A very frequent pattern of work flow arises when a data set is sequentially processed by a sequence of processes. The first process inputs the data from a peripheral device and lays down its result into an intermediate depository. The next steps forward the data under further processing to following depot places. The last step outputs the final result back to the outside world. Fig. 24 shows such a situation with two steps only. The solid lines mean calls on processes, the dashed lines indicate resource management operations, and the patterned lines hint at data flow. The flow controller apparently fits together several work units, thus actually building the entirety of an application.

7.2 Operating the system

Similar to any technical apparatus a software system, too, has to be operated by a human user. He wants to inform himself of a current state or set a system parameter to a specific value. These handlings are comparable to reading from instruments or turning tuning knobs. A third kind of operation is to trigger off an activity lasting for some time. This kind is similar to pressing a button. To realize these operating facilities one can organize a command unit according to Fig. 25.

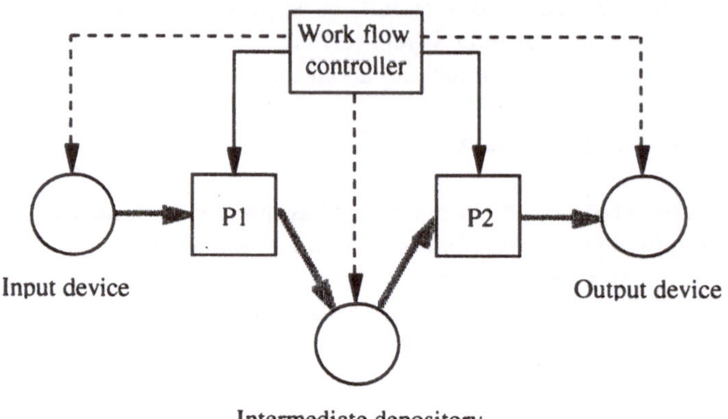

Fig. 24. Incorporation of a work flow controller

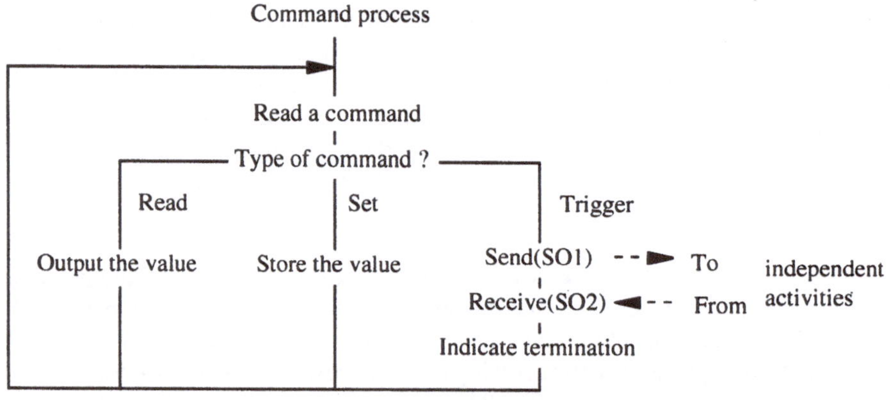

Fig. 25. Structure of a simple command unit

Reading and setting a parameter require only a short execution time and can, therefore, be placed in the body of the command unit itself. In contrast, the starting of an activity will probably be realized by sending an order message to a process followed by a synchronous receive operation to wait for the completion of the activity. The problem with such a structure is that the operability of the system is suspended for the duration of the activity. It is not even possible to manually force the activity to terminate (e.g. after a faulty handling). Unfortunately, many of our today's systems behave in such a malicious fashion.

In order to keep a system operable one needs concurrency between several operating commands. One possibility to achieve this is to organize the triggering branch of the command process as a two-phase *pipeline* as is indicated in Fig. 26. The first phase accepts arbitrarily many trigger commands and forwards them to the selected processes. The second phase expects and handles termination messages. Any other form of external concurrency (see section 5.2) could also be applied.

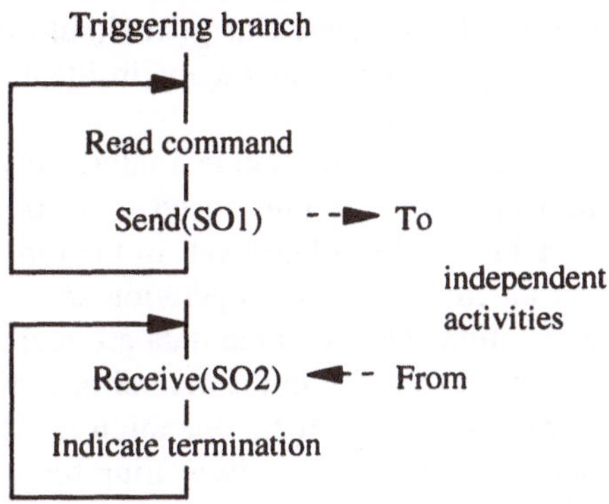

Fig. 26. Command unit as a pipeline

8 Distributed functional units

In the sections before we have mainly discussed the execution of a collection of functional units within a single physical unit comprising one or more central processors and a main working memory. This, however, may only be part of an entire hardware/software system. Considering the state of the art the system designer is confronted with the fact that a system is typically furnished with several independent peripheral processors frequently with a powerful functionality. Those processors are capable of performing dedicated tasks in parallel. For instance, a print processor may be supplied with the data to be printed on a sheet of paper and requested to decode and format the data and control the actual printing parallel to the running of the central processor. Within the notion of a family of interacting units this processing of print data may be considered as an example of a task of a specialized physical agent. Usually, such an agent is *tightly coupled* to the central processing unit, physically and logically.

Besides such an architecture, it has become very popular to physically connect several autonomous computers. This establishes the possibility to have all those computers participate in the solution of a single application. It represents a second form of an architecture for parallel execution comprising rather *loosely coupled* units. The system designer has to deal with both forms. Their realization is similar albeit different in detail. Their impact on system structure will be considered in the sequel.

8.1 Tightly coupled units

If we would not have, for instance, a dedicated print processor the task of printing had to be fulfilled by the central processor. If this were the case the print process would interact with other processes in the way discussed in section 4. Luckily, however, we can delegate this task to a separate processor and have it execute in real parallelity. Of course, this transfer doesn't eliminate the interaction necessity. However, the interaction now crosses physical unit boundaries. Our goal should be to have interaction across unit boundaries look very similar to interaction inside the central unit.

Ideally, this goal could be achieved, if the technical interface were capable of maintaining a regular communication link as introduced in section 4.3. Unfortunately, such an intelligent interface has not been build yet. In the direction from the central to the peripheral processor an emulation by cooperation and signalling can approximate this specific communication. The common data are realized as a couple of *input/output registers* (IO-registers) located in the central or the peripheral part which can be accessed by both processors. The signal operation is sufficient because in the peripheral unit there is only one process actively waiting for a demand. A simple signal can release this waiting. In the opposite direction the send operation would have to continue a waiting process and eventually preempt a lower priority process. All we can expect, however, is a technical interprocessor signal interrupting the current course of events in the central processor followed by a transfer of control to an interrupt service routine. This routine should immediately form a well defined send operation. By that, the central processor substitutes the operation the peripheral processor was not able to furnish. Fig. 27 depicts the resulting structure. Because the interrupt service routine forms a regular send operation, the interrupt should better cause an automatic invocation of the kernel (see section 3.2).

8.2 Loosely coupled units

A somewhat different, or rather extended concept must be applied, when two interacting processes are located in different computers, that is in loosely coupled processing units. Both computers see each other as a peripheral unit. However, because of the loose coupling there is no predefined cooperation area and no technical signalling. All interaction must be realized by sending messages in both directions. On the other hand, a process being invoked on a remote computer may perform an arbitrary function, not just a dedicated device handling. The process can be organized in any form, supplied with any number of parameters of any possible type. In fact, we can imagine that two processes are initially installed on the same computer being build to interact in a regular way. Then one of these processes, say the one with the server role, is moved to a different location. A proper design goal should be to keep the two processes untouched by this movement. The actual placement of a process should stay transparent to the companion process in both directions.

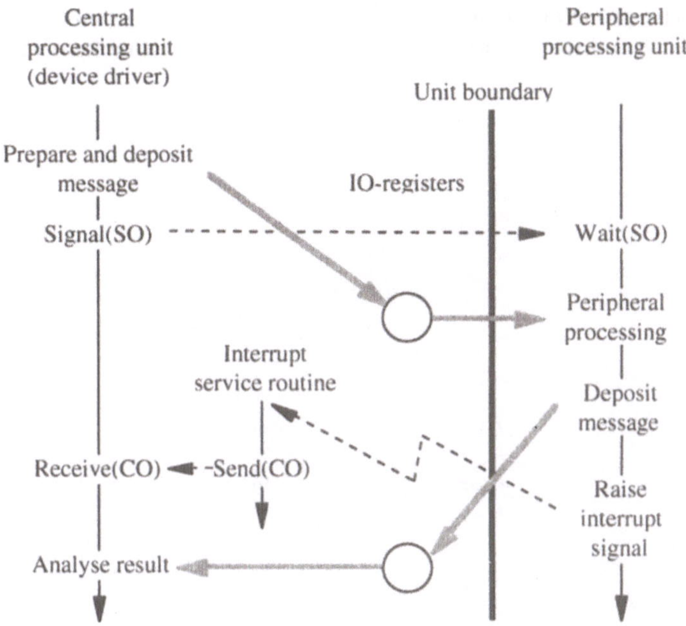

Fig. 27. Interunit interaction tightly coupled (I/O-registers in the central part)

In order to achieve this goal we mentally cut the interaction link and plug an extension lead in between, bridging the two locations. The bridge has to supply a server substitute on the client side and a client substitute on the server side which in some sense consist at least of a copy of the shell of their corresponding originals. This is a minimum the originals have to leave back when being moved apart from each other. The substitutes, sometimes called *stubs*, guarantee the original interface (cf. Fig. 28).

However, there is an additional task the substitutes have to fulfill. A request is usually accompanied by a list of parameters each one of a certain type. Since the parameters have to be transported to the remote computer they are packed into a message by the server substitute and unpacked by the client substitute. The latter has to present them to the server in the same order and in the same type as the original client would have done. If the server happens to execute on a different computer architecture, for example with a different data format (heterogeneous system), the substitutes also have to transform the data representation. Furthermore, if a matrix has to be transported the order in storage may have to be changed (e.g. from rowwise to columnwise). If many such differences have to be eliminated, the substitute units may become even larger than the original units. In those cases, distributed systems might reach their limit.

The substitutes may become even more sophisticated if the various degrees of temporal coupling of processes (see section 4.3) have to be maintained across physical unit boundaries. This problem can only be solved if communication operations are resolved into their time and space constituents (remote interprocess communication).

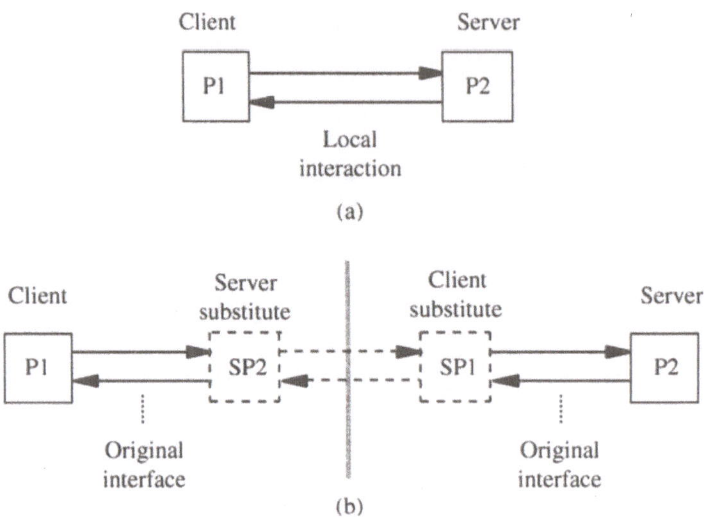

Fig. 28. (a) local and (b) remote interaction of two processes loosely coupled

Any variant of a bridging facility can be plugged in between functional units on any layer of the general system architecture shown in Fig. 2, thus giving rise to various forms of distributed systems architectures.

Literature

Burgess, R.A., Developing Your Own 32-Bit Operating System, SAMS Publishing (1995)

Davis, P.K., Operating Systems. A Systematic View, Benjamin/Cummings (1992)

Goscinski, A., Distributed Operating Systems - The Logical Design, Addison-Wesley (1991)

Milenkovic, M., Operating Systems: Concepts and Design, 2nd Ed., McGraw-Hill (1992)

Stallings, W., Operating Systems, Macmillan Publishing Comp. (1992)

Silberschatz, A., a.o., Operating System Concepts, 3rd Ed., Addison-Wesley (1991)

Sinha, P.K., Distributed Operating Systems: Concepts and Design, IEEE C.S. Press (1997)

Tanenbaum, A.S., Modern Operating Systems, Prentice Hall (1992)

Vahalia, U., UNIX Internals - The New Frontiers, Prentice Hall (1996)

Wettstein, H., Systemarchitektur, Carl Hanser (1993) (in German)

Object-oriented system development
- concepts and tools -

Wolfgang Kreutzer

1 Background and motivation

Modern computing devices and their patterns of usage continue to change dramatically at an unprecedented rate and technological capabilities as well as our expectations are very different from what they were in the past. Personal workstations and networks have largely replaced the expensive mainframe computers of the sixties and seventies and users are increasingly unwilling to tolerate tools and interfaces that are arcane and cryptic. Instead they expect technology to adapt to its users and support languages and habits which take account of their users' strengths and weaknesses. While this is a very positive and desirable development it greatly increases the resource requirements and complexity of computer software. Fortunately computers continue to become faster, smaller and cheaper, so that the difficulty of developing effective and reliable programs rather than traditional concerns with hardware efficiencies has increasingly come to dominate the cost of applications development. The term *software crisis* is often used in this context and the question of how to structure and control the "growth" of large programs has become a mayor focus of software research.

Figure 1 depicts communication between humans and computers, built on a common base, ie. a programming language, as its medium to exchange information.

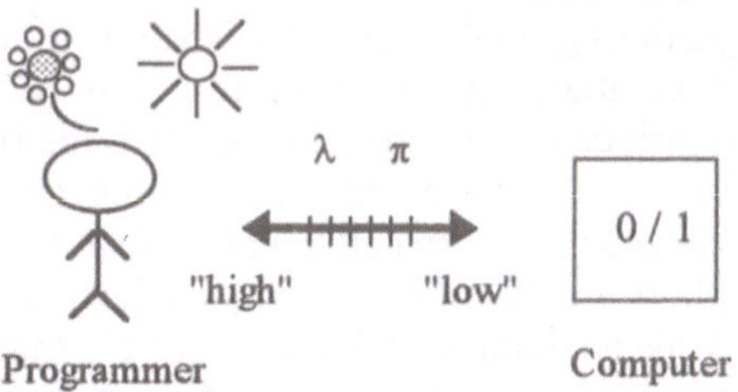

Fig.1. A model of human - computer dialogues

Any successful communication needs conventions regarding the meaning of signs and we must therefore decide which concepts we use to encode programs in. The closer such symbols correspond to a "familiar" context, the less work is needed for their encoding and decoding. Of course, if "native" frameworks differ markedly between the entities involved in a communication (ie. if they come from much different cultures), a tradeoff must be made with regard to which language to use. In the early days of computing, computers were very expensive and primitive devices, so that there was no question that communication (ie. in terms of the concepts supported by programming tools) had to occur at a level close to the hardware. The first programming tools (ie. machine code and so called assembly languages) are therefore known as *low level languages* today, simply because we consider the various forms of human communication as far "above" and superior to such primitive means of expression. Since then the most common classification of programming tools has drawn on this notion; ie. a language is "higher" level if it supports a mode of communication which is closer to the mental framework of a typical user. It must be noted that what is familiar depends very much on the user. There used to be a small group of people who had become so accustomed to using machine code that it became very familiar to them. However, the relevant point here is that humans have finite mental resources and being forced to communicate at this level constrains the complexity of the patterns of thought we are able to use. Much of the history of computing can in fact be described as an ongoing quest for reaching ever higher levels of communication, resulting in a vast number of so called *high level languages* (eg. Fortran., Algol, Cobol, Lisp, ...) and *very high level languages* (eg. Prolog, 4GLs, Miranda, ...) - with no end in sight.

Regardless of level, *all* programming languages must offer a set of concepts through which computations can be described, cast into suitable linguistic abstractions. Although most programming tools are *turing complete*, ie. they are *sufficient in principle* to describe any computable problem. In practice new languages offer new ways of viewing and thinking about particular aspects of the world - and not all are equally well suited to *all* types of applications. Thus we have programming languages which excel at numerical calculation (eg. Fortran, ...), business data processing (eg. Cobol, ...), systems programming (eg. C. ...), text processing (eg. Snobol, ...) or symbol processing tasks (eg. Lisp, ...). Of course, some languages are designed to be more "wide spectrum" than others (eg. PL/1 or Ada). In analogy to tools in a craftsperson's toolbox we need widely useful but relatively unsophisticated coding tools as well as highly specialised languages which have been optimised for a narrow range of tasks.

Most early programming tools naturally sought to extent the framework suggested by classical computer architecture (eg. the so called *von Neumann machine*), making only minor concessions to ease life for their programmers (eg. symbolic references, procedures, more sophisticated operations, control and data). This has led to many variants of the so called *imperative* style, but it has since been argued that other metaphors for describing a computation may provide better support to human problem solving techniques. Of the many alternatives which have been explored three such metaphors have lately achieved some degree of popularity. These are referred to as *logic-based, functional and object-oriented* styles of programming.

While research into conceptual foundations and the effective implementation of functional, logic and object-oriented programming styles has been conducted over many years, the wider computing community has only recently shown any interest. *Logic-based programming* became popular after its adoption by the Japanese "5th generation" effort (an ambitious attempt to build computers well suited to logical inference in support of so called expert systems). *Functional programming* styles are still rarely used outside universities and research institutions. *Object-oriented programming*, however, has now "hit the streets". The main reason for its mushrooming popularity is due to a growing unease in the software industry that classical methods of program design, as characterised by structured analysis and design, imperative programming tools and relational databases, seem to offer little hope for solving the complexity problem. The state of software engineering today is in fact reminiscent to what the philosopher of science Thomas Kuhn has identified as a *paradigm shift*, ie. a time of crisis in which an established methodological framework (ie. imperative programming) is about to be overthrown by a new one (ie. object-oriented design). Kuhn (1970) is a science historian who, in the early sixties, wrote "The Structure of Scientific Revolutions" in which he discusses two qualitatively different types of Science Normal science is anchored to a paradigm, a mental, social and methodological framework which determines what background assumption can safely be made, what questions should legitimately be asked, what types of methods we are allowed to use, and how we can recognise valid answers. As long as things go well a paradigm remains largely unchallenged (and heretics are labelled as crackpots). If, however, too many things go wrong (ie. we can find no acceptable answers to urgent questions) the paradigm starts to topple and exceptional people will propose new frameworks, one of which may become a new paradigm. Notable examples for such paradigm shifts are the emergence of relativistic physics, quantum mechanics, or the structured programming movement. Generally, however, one must be careful not to overstate the capabilities of any new technology, a restraint which the computing culture has found difficult to accept. Object-orientation can not be the answer to *all* our problems (nor will any other single method), but it will and has already become a very valuable and widely used tool for tackling software complexity and there are good reasons to be confident that its growth in popularity will continue.

2 General concepts

Object orientation is not a novel idea. While the term was invented by Alan Kay (1977) in connection with his work on Smalltalk, object orientation's most central concepts can be traced to an early graphics system called Sketchpad (Sutherland 1963) and a simulation language called Simula67 (Dahl and Nygaard 1966). The underlying ideas have an even more time-honoured history if we do not constrain our search for their intellectual ancestors to computing science itself. In fact, the object-oriented method fairly accurately reflects the practice of what is widely known as the *scientific method* of problem solving. Given this pedigree it may seem rather surprising it has largely been ignored by main-stream computing for over 20 years. One of the reasons for this obscurity can doubtlessly be found in its relatively large computing and memory requirements when compared to conventional programming techniques. If we remember that imperative programming styles grew out of formalisms based quite directly on the so called von Neumann style of computer design (ie. they are basically "higher level" extensions of the concept of Turing machines) it should not be surprising that tools based on different formalisms such as predicate logic for logic based languages and lambda calculus for functional languages will need to pay a performance penalty. Since they were therefore always seen as wasteful in terms of computing resources while human effectiveness was largely ignored or dismissed as much less important, the use of alternative programming styles has long been considered too expensive and even frivolous for practical applications. As a result their use and development was largely restricted to the research community, a perception which only changed in the mid 1980s, prompted by two developments which Balci and Nance (1988) refer to as *applications pull* and *technology push*. The first of these terms reflects the fact that applications have now become so complicated that we need all the help we can muster to build them at all - never mind the expense. The second term indicates that we are now in the fortunate position of being able to "throw" more hardware at the problem than possible or economical before. Computer memory in particular has become very cheap and this has had a tremendous impact on the economics of different styles of software development.

Phrases such as *software factories, rapid prototyping and reusable components* reflect the desire to base software construction on more productive, reliable and predictable methods, similar to those used in the engineering community. Although object-oriented techniques may give better assistance than more conventional methods none of these goals can be achieved easily. Designing programs is a difficult task, which has not responded well to any attempts to replace "art" with more method. There is in fact growing evidence that this may be inherent in the subject and not just a reflection of failures in understanding (Blum 1996).

While object-oriented programming has entered main-stream software development only recently, it has for many years been used as a successful and popular tool in simulation modelling (Kreutzer 1986), a task for which it is particularly well suited. Users of object-oriented programming tools can therefore draw on a long history which transfer quite well to other types of software development.

We have already mentioned that object-orientation shares the philosophical basis of what has become known as the *scientific method.* Both strategies seek to build models of reality with a high degree of descriptive and predictive power. Through the works of many philosophers and scientists we know that it is quite impossible to derive any objectively "correct" understanding of the world from our senses, as Dennet (1991, page 101) rightly observes: "... wherever there is a conscious mind there is a point of view. A conscious mind is an observer who takes in a limited *subset* of all the information there is ..."

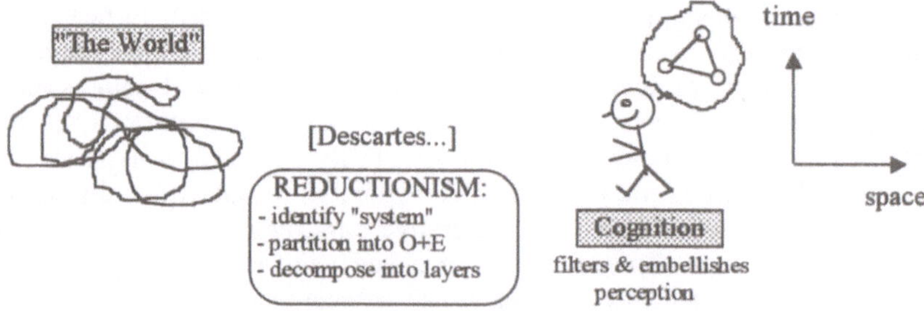

Fig.2. Making sense of the world

The elusive notion of *understanding* must therefore be viewed as a process of filtering and interpretation of information relative to a taken point of view. If we accept the underlying thesis that we can not help but use preconceived concepts to process raw sense impressions, it seems plausible to assume that a spatio-temporal network anchored in objects and events serves as our most basic frame of reference. Coad and Yourdon (1991) express this quite well: "Object orientation is based upon concepts that we learned in Kindergarten: objects & attributes & behaviours, classes & members, wholes & parts, ..."

Although we can never *guarantee* to be in possession of provably correct models of the world we still try our best to find useful ones, ie. models which are helpful for explanation and prediction. This observation stresses the importance of abstraction, ie. simplification, as a crucial foundation for model building, and the strategy known as *reductionism* can help us in building abstractions systematically. Although reductionism is now well established in western society as *the* method of rational inquiry, this has by no means always been the case and the so called enlightenment in the 17th century may be seen as the "paradigm shift" responsible for its almost universal adoption. Reductionism is foremost a strategy for controlling complexity. In order to analyse a complex phenomenon it asks us to identify and isolate a *system*, as a collection of objects and events which are relevant to a given task and which are spatially, temporally or causally related. Of course, by singling out any such objects from others we must throw information away, an act which we justify by the often tacit assumption that it will not be *relevant* to the problem we set out to solve. Reduction is a recursive technique and can again be applied to each of the simpler objects we arrive at (ie. by considering each of them as decomposable systems in their own right), until a level is reached at which elementary objects and their interactions are simple enough for us to understand. Of course the success of this strategy critically depends on the assumption that in spite of our desire to simplify we have managed not to abstract the problem away. It therefore works best for simple problems in which only a few factors are seen as important. While less successful in the biological and social sciences reductionism has reigned unchallenged in the natural sciences. More recent investigations into so called chaotic systems, however, have questioned its adequacy even there.

Object-oriented styles of systems analysis and program development support a reductionist strategy well. From this perspective writing a program becomes building a model of a domain. Such models contain objects, relationships among these, and processes in which they engage in. Objects with similar properties and behaviour can be grouped into classes, whose descriptions encapsulate all properties and actions that we consider relevant for our purpose. Relationships bind objects to structures, which then play their role in a process. In object-based software processes spawn computations, described by procedures framed as a series of messages sent among objects.

Viewed in the temporal dimension such computations unfold as a sequence of all of the participating objects' changes in state.

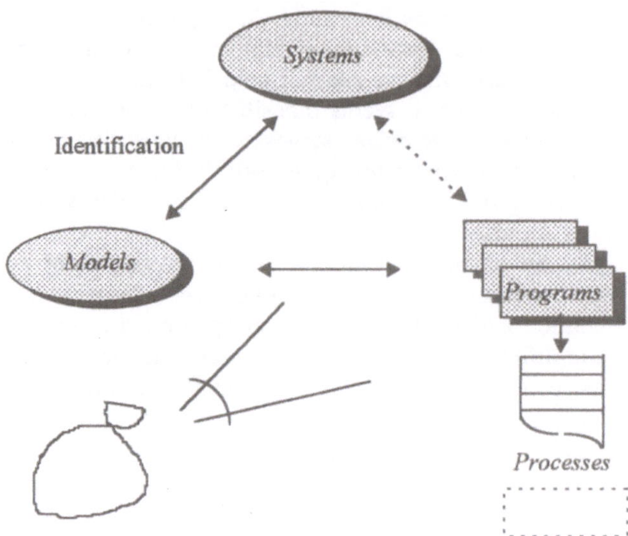

Fig.3. A model of the modelling process

While any programming tool can in principle be used to transform an object-oriented model into an executable program, the so called *conceptual distance*, defined as a measure of mapping one set of concepts into another, may become large if low level tools (eg. languages like Fortran, C or Pascal) are employed for its implementation. Here it would be classed as "small" if the modelling constructs were part of the linguistic abstractions provided by the programming language, and "large" if it offered only some general but unrelated data structures and operations. To safely map models into a program it needs to be small, otherwise the complexity of the mapping may quickly become unmanageable. This means that the concepts we use (ie. at the modelling and programming level) should not differ too much from each another, a requirement which calls for a tool which either already offers an appropriate set of abstractions, or which makes it easy for suitable concepts to be defined. The second alternative is exactly what object-oriented programming styles can provide.

To justify our analogy between model and software construction the model in Figure 3 illustrates and compares the main phases of their development cycles.

In simulation these phases are commonly referred to as system identification, model design, model validation, model implementation, program verification, and experimentation. The lifecycle of object-oriented software shows all of these phases as well. The names, however, have changed, and the literature typically distinguishes between *object-oriented analysis*, *object-oriented design*, and *object-oriented programming*.

From what we have written so far it should be obvious that *objects* form the core of object-oriented programs. They have an identity, some properties and a set of actions they can perform on demand. Demands are carried by *messages*, ie. whenever we prompt objects for actions we can think of invoking a "contract", the details of whose execution we need not worry about. This has the potential to reduce mental complexity by keeping information "local", ie. within an as tightly constrained context as possible. This notion of *encapsulation* is again by no means a new one, and in computing its merits have long been promoted under a variety of labels: eg. abstract data types, working sets ... By insisting on strict encapsulation of *all* information relevant to an object (ie. state *and* behaviour), object-orientation carries this idea further and objects now become autonomous entities with which we communicate through a *message protocol* or *signature*. Applying the principle of *information hiding* we can limit complexity by keeping this interface as narrow as possible, separating an object's *specification* (as defined by externally visible services) from its implementation (as described by internal properties and methods).

Encapsulation also elegantly supports the concept of *polymorphism*, meaning that the effect of an action can take "many forms", depending on the context in which it is performed, in that we can name actions without undue concern about the context in which they may be applied; ie. the kind of object to which messages are sent. This is achieved by delegating the association of messages with an implementation to the receiver. For example, we can define a generic *show* message which behaves differently over a wide range of objects (eg. text, images, figures, movies, sound, ...). While all of these objects' responses will result in some kind of for display, the results (ie. what will appear on a screen, printer, or through a speaker) as well as the code may differ widely.

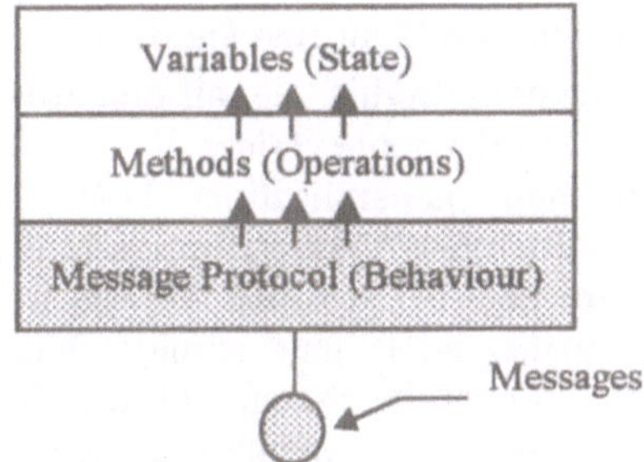

Fig.4. An encapsulated object

Composition (ie. creating large structures from small ones) is another technique whose use for reducing complexity is well established. Object orientation supports this idea by assembling a program in layers, where the components of complex objects are themselves simpler objects - ultimately grounded in a given language's linguistic abstractions.

Fig.5. Wholes and parts

Figure 5 shows an example decomposition of a "whole" car into some "parts". This type of abstraction is well supported by most modern programming languages.

In addition, however, object oriented tools provide classification and generalisation to foster *abstraction*, ie. the removal and aggregation of superfluous detail. *Classes* are templates by which sets of similar objects (remember that the notion of similarity requires a point of view (ie. a purpose) - similarity is always in the eye of the beholder !) can be described and they can also be used as *generators* to create objects of a given category - eg. an *account* class in a banking program could be responsible for creating accounts on demand. Either inheritance or delegation can serve this purpose, but weaving classes into inheritance hierarchies is the most common strategy used by object-oriented programming tools. *Inheritance* reflects the fact that some classes differ from others in only a small number of aspects. To improve flexibility and mental economy we can therefore factor all commonalities into "superclasses", from which they are then inherited by all of their variants. This multi-layered classification is prevalent in all sciences; well known examples are Linne's systematisation of the animal world or the periodic system of elements. While such schemes for classification are often arranged into hierarchies they can also form lattices or networks. In object oriented development this is reflected in single or multiple inheritance, but because of semantic complexities (eg. what to do when aspects with the same name are inherited along different paths ?) multiple inheritance is not supported by all tools.

Figure 6 shows an example where the concepts of apple tree and apple orchard are introduced as specialisations of fruit tree and fruit orchard. Note how the notion of composition runs orthogonally to this - ie. all orchards *contain* fruit trees, but in apple orchards they must be of the apple variety.

Both composition and generalisation have specific advantages and disadvantages. Whole-part *composition* can be programmed quite easily, eg. by using variables whose values are objects. Note that we tend to refer to these as "links" or "pointers" at the implementation level. This strategy has the advantage that such relationships are dynamic and can be changed on the fly (ie. at program execution time). On the negative side we need to mentally trace objects' history to make sure what associations will currently hold. *Classification* is better suited to define enduring and less dynamic relationships, which it makes easier to reason about.

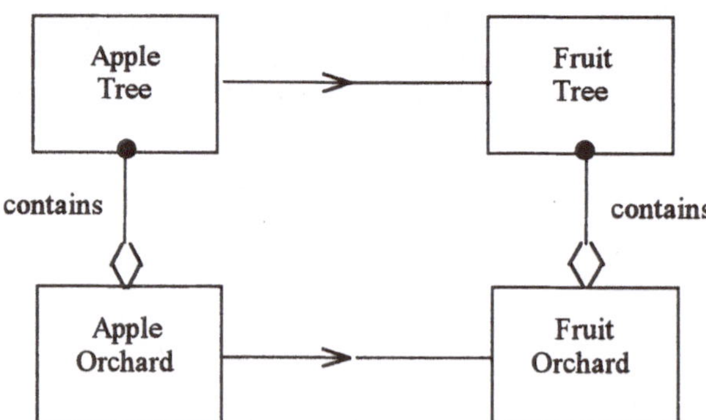

Fig.6. Decomposition & inheritance in an apple orchard

However, few general purpose programming tools support it directly and we trade easier understanding against lost flexibility. On the positive side inheritance provides us with a convenient means of customising objects by specialising and redefining canonical structures and behaviours, which is an important prerequisite for constructing reusable library classes. Both strategies impose a performance penalty caused by indirect references - ie. to refer to an object we may need to traverse paths with multiple segments.

3 Object-oriented analysis and design

Once they are used in large scale program development object-oriented methods can no longer cater for only the needs of a single user and more sophisticated tools to support analysis and design as well as methods for managing programmer teams are required. As a consequence research in these areas has seen an explosive growth over recent years.

Object-oriented development does not sit well with classical methods of software construction and can be quite different from the recommendations given by *the Structured Programming* and *Structured Design* traditions which have dominated main stream computing since the sixties and seventies. Since the eighties these methods have been augmented by CASE (ie. Computer Aided Software Engineering) tools and a whole industry has emerged to support their ideals. Unfortunately the tradition of *top down functional design* has not been able to live up to all of its promises.

In introductory programming classes students usually start from precise specifications and use formal means of deduction to arrive at a working program. The ideal of this method is for a program and its *proof* to evolve "hand in hand". Although the power of this style of *top down* development has been convincingly advocated and demonstrated by well known computer scientists using small and much cited examples, it is only suitable for problems whose properties are so well understood that invariant and rigorous specifications are available prior to the coding stage. By asking us to avoid applications with ill-defined specifications and first gain a precise understanding of the space of a problem's solutions *before* we embark on a program, this school of thought moves most interesting problems beyond our reach. In many areas which are characterised by a high degree of mental complexity there is a dire need for some guidance on how such an understanding can ever be reached. A possible solution to this dilemma suggests itself when we look at how other disciplines (eg. engineering or architecture) address a design. Often a bottom up view of problem solving (ie. modifying solutions which are known to work) is prevalent there. In analogy to this strategy we would start with a collection of program components and compose applications by combination and customisation. This style fits an object oriented perspective well - systems are built as combinations of autonomous elements (ie. classes), where each of these represents an abstraction of interest, in terms of its properties and behaviours.

Abstraction, encapsulation and layering can do much to help master complexity, but we can not rely on these methods alone. In fact, finding the *right* representation for a problem (whose description may be quite fuzzy at the start) is a difficult task and experimentation is often the only way to make progress. Tasks of this nature should be considered as design rather than implementation problems and *program* and specification should evolve hand in hand. The method of *exploratory development* of prototypes recognises this need and is therefore an indispensable tool for building any complex pieces of software· Instead of following a waterfall model's prescription of strictly top-down decomposition we proceed in two directions at once - from both, top (eg. by identifying relevant classes of objects, ...) and bottom (eg. by adapting existing components which are useful). This process is highly dynamic in nature and analysis, design, implementation and experimentation with prototypes tend to be tightly intertwined. While this method of *rapid prototyping* has sometimes been be-devilled as only a slightly more dignified form of hacking it has the potential for finding working solutions quickly, which can then be further refined and polished.

At this stage it is important to stress that classical and exploratory styles must not be seen as exclusive, since the second serves to generate the insights on which the precise specifications needed for the first can be based. While exploratory development styles can boost individual programmers' productivity more formal frameworks are needed to support programmer teams. Lure by financial gain a large number of prescriptions for object-oriented analysis and are vigorously competing with each other. Although it is too early to predict which method will come to dominate one can detect many similarities among them.

Deciding what should be included in a conceptual model falls within the realm of *analysis*, while commitments to a particular kind of computer-based representation belongs to *design*. The results of design can then serve as specifications from which implementation can proceed.

Typical problems which must be solved during *analysis* lead to questions such as the following:

(1) What *classes* do we need ? (eg. identify relevant concepts which should be part of a model)

(2) What *functionality* must be supported by the model (eg. identify each class' responsibilities).

(3) How should classes be *related* to each other ? (eg. describe compositions & generalisations).

(4) How do objects *cooperate* in an application (eg. describe scenarios and event traces).

Although some guidance on how to approach questions like these can be given one must never forget to remember that the results of a modelling exercise depend strongly on the conceptual framework used by the modeller, and that different observers will classify reality in different ways. Finding good answers to the above questions needs experimentation and can greatly be aided by relevant experience. To answer the first two questions it is often suggested that one should study the language of a domain. This can be done through the study of journals and textbooks or by talking to experts. At the very least the outcome of this exercise should be an improved understanding and a *data dictionary*, of key terms and phrases. The next step is classification, which demands that we find similarities, a task which presupposes choosing a point of view and level of abstraction, so that we can group things that show common structure or behaviour. Creating a class can be justified when relevant data is (ie. there is enough functionality and variation) or when there are opportunities for reuse and extension. For example, information about a vehicle's speed will rarely be elevated to classhood but is normally better stored as an attribute. Information about files (eg. in a design for an operating system), however, will often be complex enough and have sufficiently varied functionality to warrant a class of its own (Moessenboeck 1995).

Protocol analysis (ie. stepping through a collection of typical *use cases*) can be a helpful and role playing has also been found productive. One particularly popular method was invented by Beck and Cunningham (1989), based on their Smalltalk experience. It uses what has since become known as *CRC cards*. This term stands for "Classes - Responsibilities - Collaborations" and indicates the three types of information participants in a brainstorming session are expected to write on a set of index cards. During such sessions participants suggest and criticise proposals for classes, responsibilities and collaborations, followed by actively playing the roles of such objects, simulating "what happens when ..." types of questions. Clients (ie. end users) as well as designers and implementors participate, so that the essential functionality and "look & feel" of an application can be tried out and discussed. Often such sessions uncover inadequacies in design and may even suggest unanticipated types of usage. Eventually the information recorded on these cards will serve as a blueprint for an application's structure and can also help documenting the software later. Once we have derived what classes we may require we arrange them into whole-part and generalization-specialization relationships. Typically the results are documented in simple classification graphs and variants of the entity relationship diagrams commonly used in database technology.

Figure 7 shows a class relationship diagram for a small personal finance application, using Coad and Yourdon's (1991) symbols for *subject*, *structure* and *attribute* layers. The diagram shows five classes, together with attributes and relationships. We have *Wallet* and *PiggyBank* as specialisations (ie. subclasses) of *Container*, and different relationships with suitable cardinalities among them. The so called *services* layer (ie. the classes' methods) is not shown.

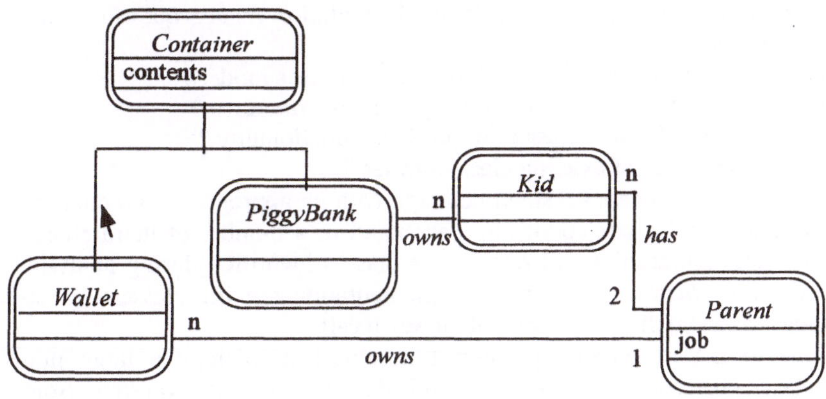

Fig.7. A Coad-Yourdon class/relationship diagram

Since they capture an application's structure these types of document what is known as the *static model* of an application. To specify an application's behaviour we need to prescribe a *dynamic model*. Again, many types of representations have been proposed for documenting dynamic models, ranging from simple event traces to state transitions diagrams, life cycle diagrams and petri nets. Figure 8 shows an *event trace* of the dynamics of pocket money collection as a sequence of messages passed among objects. Of course, this is a vanilla scenario on which numerous variations exist (eg. what happens if parents are broke ?, what if the kid recklessly spends all the money?).

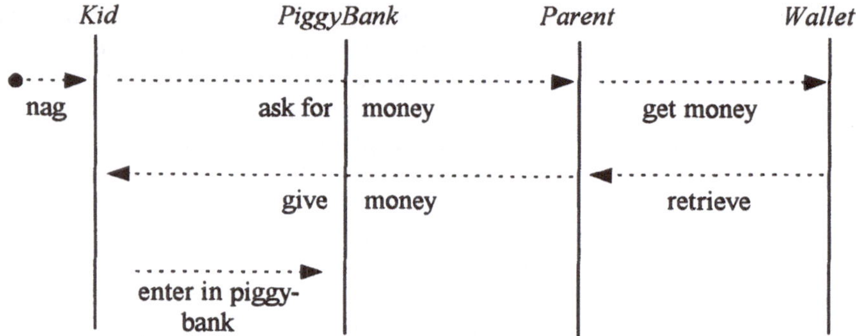

Fig.8. Event trace of a financial transaction

An additional means for depicting a model's dynamics could be the use of a state transition diagram - STD (Rumbaugh et al. 1991). STDs are helpful to summarise important entities' (classes) behaviour patterns, whereas event charts are more useful to show interactions.

While the results of object-oriented analysis provide a conceptual model of the problem domain, this phase is followed by *object-oriented design*, which serves to answer more technology-specific and. implementation-related questions. We might, for example, ask

(1) how many processes are needed to implement this model ?
(2) do we need additional classes to support a user interface ?
(3) do we need additional classes for database functionality ?
(4) can we make use of existing class libraries ?

At the end of this process we should emerge with an appropriate and complete set of class descriptions for our application, arranged in a number of hierarchies. Higher levels of these hierarchies will refer to the classes identified during analysis (the so called *domain classes*), while more implementation specific classes (eg. so *called interface classes*) will tend to populate lower levels.

Due to the mushrooming interest in object technology a large number of proprietary object-oriented analysis and design methods, varying from fairly conventional modifications to traditional methods (eg. the methods of Shlaer/Mellor 1992 and Martin/Odell 1992) to more radical proposals (eg. Jacobson et al.'s *Objectory* (1992) and the responsibility-driven design method described by Wirfs-Brock et al. 1990) have appeared on the market. Generally all of these suggest a phase model, some heuristic for completing the tasks defined for each phase, a set of notations for documenting the emerging conceptual model, and possibly some computer-based tools. Berard (1993) uses an interesting classification of methods as three-part or one-part approaches, and claim that the three-part approaches (which clearly separate between analysis, design and implementation) evolved from extensions to traditional methods of structured analysis and design. All these methods distinguish between static, dynamic and architectural (ie. how parts of a program are grouped into modules and how these fit together) views of a software system and document them in some variant of entity relationship diagrams (for the static model), state transition diagrams (for the dynamic model), and data flow diagrams (for an architectural perspective). Developers of such methods can typically draw on substantial experience with traditional methods. Methods proposed by Shlaer and Mellor (1992), Martin and Odell (1992), Booch (1990), Rumbaugh et al (1991), Coad and Yourdon (1991), and others fall into this category. The inventors of what Berard refers to as one-part approaches instead started with design concepts first, followed by additional analysis tools later. Generally they stress the perspective of a system's user more strongly (eg. by analysing so called *use cases*) and characterise the interaction of objects in terms of contracts and responsibilities towards others. These developers typically draw on extensive experience in using some object-oriented language (eg. Smalltalk) and are very much aware of the importance of mixing both bottom up and top down approaches. Methods proposed by Wirfs-Brock et al. (1990), Jacobson et al (1992), and Reenskaugh et al. (1996) are good examples for proposals with these characteristics. More recently, however, we can observe some convergence of methods - eg. the *Unified Method* of Jacobson, Booch and Rumbaugh (Quatrani 1997).

4 Object-oriented programming tools

Good notation should simplify and suggest rather than hinder the expression of ideas. To foster this goal it must express familiar concepts as directly and non-intrusively as possible. In section 1 we argued in favour of "higher level" tools in order to boost the productivity of computer augmented tools. A perfect tool would be transparent, so that we would feel as if we were working directly on a problem. Unfortunately, however, such perfection remains an elusive goal towards which today's programming languages still have a long way to travel. Although we now know much more about computer based problem solving and program development and although we now have much better hardware to support user-friendlier styles of interaction there is still much resistance to change cherished habits (Gabriel 1996). Novel ideas are only slowly accepted and make very gradual impacts on professional practice. As a result many commercial tools try to steer a difficult compromise between innovation and conservation, in that they regularly offer small chunks of modest improvements while trying their best to stay compatible with the horrors of the past. Unfortunately this breeds inelegant, baroque and cludgy systems which rarely work well. Some object-oriented programming tools have fallen prey to this trend and their designers would do well to remember A. de Saint-Exupery's well chosen words of caution: "Perfection is reached not when there is no longer anything to add, but when there is no longer anything to take away".

It is a common mistake to try learning a programming language in isolation, ie. by studying only its syntax and semantics. Using a new tool while perpetuating old habits will invariably be disappointing, since history, programming environments, and typical idioms of usage deserve much attention. The widespread reluctance to invest in learning a whole new style can only partly be blamed on the time and effort this process requires. Often it is also a defence on the part of the user, since it is hard confront the fact that hard-won skills may have become obsolete.

Based on its relatively long history the subarea of object-oriented programming tools shows a much higher degree of maturity than any other. From a large number of competing designs a relatively small number of languages have emerged to dominate commercial practice today. Among these languages Common Lisp Object System (CLOS), Smalltalk, Eiffel, C++ and particularly Java have proved most successful; each catering for its own class of applications. CLOS (Keene 1989) is most popular as a tool for programming artificial intelligence techniques (eg. expert systems). Smalltalk (Goldberg and Robson 1983) is an excellent prototyping language, particularly if user interfaces and graphical representations are involved. In addition it is also an excellent choice for learning and teaching good object-oriented style. While Eiffel (Meyer 1992) is a good teaching tool, it is focussed on large projects where software reliability is of particular concern. Here it competes directly with C++ (Stroustrup 1994), which is a much more pragmatic language with less solid formal foundations, but with particular strengths in supporting efficient "low level" implementation techniques (ie. as a tool for writing programs close to the hardware). Java's (Arnold and Gosling 1996) popularity has been fuelled by its ability to make the World Wide Web come alive. In terms of numbers of users C++ still dominates all others, although Java's popularity rises quickly.

Figure 9 gives a brief overview of the historical development of some of the many object-oriented programming languages on the market today.

There are many ways in which object-oriented programming languages can be classified, but two such schemes have proved particularly popular. The distinction between so called *pure* and *hybrid* languages hinges on the question whether a language has been designed from the ground up to offer only object-oriented features (which means that it is impossible to write programs in any other styles), or whether object-oriented extensions were grafted onto an existing imperative, logic or functional base. From this perspective Smalltalk, Java and Eiffel may be considered as "pure", while C++, CLOS, Ada 95 and others fall into the "hybrid" category. Both families have their strengths and weaknesses. Pure languages are better suited for teaching good style, since clinging to obsolete habits is impossible. Hybrid languages on the other hand make it easier to reuse existing experience where a mixture of styles seems advisable. Orthogonally to this classification the terms *static* and *dynamic* are used to distinguish programming tools in a different dimension. Static languages require the structure of all classes and relationships to be frozen at the time a program or module is compiled. Dynamic languages, on the other hand, allow us to change a program's structure "on the fly". One can, for example, easily write programs which build new or restructure existing classes at execution time. C++, Eiffel, Ada 95 and other languages of this ilk are static in this sense, while languages like Smalltalk and CLOS can be classed as dynamic. Advantages relate to efficiency and reliability for static versus flexibility for dynamic languages. In terms of their applications, static languages should be used for long-lived programs with well known requirements while dynamic languages excel as prototyping tools; where requirements may initially be fuzzy. Lately, however, this distinction has been challenged by faster hardware and more efficient implementation techniques. Also there is much debate whether the processor cycles saved through compiling outweigh the more tedious development cycle associated with static languages. Dynamic languages typically have excellent instrumentation tools, which can help pinpoint bottlenecks on which one can then focus attention.

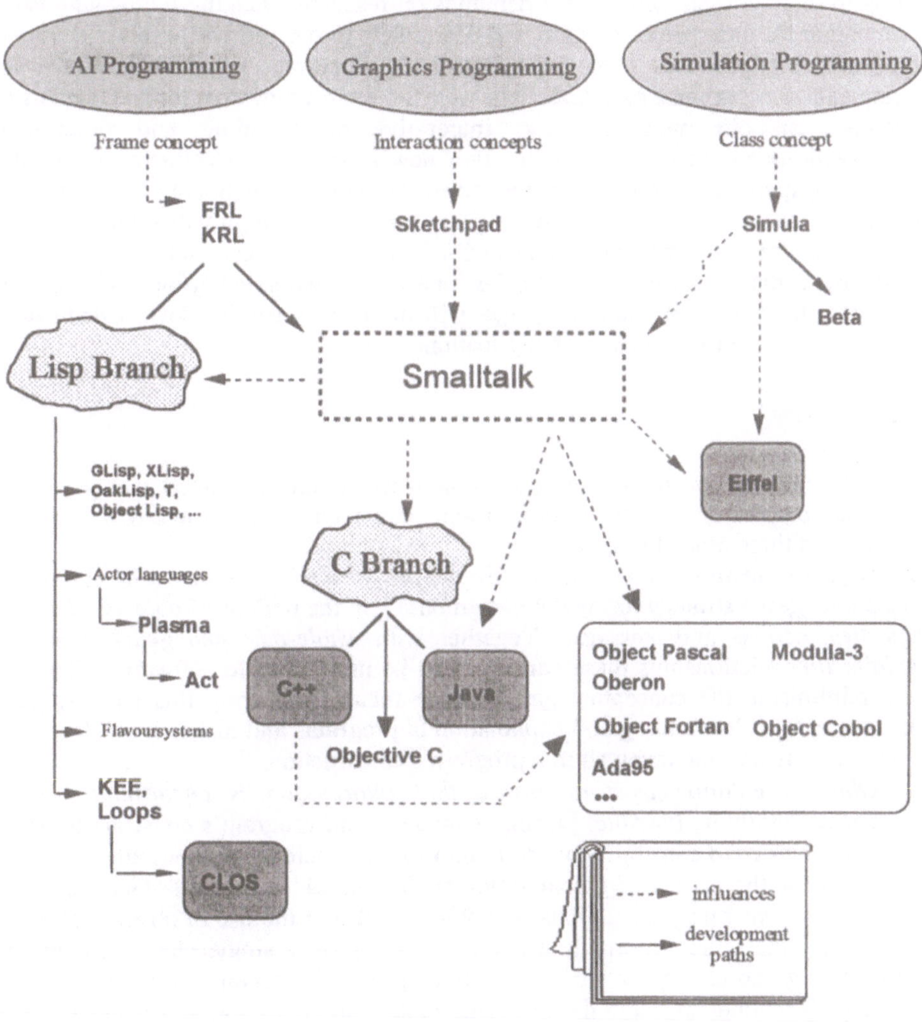

Fig.9. The OOPL landscape

A programming tool's success - or lack of it - can not be explained solely in terms of technical merits. Many other factors have often more dominant roles to play. According to Gabriel (1996) we should be aware that programming languages and other tools are accepted and evolve through a social process, which is strongly influenced by habits and values. Successful languages must be perceived as familiar and must not require high levels of mathematical sophistication. They must also have simple performance models and acceptable resource requirements. Few people will make the effort to learn how to use a new tool unless they run into problems resisting the tools they are already familiar with (old habits are hard to break !). This tendency favours incremental over radical change. While today's programmers are willing to tolerate tools with higher demands on resources, making a language very much bigger for little perceived gain will be unlikely to succeed where fewer changes along with big promises may. This fact is an important reason behind C++'s phenomenal success, and may bode less well for more innovative tools. Of course, sometimes an old paradigm (e.g. imperative programming and functional decomposition) proves unable to cope with new problems and collapses under its own weight, giving new frameworks (eg. object-oriented or functional programming) a chance to take over. After a number of years the new paradigm will establish itself and become a new standard, now to be challenged by younger contenders.

While a discussion of any particular language's details is beyond the scope of this paper all of the mentioned languages will most likely survive, attract dedicated users and carve their own niches of applications.

5 Summary

In this article we have made a case for object-orientation as a good metaphor for mastering complexity in software design and implementation. To justify this claim we identified three crucial properties:

(1) *Object-orientation is a "natural" framework for modelling the world.* Its basic concepts suggest a strategy for model design based on the notions of *objects, classes, properties, actions* and *messages*. Together with *whole-part and generalization-specialization* relationships these concepts can be used to build software in layers, while minimising the conceptual gap between them. Employing this strategy can help us to ease understanding and explanation of programs and makes it much easier for end-users to become involved in a program's construction.

(2) *Object-orientation tries to ensure that information is encapsulated and localised as tightly as possible.* During a model's and program's construction this strategy enables us to constrain our attention to small contexts, without any need for thinking about the effects of global states or "non-local" influences. Good object-oriented programs rarely need global variables at all and the use of abstract classes ensures that "the same" information is only ever kept in a single place. This eases understanding and change as well as the tracing of error conditions; which can in turn simplify maintenance and development. Object-oriented programs achieve their locality by *encapsulating* an object's state and behaviour in an entity which can only be accessed through a narrow interface. A beneficial side-effect of this strategy is *polymorphism*, which ensures that we can choose message independently of the type

of object to which they apply. This can make models more readable and greatly eases the task of integrating them into new contexts (ie. different programs).

(3) Compared to conventional procedure or module libraries we gain *flexibility*. Encapsulation, the fact that assemblies of objects can be composed and decomposed into whole-part relationships, and inheritance's support of *generalisation and specialisation* make it easier to build collections of *reusable components* (so called *class libraries* and *frameworks*). Since hardly any global context need to be assumed classes can simply be treated as black box components whose behaviour can be changed, restricted or extended by subclassing; without touching the original implementation. This incremental extensibility offers great advantages for the development of reusable software. The *patterns* community (Gabriel 1996) has made good use of these ideas.

In summary there are good reasons to believe that object-orientation will continue to gain popularity and remain an important new approach for software construction. There are of course costs associated with it, but in the light of technological advances these will become easier and easier to pay. After all, we need better tools to help us cope with the complexity of software construction, otherwise faster hardware will only allow us to make our mistakes at a faster rate.

What is still needed is further research in some areas, particularly with regard to formal foundations, the integration of object oriented programming languages with other resources (eg. databases and networks), better methods of analysis and design, the development and administration of class libraries and frameworks, and suitable techniques for managing teams of programmers working within this culture. Above all, a new paradigm needs a shift in values. Current development is too narrowly focussed on short term gains. To be successful with an object-oriented approach we need to adopt a more long-term and investment-based perspective. Design, construction, use and refinement of a class library, for example, requires considerable effort; which will only pay for itself over time. Empirical studies have shown that the hoped for productivity advantages of object-orientation will not manifest themselves immediately. In the longer term, however, the experiences gained, and the tools and components developed within this framework will make the design and implementation of future applications easier, faster, more reliable and more productive.

References

Arnold, K. and Gosling, J.: (1996) The Java Programming Language, Reading: Addison-Wesley.

Balci, O. and Nance, R.E.: (1988) Simulation Model Development: The Multidimensionality of the Computing Technology Pull. Technical Report, Virginia Polytechnic Inst. and State University, Number TR-88-26, November 1988.

Beck, K. and Cunningham, W. (1989): A Laboratory for Teaching Object-Oriented Thinking Proceedings of the OOPSLA '89 Conference on Object-oriented

Programming Systems, Languages and Applications, pp. 1-6, October 1989.

Berard, E. (1993): Essays on object-oriented software engineering, Vol. 1, Englewood Cliffs: Prentice-Hall.

Blum, B. (1996): Beyond Programming: To a New Era of Design. New York: Oxford University Press.

Booch, G. (1990): Object-Oriented Design with Applications, Benjamin-Cummings.

Coad, P. and Yourdon, E. (1991): Object-Oriented Analysis, Englewood Cliffs: Prentice Hall.

Dahl, O.J. and Nygaard, K. (1966): SIMULA, an ALGOL-based simulation language. Communications of the ACM, 9(9), pp. 671-678, September 1966.

Dennet, D.C. (1991): Consciousness Explained. Boston: Little, Brown & Co.

Gabriel. R. (1996): Patterns of Software. New York: Oxford University Press.

Goldberg, A. and Robson, D. (1983): Smalltalk-80 - The Language and its Implementation. Reading: Addison-Wesley.

Jacobson, I. et al. (1992): Object-Oriented Software Engineering: A Use Case Driven Approach, Reading: Addison-Wesley.

Kay, A. (1977): Microelectronics and the Personal Computer. Sci. Am., pp. 231-244, September 1977.

Keene, S. (1989): Object-Oriented Programming in Common Lisp. Reading: Addison-Wesley.

Kreutzer, W. (1986): System Simulation Programming Styles and Languages, Reading: Addison-Wesley.

Kuhn, T. (1970): The Structure of Scientific Revolutions. 2nd. ed., Chicago: Chicago University Press.

Martin, J. and Odell, J. (1992): Object-Oriented Analysis & Design, Englewood Cliffs: Prentice-Hall.

Meyer, B. (1992) Eiffel: The Language, Englewood Cliffs: Prentice Hall.

Moessenboeck, H. (1995): Object-Oriented Programming in Oberon-2, p. 278, New York: Springer.

Quatrani, T. (1997): Visual Modeling with the UML: A Rational Approach. Reading: Addison-Wesley.

Reenskaug T (1996): Working with Objects - The Ooram Software Engineering Method. Reading: Addison-Wesley.

Rumbaugh, J. et al. (1991): Object-oriented Modeling and Design, Englewood Cliffs: Prentice Hall.

Shlaer, S. and Mellor, S. (1992): Object Life Cycles: Modeling the World in States, Yourdon Press.

Stroustrup. B. (1994): The Design and Evolution of C++. Reading: Addison-Wesley.

Sutherland, I. (1963) Sketchpad: A Man-Machine Graphical Communication System. Proceedings AFIPS Spring Joint Computer Conference, Vol. 23, pp. 329-346, May 1963.

Wirfs-Brock, R. (1990): Designing Object-Oriented Software, Englewood Cliffs: Prentice-Hall.

Model-based software engineering for interactive systems

Christian Märtin

1 Introduction

Software solutions for many target domains (e.g. business, financial, CAD, scientific visualization, Internet) are typically organized as interactive systems. This article discusses the design of interactive software systems in general and presents a model-based environment for computer-aided design of such systems, the *A*pplication *Modeling E*nvironment *(AME)*.

One natural way for modeling, designing and implementing interactive software systems is object-orientation. Important object-oriented concepts are therefore introduced in this chapter. Each complex software design process has to be divided into smaller design activities. A software life cycle assigns specific design tasks to any of these activities, defines the software product, resulting from each activity and specifies the design workflow structure. The importance of life cycles for software engineering in general and the specific life cycle requirements for object-oriented development are also briefly discussed in this chapter.

Chapter 2 defines the structure of interactive software systems and discusses the various abstraction levels, present in human-computer interaction. An interactive software system also includes non-interactive components. The integration of design issues for the remaining parts into the overall software design life cycle is briefly discussed.

The object-oriented scheme used by AME for representing interactive application models in all their refinement stages is discussed in chapter 3. It is also shown, how the functionality and HCI-related knowledge for refining models are organized and maintained.

Chapter 4 illustrates AME's life cycle activities and shows, how applications are developed with the available tools. It is demonstrated, how an abstract model can be gradually refined into a running interactive application.

Some related or alternative approaches for the computer-aided modeling and design of interactive software-systems are discussed in chapter 5.

1.1 Object-orientation as a modeling approach

The proposed software design process starts with the definition of an abstract object-oriented model of the structural, functional and dynamic features of the planned application. In order to support a number of software automation techniques, a flexible interpretation of object-orientation, inspired by object-oriented analysis and design methodologies (Rumbaugh et al. 1991; Coad, Yourdon 1991) and by the frame concept (Fikes, Kehler 1985), serves as the basis for system modeling.

Object-orientation in general means that a software system at a given abstraction level can be described by a set of discrete, related classes and their objects. A *class* is the basic unit of an object-oriented software system. It defines the features of an important notion type of the target domain: its data structure and its functionality. The data structure of a class is defined by typed class attributes and default attribute values. The functionality of a class is defined by typed class operations, i.e. the operation names, their calling parameters and the type of their return value. On a more detailed design level some symbolic description of the functional behavior of operations can be given.

Classes can be specialized by adding compatible new attributes and operations, modifying the types of attributes and operations in a consistent way and providing specific operation implementations (methods). Such specialization classes are called subclasses. A subclass inherits all features from its superclass, but may delete attributes and methods, not needed. Two methods with the same name, which are located in different subclasses of a common superclass, may have different implementations and therefore behave in a different way (polymorphism). If a class inherits attributes and operations/methods from two or more superclasses (multiple inheritance), a careful conflict detection or resolution strategy is needed for cases, where attributes or methods with the same name, but different specification details, are inherited to the common subclass.

An *object* is a uniquely identifiable instance of a class. Each class definition is located in a class descriptor object, which may be handled like any other object. Therefore, there is a very smooth transition from classes to their objects. The main difference between them is the inheritance relation, which is only available for classes. Objects can be dynamically created and deleted at runtime. The same is true for classes. All objects of a class own the attributes and methods specified by the class, but will typically assign values to attributes individually. The set of all attribute values of an object at a given time defines the state of this object. Objects may use their own methods for attribute data manipulation or attribute queries. They also may act as clients of an object of another class (supplier) and use the supplier methods by sending a specified message to the supplier object. The client-supplier relation between objects of different classes may be implemented by message communication. In addition, some other relations for defining structural and semantic properties of the system, like associations between classes or the construction of aggregation classes from simpler classes can be introduced.

Object-oriented techniques for the development activities analysis, design and implementation enable an easy transition from abstract to concrete stages of a model representation. In AME all modeling stages are represented by so called application system objects. The most abstract application model is an object-oriented analysis model of the system. The most concrete model is the object-oriented program source code of the application, which can be compiled into a running software application. AME´s object-oriented representation scheme is discussed in chapter 3.

1.2 Software life cycle

The presented software engineering approach follows a computer-aided development life cycle, in which the application model is gradually refined into the interactive target system with all its aspects concerning structure, functionality and usability. The life cycle covers the development of both domain-oriented and interactive components of the target application. A life cycle may generally include activities A_i for *planning, analysis, global design, detailed design, implementation, validation & verification,* and *maintenance.* AME's life cycle activities are discussed in detail in chapter 4. Various kinds of *knowledge resources* including design methods, design rules, reusable software patterns, user preferences, features of the target environment and the actual designers' experience are exploited during AME's life cycle activities in order to achieve high software quality, good usability and to improve developers' productivity.

In AME a computer-aided software engineering (CASE) approach is applied. This means that specialized software tools are assigned to most life cycle activities. Some of these tools are themselves interactive software systems. Other tools are automated batch facilities. However, the result, produced by any tool, can always be interactively modified by the designer. This is an important feature of a flexible design system. For instance, evaluation and simulation of software products resulting from a given activity might not show the expected behavior, due to faulty specifications provided by an earlier activity. In other cases system requirements might suddenly change during development. The designer therefore needs active control over all interactive and batch tools involved.

The *waterfall model,* which was used as a pre-object-oriented life cycle approach for many very large-scale systems during the last 15 years (Boehm 1981) can be characterized by its clearly separated sequential phases and exactly defined documentation or software products resulting from each phase. For each phase different developer groups were responsible. Large personnel resources were needed. It caused enormous work efforts, when system requirements were modified during development periods which could last for several years.

An advanced life cycle approach for the development of medium-scale, object-oriented software systems has to provide support for prototyping part of the target application, e.g. the user interface. It must also be possible to fully or partly iterate the life cycle and to simulate the outcome of each life cycle iteration, e.g. functional or structural parts of the user interface. It must be possible to test and verify intermediate results in the development environment. Although the individual activities of the waterfall model will also be found in such a more flexible approach, the steps of an object-oriented life cycle cannot be separated as clearly as with the waterfall model. Life cycle activities typically overlap. In many cases the people responsible for the various activities are the same over the whole life cycle. Some recent ideas for object-oriented software life cycles were presented by (Meyer 1995). In his view, large projects should rather be divided into different clusters, which may be developed in parallel. A cluster consists of a group of semantically related classes. Each cluster maintains its own life cycle. Each team is responsible for one cluster during its whole life cycle.

Such ideas for restructuring the life cycle form a sound basis for interactive system design. However it is questionable, whether they are also appropriate for automating as many parts of the design process as possible. There is good reason for such doubts. When we compare software and hardware design, we find that even the most advanced hardware design processes show similarities to the waterfall model. It is too expensive and time-consuming to apply several iteration cycles on the design of complex VLSI chips. It is vital for the success of a hardware project that there exist highly trusted specifications or models of the system structure, logic and functional behavior. Automation methods and cooperative tools are available for transforming high-level structural and behavioral specifications step by step into more low-level specifications, synthesizing parts of the chip hardware and generating the chip floorplan and layout.

If we wish similar productivity and quality enhancements in the software area, we can neither exactly copy the hardware-oriented view, nor use Meyer's clustering approach without any modifications. We will have to combine the benefits of both views. It will not be possible to raise productivity and quality with pure production-oriented software factory methods. We need to study the individual life cycle activities for such tasks which may be easily automated and for other tasks, which are still best understood by experienced human designers. We will then have to define cooperative design interfaces between these different task groups.

It can be shown that object-oriented development leads to systems with reusable components. Object-oriented approaches also support distributed as well as interactive software architectures in a very natural way. It is therefore necessary, to join the forces of object-oriented systems and of classic life cycle approaches with their high potential of automation. AME's life cycle was developed with these goals in mind.

2 Interactive software systems

2.1 Structure of interactive software systems

The structure of an interactive software system *(ISS)* is shown in Fig. 1. Its main components are a user interface *(UI)*, an application kernel *(AK)* and a communication interface between UI and AK. If UI and AK do need not reside on the same workstation, a client-server relation between AK (client) and UI (server) exists.

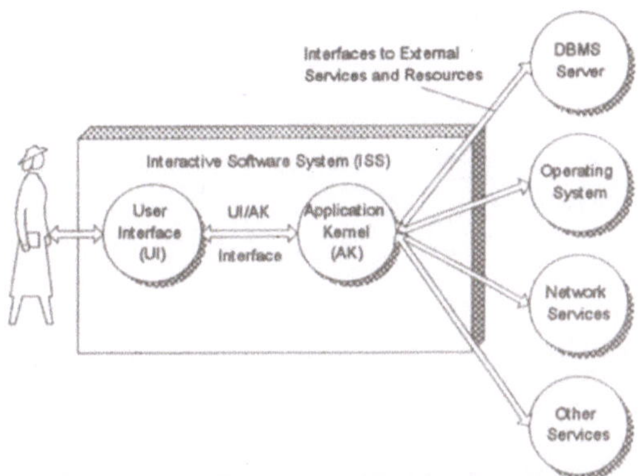

Fig. 1. Logical structure of interactive software systems

User Interface. The UI is in itself divided into several logical levels which may be organized as software layers or cooperating processes. The UI levels handle all aspects of user-system interaction from low-level input/output-device management to conceptual human-computer dialogue. The UI structure is further discussed in 2.2.

Application Kernel. The AK (Fig. 2) contains the domain-dependent functionality and data structures of the ISS. The AK may be subdivided into several logical parts. The core functionality and data structures *AKC* should be independent of any UI aspects. Data and functionality of the application, which have to be represented in the UI and mirror application-specific interactive requirements, will reside in the interactive part of the AK *(AKI)*. Domain-specific interactive resources for the application user interface, like non-standard graphical visualization or animation techniques also reside in the AKI.

An ISS may include additional interfaces to databases, operating systems, network services and to other interactive software systems in a distributed environment. There is a certain symmetry between the AK and the UI on the one side, and the AK and these other services on the other side. Application-specific functional requirements to these services (e.g. the transformation of object-oriented data into relational database tables or the mapping of remote procedure calls to low-level communication protocols) will reside in service-specific AK parts *(AKS)*. Recently several standard solutions for distributed connectivity services have become available mainly in the form of object-oriented communication environments. These systems are called "middleware" (Bernstein 1996).

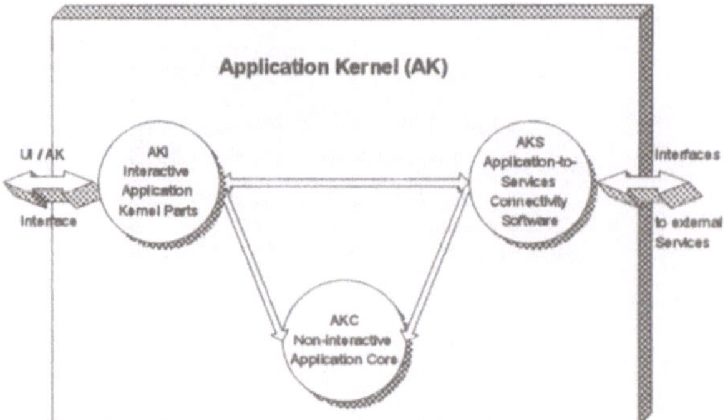

Fig. 2. **Structure of the application kernel (AK)**

The design of the AKS and their interaction with the rest of the ISS with or without the usage of middleware is not explicitly covered here. Note, however, that communication between AKS and non UI-services in a distributed environment is similar to interaction between AK and UI in as far as in both cases messages of some defined type are dynamically passed between loosely coupled systems in order to establish communication at runtime. There exist interface design languages *(IDLs)* and service management components for developing standardized interaction between the elements of a distributed system. A common user interface, shared by all components, may also be included in such a distributed system. Middleware also allows an easier implementation of computer-supported collaborative work (CSCW) applications, where several users on different workstations cooperatively work on the solution of some problem. In this case, UI coherency must be guaranteed, by synchronizing parallel interactions of different users with identical parts of shared application data structures. Note, that the terms *interactive*, *interaction* and *interactive system* as discussed in this article, are limited to communication of users with the UI and of the UI with the AK.

2.2 Abstraction levels in human-computer interaction

The theory and practice of user-system interaction and user-centered design of interactive systems are covered by the field of *h*uman-*c*omputer *i*nteraction *(HCI)* (Salvendy 1997, Shneiderman 1997). Interaction between human users and an ISS concerns various logical levels of communication. At each logical level certain properties of the ISS are specified. It depends on the available design tools and the target environment whether these properties are specified during development time or dynamically at runtime. Specific methods and tools for UI design and management can be assigned to each level (Märtin 1996a):

Discourse. The highest level specifies pure content-based dialogue (discourse) between a user and a system in a non-computerized way, e.g. as natural language utterances. Artificial intelligence or pattern analysis methods and techniques are needed to interpret or generate such dialogues. Knowledge and methods from the application domain are needed to implement the discourse level. It is an important part of a complex ISS, but it is not a standard part of the UI. It makes sense, to technically separate functionality of this level from pure domain functionality. Reusability of discourse functionality for similar domains may thus be increased.

Dialogue. The dialogue level defines the way, in which communication requirements of an AK are transmitted to the UI and vice-versa. It includes three sub-levels: abstract dialogue, semi-abstract dialogue and concrete dialogue. The more abstract the chosen level, the less an application designer needs to know about specific UI design techniques, presentational, and usabiltiy issues. This specific knowledge will be provided either by automated tools or by experienced human *dialogue designers,* who interactively refine the dialogue specifications produced by *application designers.* The three sub-levels are discussed in the following.

Abstract dialogue requirements are expressed in a way that neither specifies the dialogue medium (e.g. natural language, 3D-animation, text), nor the interaction technique (e.g. graphical button, function key, textual command input). Only the data contents, the data types, data cardinalities and possibly some meta-information about the dialogue data are exchanged between the AK and the UI. These data are specified as features of AK classes. At this level the sequence of the dynamic interactions between users and the target ISS and the possible responses of the ISS to input/output events are not directly specified. Such dynamic information can however be partly derived from the dynamic properties of the domain classes found in the AK and their dynamic interrelationships. In other terms: a problem domain model is needed for this purpose. Another approach for gathering information about the dynamic behavior and the needed interaction techniques of the target system under construction would be the analysis of the task structures to be supported by the ISS under construction. A task model could be provided for specifying this information (Elwert, Forbrig 1997).

Based on the abstract dialogue specification and requirements and constraints of the design tools and the target environment, dialogue design decisions are delegated either to experienced human UI designers or ISS design tools. Dialogue management tools, which dynamically apply ISS design knowledge at runtime, may also provide UI design decisions. When selecting dialogue media and interaction techniques for a given set of abstract dialogue requirements, in particular the human factors issues related to the application domain have to be considered. In order to improve usability, individual user models, representing different experience levels or dialogue preferences, may be provided by design systems. When using runtime dialogue management tools, the UI may even be dynamically adapted to the varying needs or wishes of differing users or provide contextual help information for complex user tasks.

Semi-abstract dialogue leaves either the medium or the interaction technique unspecified. In many cases the selection of a dialogue medium is constrained by the fact that one of the standard window-oriented graphical *UI*s *(GUIs)* will be used in the target environment. The primary medium will then be the two-dimensional, pixel-oriented graphical color display. Even then it is still not trivial, to look for an appropriate interaction technique for expressing one specific dialogue requirement.

Concrete dialogue specifies both a dialogue medium and an interaction technique for a given dialogue requirement (e.g. a scanned bitmap form for income tax preparation for entering individual data). All dialogue sub-levels are still independent of the underlying input/output levels and the UI tools located there. It is not necessary to know the specific target environment, when developing the dialogue interface. However, it may be helpful to know the scope of available target enviroments and their features during dialogue design. The interaction techniques specified during dialogue design are often called *a*bstract *i*nteraction *o*bjects *(AIOs)* because they are mapped to graphical objects (controls) or *c*oncrete *i*nteraction *o*bjects *(CIOs)* of a specific GUI toolkit (e.g. buttons, pulldown- and popup-menus, input fields, bitmaps, slides, meters) during input/output design. At this level the detailed interaction sequence or the behavior of the ISS, when a defined event occurs, can be specified using a dialogue specification language or some graphical specification tool. Such tools are typically based on state/transition approaches like Petri-nets (Janssen et al. 1993).

Input/Output. The four stages of the input/output level *(toolkit, I/O-server, device driver, device hardware)* link the specification results produced at the dialogue level to the tools and features of the runtime target environment. The I/O levels are usually accessed by application developers. Some recent approaches hide the input/output levels from application developers and let intelligent tools, equipped with UI design knowledge, access resources at the input/output level. For business-oriented application domains with their typical UI requirements a lot of design time can thus be saved. At the same time high-quality UIs conforming to widely accepted style and presentation standards can be produced. The most flexible approach is provided by cooperative design environments, which allow the interchange and refinement of design results between developers and automated tools.

Toolkit level. Within an interactive software system, interaction between users and application functionality is typically mediated by a graphical user interface toolkit or a toolkit which supports additional media like speech input or video output. At the toolkit level dialogue requirements are specified for one specific toolkit. A GUI toolkit supports a standardized style (e.g. Microsoft Windows or OSF/Motif) for presenting specific application functionality to users. The GUI of an interactive application is composed of one or more interrelated windows. A window contains CIOs for input and/or output. The toolkit allows the designer to specify the exact structure of windows and interaction elements and part of the GUI behavior.

The functionality of a toolkit is typically accessed via an interface to a standard programming language. Toolkit users should be HCI experts, because the toolkit cannot control usability of the resulting GUI structure, layout and behavior. At the toolkit level, graphical editors for GUI design are sometimes provided. They allow designers to construct GUIs interactively. Toolkits alternatively provide low-level specification languages for describing a user interface, its elements and its behavior, e.g. how the direct manipulation of a button with a mouse or a similar input device is mapped to a function call in the AK.

I/O-server level. At this level I/O commands and graphical primitives are specified by using library functions, which are directly mapped to a basic window system protocol, like X11. This step is typically done automatically by translating toolkit functionality into sequences of I/O-server activations. The remaining I/O levels are very hardware-dependent and device-specifc. They are not further discussed here.

3 Representing models of interactive software systems in AME

A model-based ISS representation scheme has to mirror the ISS model before and after all refinement activities defined by the software engineering life cycle. In order to raise overall software productivity and allow to build software systems with standardized look and feel characteristics, it seems reasonable to strive for a high degree of design automation during all life cycle activities. The *A*pplication *Modeling E*nvironment *(AME),* which is discussed in this and the following chapter, both provides a common representation mechanism for defining ISS models in all their development states and an iterative life cycle which provides a set of interactive and automatic tools for transforming an ISS model from one development state into the next (Märtin 1996a,b).

3.1 ISS models and resources

ISS model specifications are represented by an object-oriented scheme that can be defined as a set of classes $M_{Model} = \{M_{ASO}, M_{UI}, M_{Media}, M_{Data}, M_{User}, M_{Env}, M_{Admin}\}$. The classes of M_{Model} provide the expressional power for modeling ISSs at all development stages. They also are the basis for defining variable knowledge resources which can be exploited by AME tools for the different life cycle activities A_i. AME tools also use built-in structural and content-based design knowledge which is not represented in the ISS model specifications.

Any ISS model specification $S_{i,j}$, $0 < i < n$, can be represented by using or referencing objects and subclasses of the classes in M_{Model}. A model specification $S_{0,j}$ is an abstract problem description of the target application and can e.g. be expressed as a textual description, whereas $S_{n,j}$ represents the programming language source code, which is produced by the final life cycle activity A_n. The index j denotes the current life cycle iteration (see **4.1**). The classes in M_{Model} and their role in a model specification are now discussed in more detail.

M_{ASO}. Objects of this class are used to define the structure, relations and possible contents for the ISS representation of domain classes and their objects (*A*pplication *S*ystem *O*bjects: *ASO*). M_{ASO} provides a generic infrastructure for representing all necessary features of individual application domain classes throughout the whole life cycle.

Classes in an application domain model differ in the number and types of their attributes and operations/methods. In order to provide the necessary flexibility for automation purposes M_{ASO} offers a fixed set of standard attributes (slots) to which the individual configuration of attributes and operations of domain model classes is mapped as M_{ASO}-slot values or variable length slot value lists. An overview of M_{ASO}-slots is given in Fig. 3. Each application domain class C_i is represented by an object $c_i \in M_{ASO}$ (i.e. c_i is of type M_{ASO}). A slot s_k of object c_i is denoted as $c_i{:}s_k$. The number of elements in a slot value list is denoted as $\#c_i{:}s_k$. A method m_k of object c_i is denoted as $c_i.m_k$.

Application System Object (ASO)

actions: multiple text
action_types: multiple text
as_action: multiple text
as_components: multiple object
as_construction: multiple text
as_content: text
as_frame: text
as_description: multiple text
as_name: text
as_parent_profile: text
as_presentation: multiple text
as_type: text
Association: multiple object
Association_types: multiple text
attr: multiple text
contents: multiple text
content_types: multiple text
data_type: text
data_length: integer
de_instance: object
dialog_construction: multiple text
dialog_medium: text | multiple text
dialog_object: object | multiple object
dialog_preference: object
dialog_presentation: multiple text
WholePartRelationsFrom: multiple object
WholePartFrom_types: multiple text
WholePartRelationsTo: multiple object
WholePartTo_types: multiple text
GenSpecRelationsFrom: multiple object
GenSpecRelationsTo: multiple object
InstanceCounter: integer
MessageLinksFrom: multiple object
MessageFrom_names: multiple object
MessageFrom_priorities: multiple text
MessageFrom_types: multiple text
MessageLinksTo: multiple object
MessageTo_names: multiple text
MessageTo_priorities: multiple text
MessageTo_types: multiple text
name: text
prototype: object
semantic_neighbors: multiple object
sub_level: boolean
sub_object: multiple object
sub_level_conn_type: multiple text
synthetic: boolean
value: text | multiple text
visible: boolean

MakeDialogObject
Behavior
MakeLayout

Fig. 3. M_{ASO} class attributes and operations.

Any two c_i, $c_j \in M_{ASO}$ own the same attribute slots, but typically have different slot values or value lists. Generally c_i:contents $\neq$ c_j:contents, #c_i:contents $\neq$ #c_j:contents, c_i:content_types $\neq$ c_j:content_types and #c_i:content_types $\neq$ #c_j:content_types, because the attributes of C_i and C_j become the values of slots c_i:contents and c_j:contents. Attribute types are mapped to c_i:content_types and c_j:content_types. Operations are mapped to slots c_i:actions and c_j:actions. Their types are mapped to c_i:action_types and c_j:action_types. Typically c_i:actions $\neq$ c_j:actions and #c_i:actions $\neq$ #c_j:actions, c_i:action_types $\neq$ c_j:action_types and #c_i:action_types $\neq$ #c_j:action_types. One representation class with a fixed set of attributes and operations therefore suffices for expressing all possible variations in attribute and operation type and count in domain classes. The relevant slots are list valued slots. In an object c_i the type information for a content or action element is located in the relevant type slot under the same index as the content or action element.

After the early life cycle activities many features of the target ISS are still unknown or unspecified. *NULL* values in M_{ASO}-slots denote unspecified class features. Tool-based refinement of the ISS model over the different life cycle activities gradually leads to more M_{ASO}-slots with defined values, i.e. specified features of ISS-model classes and their objects. As discussed earlier, the interactive constituents of any ISS can be modeled on different levels of abstraction. The M_{ASO}-representation objects mirror the abstraction level by either providing values of the slots for such important ISS features as the AIO type (c_i:dialog_object) or the selected target medium (c_i:dialog_medium) or by leaving these slots unspecified. Reuse of an existing UI component prototype by objects of a new domain class is supported by an appropriate slot (c_i:prototype). If explicitly no dialog object is specified during the life cycle, the c_i is considered a *domain* class or object that may represent a resource or domain function provided by some AKS or AKC component.

Another requirement for representing ISSs at all their life cycle stages is met by M_{ASO}-objects: In many design situations a single domain analysis class is mapped to a whole set of related classes, later in the life cycle. Specific M_{ASO}-slots maintain the links to such design or implementation classes, which are created during design time. A typical example for this situation is some domain class C_1 for a business object, e.g. a tax preparation form with attributes for all the required address and tax data. C_1 is represented by an object $c_1 \in M_{ASO}$. During design, such a complex class will be automatically expanded to an aggregation of classes $\{C_1, C_{11}, C_{12}, ..., C_{1n}\}$, resulting in a set of representation objects $\{c_1, c_{11}, c_{12}, ..., c_{1n}\} \subset M_{ASO}$, where c_1 now represents a container window and the other objects represent entry fields for numerical or string values. The necessary aggregational relations are generated by adding the list of values $c_{11}, c_{12}, ..., c_{1n}$ to the slot c_1:WholePartRelationsTo. Vice versa, each c_{1i} automatically becomes a part of c_1 by adding c_1 as a value to slot c_{1i}:WholePartRelationsFrom. For specifying the cardinality of an aggregation between two classes (e.g. one class c_i may have n parts of the same type c_j), additional type slots are provided.

Other relations between classes can be considered in a similar way: Inheritance between a superclass C_i and a subclass C_j is represented by the following slot entries: c_i:GenSpecRelationsTo = c_j and c_j:GenSpecRelationsFrom = c_i. Associations and their types (cardinalities, roles) are represented by the slots c_i:Association and c_i:Association_types. Associations may be used for defining some semantic neighborhood between classes. Depending on the specific association type, life cycle tools can for example generate layout neighborhoods between ISS interaction objects or create data transfer interfaces between objects of associated classes.

In order to express dynamic relationships between objects of two classes, message channels between objects of ISS classes can be created. A set of slots represent message-based communication between ISS components at all abstraction levels and during the whole life cycle. Message channels and the specification characteristics of class operations are exploited by AME tools for generating client-supplier relationships and interactive behavior of ISS objects at runtime. Several additional slots are provided for representing administrative data and managing class features that are dynamically adapated to the needs of the target system during the life cycle (Märtin 1996a).

M_{ASO}-objects by default own three construction methods: c_i.MakeDialogObject, c_i.Behavior, c_i.MakeLayout. These methods are activated by AME tools in order to find appropriate AIOs for any given c_i, to provide specifications for mapping c_i-

features to the CIOs of the target ISS and to generate layout proposals for ISS windows and dialog boxes.

M_{UI}. This extensible resource class hierarchy provides four subclasses {U_{Window}, $U_{Control}$, U_{Menu}, $U_{DomainUI}$} for modeling AIOs and their related interaction techniques and provides prototypical CIOs. M_{UI}-classes specify the features of multimedia interaction objects which are available for automatic or interactive assignment to M_{ASO}-objects. Subclasses of U_{Window} specify all available window types (e.g. application main window, window, standard dialog box, individual dialog box). $U_{Control}$ offers a large set of interaction object classes like editors, edit fields, buttons, button groups, voice input classes, video output classes, etc. U_{Menu} supports various menu classes including pop-up menu, pulldown menu, icon menu, voice command menu, etc. $U_{DomainUI}$ allows the integration of domain specific interactive classes which are not part of typical user interfaces but are necessary for modeling specific interactive application functionality like 3D-modeling or animation. Not all M_{UI}-classes are supported by concrete interaction objects in typical target environments.

M_{Media}. This extensible resource class hierarchy specifies a set of media (e.g. visual, speech, natural language, sound, video, etc.) and their properties. Each class of this hierarchy contains information about the specific requirements of one medium and the relations between the medium and available AIOs for this medium. Not all M_{Media}-classes are supported by typical target environments. Media information can be exploited by life cycle tools in order to optimize the search for appropriate AIOs for given M_{ASO}-objects. M_{Media}-classes also allow to specify the functionality for including new media into target ISSs or supporting multimodal dialogue.

M_{Data}. This extensible class hierarchy defines the data types available for AME applications. Data type information is useful for constraining AIO-search to relevant M_{UI}-classes, when M_{ASO}-objects specify a certain data type (e.g. string) in their slot c_i:data_type. Data types which are reserved for non-interactive use are called unexploitable data types. Complex data types (from simple office documents like letters or memos to complex business objects for income tax preparation) are available as M_{data}-classes. They may serve as content templates for the rapid automatic generation of ISSs. This was demonstrated by earlier system CT-UIMS (Märtin 1990) and is also supported by a specific AME tool.

M_{User}. Any ISS developed and generated with the aid of AME tools can automatically be adapted to the specific needs and requirements of an individual user or a predefined user group. A user profile can be interactively constructed by a default designer-system dialogue for gathering information about a target user's or a user group's preferred interaction techniques, interactive media, colors, presentational settings and computer skills. This information is stored in a user profile object $p_{User} \in M_{User}$. Designers may also provide user or group specific *rule* or *function based problem solvers* for ISS customization and make them available by their names in specific p_{User}-slots.

M_{env}. An ISS can be adapted to specific target environments. Environment profiles $p_{Env} \in M_{Env}$ are constructed like user profile objects. They contain detailed information about ISS properties that depend on hardware and operating system features. Such profiles also hold information on available CIO-hierarchies and interaction media. Environment dependent *construction rule groups* can also be made available by naming them in specific p_{Env}-slots.

$\mathbf{M_{Admin}}$. During the life cycle subclasses and objects of this class provide administrational functionality and serve as temporary data stores, resource data bases and interfaces to the ISS target environment. The subclass $M_{Synonym} \subset M_{Admin}$, for instance, provides objects with synonym lists for mapping method names, specified during analysis, to default names of standard menu entries in target window systems. ISS prototypes are constructed before runtime for evaluation purposes and interactive modification. They use auxiliary subclasses and objects of M_{Admin}.

3.2 Patterns

Recently, software patterns have been widely discussed as a useful means for raising software quality and productivity (Gamma et al. 1996, Schmidt 1996). Software patterns are reusable object-oriented standard solutions for some generic or domain-dependent analysis or design problems. Such patterns are typically defined in a semi-formal way. They can be tailored to specific application contexts by developers, e.g. by renaming classes, objects, attributes, methods, and association roles. AME uses structural patterns as one knowledge category for automatically generating detailed design models from abstract analysis models. The key idea in using patterns for software automation is to search for high-level intra-class and/or relational inter-class patterns in abstract models, and to automatically expand them to less abstract design patterns, which specify part of a detailed solution model. For example, one can imagine that communication between objects of different classes is only specified at the domain level. Later, typical communication patterns are recognized and automatically translated into the detailed design specification of user system dialogue and inter-object communication. The functionality for recognizing and expanding structural patterns is located in AME's tools for OOA to OOD refinement (see **4.4**).

3.3 Rule- and resource-function-based problem solvers

During all life cycle activities any class or object $c_i \in M_{ASO}$ is accessible for tool interaction. Most of the available tools are implemented in an object-oriented fashion. However, for allowing greater design flexibility, public access to all attributes and methods of M_{Model}-classes and objects is possible. This allows the implementation of rule and function based problem solvers for adpating ISS models to environmental and user-specific requirements. AME tools uses rule cascades which are activated by method code and which interpret slot contents of a given $c_i \in M_{ASO}$ in order to find an appropriate AIO for the represented domain class. Problem solvers that combine forward-chained rules of the form *<rule_name> <rule_priority> IF <condition> THEN <action>* may be used to solve detailed design problems like coloring the user interface, selecting presentational attributes, and specific layout problems. Rule conditions inspect the slots of one or several $x_i \in M_{ASO} \cup M_{UI} \cup M_{media} \cup M_{data}$ for specific value configurations. Rule actions modify the features of such objects in the specified way and may alter the conditions under which other rules can become active. Problem solvers can be added to AME as standard resources or developer-specific resources. They are used by life cycle tools in addition to default methods, whenever the names of one or more problem solvers are specified as values of p_{Env}- or p_{User}-slots. Instead of rule based problem solvers, function-based problem solvers may be activated by tool methods in a similar way.

4 Life cycle activities and tools

4.1 Generic structure of the ISS development life cycle

A generic software engineering life cycle can be defined as a non-necessarily finite sequence of activity instances ($A_{1,1}$, ..., $A_{k,1}$, ..., $A_{i,j}$, ..., $A_{l,m}$, ...), where the first index denotes a given life cycle activity A_i, $1 \leq i \leq n$. The second index denotes the number of a life cycle iteration. An activity instance $A_{i,j}$ is the j-th occurence of an activity A_i. A life cycle iteration is defined as the sequence of *one or more contiguous activity instances* $A_{i,j}$, $A_{i+1,j}$, ..., $A_{i+d,j}$, where $1 \leq i \wedge i+d \leq n$. Activity instances may partly overlap in time, but $A_{i+1,j}$ will not start before and deliver its partial result not later than $A_{i,j}$. Because AME uses an object-oriented modeling approach, this can be interpreted in the following way: Each activity instance consumes a variable amount of time. The design process, however, does not always proceed with the same speed for all classes and objects in a model. Some classes will be specified in detail before the designer or automated tools will focus on the rest of the system. Tools or designers therefore must be able to recognize the current state of each class or object in the model. During the same time interval they sometimes have to apply various design tasks from different activity instances to different groups of classes or objects.

Each activity A_i is characterized by the availability of a set of specific interactive or batch-oriented tools for transforming an input model specification for an ISS to an output model specification. An activity instance $A_{i,j}$ transforms an input model specification $S_{i-1,j}$ to an output model specification $S_{i,j}$. Each output model specification is less abstract than the respective input model specification and provides a more detailed description of the target ISS. $S_{i,j}$ is also input model specification for the next activity instance within life cycle iteration j. $S_{i+d,j}$ is the final result of the current life cycle iteration.

Whenever a new life cycle iteration j+1 is started, the first activity instance of this new iteration, $A_{i,j+1}$, uses the most current or an older input model specification $S_{i-1,q}$, which is available for this activity and was produced by some earlier life cycle iteration, i.e. $1 \leq q \leq j$. This is equivalent to applying an activity instance $A_{i,j+1}$ to a model specification $S_{i-1,j}$, where $S_{i-1,j} = S_{i-1,q}$. In other words: model specifications of any given abstraction level that were produced by activity instances of earlier life cycle iterations, have to be stored for later reuse in a *model repository*.

Activity instance $A_{i,j+1}$ typically applies the tools of activity A_i to $S_{i-1,q}$ in a different way as activity instance $A_{i,q}$, e.g. by adding some new details or extensions to the model. The output specification $S_{i,j+1}$ will therefore be different from $S_{i-1,q}$ and will hopefully represent some progress in the design process towards the target ISS. An example of an iterated life cycle is shown in Fig. 4.

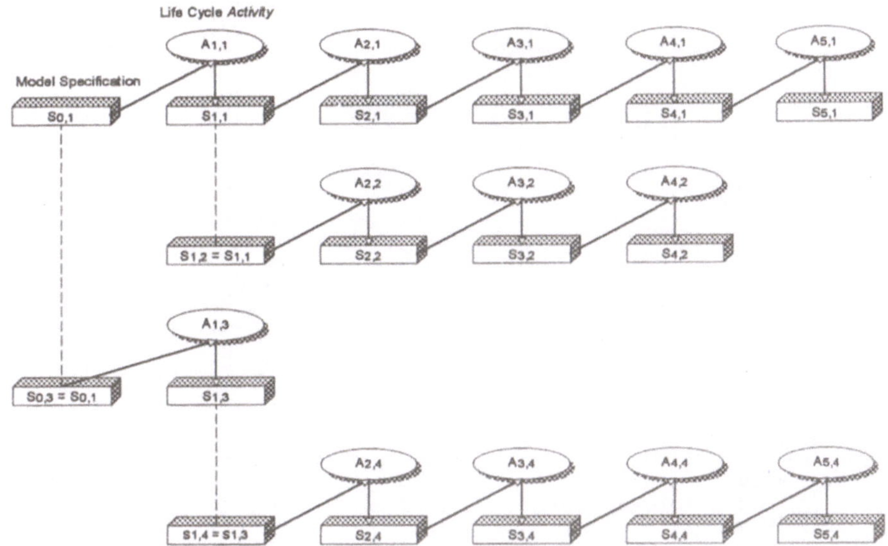

Fig. 4. Example of an iterated life cycle.

4.2 Tool implementation of AME´s Life Cycle activities

The current AME prototype provides tools that support the life cycle activities shown in figure 5. The life cycle distinguishes between six activities, all of which are supported by *interactive* tools. Methods and tools for designing non-interactive ISS components or middleware are not covered by the prototype. They may, however, be integrated into the life cycle, without modification existing AME tools.

For activities A_2 to A_6 there also exist tools for the *automatic refinement* of an input specification S_{i-1} to an output specification S_i. The life cycle can be iterated as discussed in **4.1**. There are no explicit activities for validation, verification or maintenance. Rather, the designer may use the interactive tools of an activity instance $A_{i,x}$ to inspect the results produced by $A_{i-1,x}$ in detail, and possibly improve them. For maintenance or redesign of a target ISS produced by $A_{6,x}$, a new iteration of the life cycle is started on the basis of some appropriate activity instance $A_{i,\,x+1}$ and a specification $S_{i-1,\,q}$.

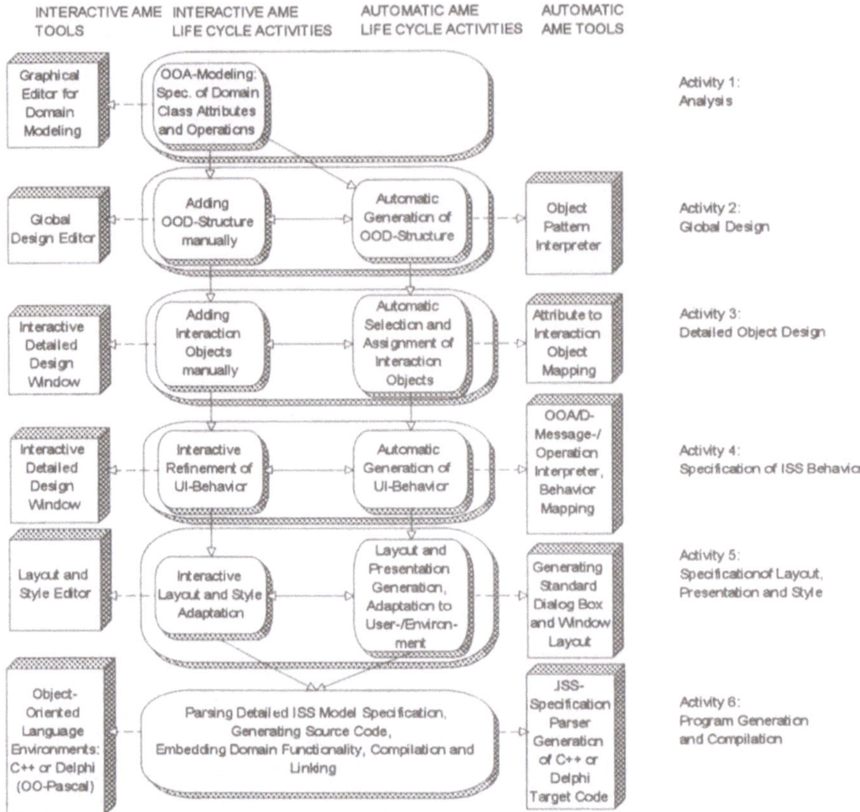

Fig. 5. Overview of life cycle activities and tools in the current AME prototype.

4.3 Activity A_1: Analysis

Based on an informal *problem domain description* $S_{0,j}$, an object-oriented analysis (OOA) model $S_{1,j}$ is created by the designer. $S_{1,j}$ does not include any information about how domain classes are represented in the target user interface. OOA classes are specified by their attributes (name, data type, starting value) and operations/methods (name, calling parameters, data types of calling parameters, return value, return type). Available inter-class relations are generalization /specialization (including multiple inheritance), aggregation, association, and dynamic link (link name, message contents, type and priority).

In the latest AME prototype the designer uses *ODE*, a graphical OOA/D editor to specify the domain object model of the application. Fig. 6 shows an *ODE* screen during OOA modeling of an example application.

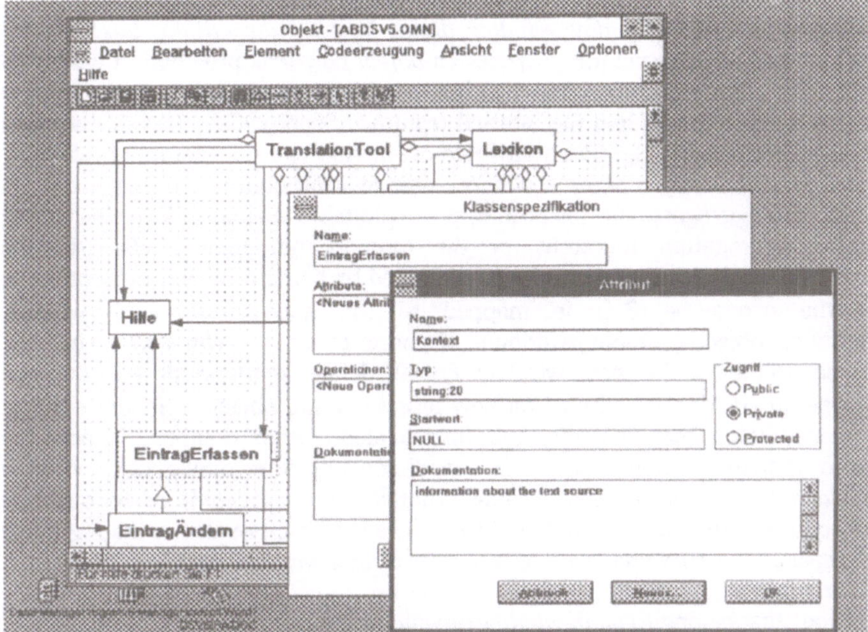

Fig. 6. Modeling domain attributes of an OOA class with ODE

The OOA model is stored as a model specification $S_{1,j} = \{c\_ooa_k \mid c\_ooa_k \in M_{ASO} \wedge 1 \leq k \leq m\_ooa\}$, which includes a representation object for each of the m_ooa classes in the created OOA model. OOA class names are stored in the slot c_ooa_k:name. Attributes, attribute types, methods and method types are mapped to the slots c_ooa_k:contents, c_ooa_k:content_types, c_ooa_k:actions, c_ooa_k:action_types as value lists. Inter-class relations between any two classes C_k and C_l are translated into directed pointers between objects c_ooa_k, $c\_ooa_l \in S_{1,j}$ and stored in the appropriate slots of c_ooa_k and c_ooa_l.

4.4 Activity A_2: Global design

This activity specifies the object-oriented design (OOD) structure of the application. The M_{ASO}-representation of the OOA model is automatically expanded to an OOD model, also represented by M_{ASO}-objects, which includes the window structure and the menu and command hierarchy of the application.

To provide global design automatically the system needs some basic information about the target runtime environment at this early stage. The AME prototype uses structural knowledge about the target window environment (e.g. about the available standard menus *File, Edit, Help, View* etc. in MS-Windows and their default menu items as well as their synonyms). Textual pattern matching techniques are used to map the method names of OOA classes (c_i:actions) to the synonymous items of standard- or application-specific pulldown-menus in the Windows environment. It is not easy to automatize this task, because standard Windows applications provide pulldown-menus only for the main window of an application.

Therefore, the matching algorithm has to know which OOA class will be mapped to which window type *(main window, window, dialog box, standard dialog box)*.

For this and other structural purposes an *object parser* is provided. It examines and exploits the generalization/specialization, association and aggregation structure of the OOA domain model and the internal features of each OOA class on the basis of the OOA specification $S_{1,j}$. Each attribute or method can only be inherited once. The parser automatically assigns a window type to each complex object, i.e. each OOA class with aggregated classes or multiple exploitable attributes. The class at the top of the aggregation hierarchy or the topmost non-generic class in the generalization/specialization hierarchy (represented by c_main, which is identical to one of the $c\_ooa_i \in S_{1,j}$) is mapped to the application main window (c_main:dialog_object = main_window). If more candidate windows exist, the designer has to choose the main window. For all default main window menus and possibly one application specific menu new representation objects $c\_menu_i \in M_{ASO}$ with c_menu_i:dialog_box = menu, are instanciated. Other complex objects are mapped to individual dialog boxes (c_ooa_i:dialog_object = dialog_box), if they contain *Cancel* and *OK* methods (name synonyms are accepted) or to ordinary windows (c_ooa_i:dialog_box = window), if not. The cardinality-value of aggregations is used to specify whether one or more instances of this window class may be created at the same time.

Methods that belong to *main window* or *window* objects are mapped to pulldown menu entries of the main window. Window methods for which no synonyms can be found are mapped to an application-specific pulldown menu. If the OOA classes contain methods for standard services (e.g. *Print, Find, Replace, Open, Close),* these methods are mapped to the corresponding menu items and linked with standardized objects $c\_stdlgbox_i \in M_{ASO}$, which represent the related dialog box or a pure default action.

To resolve name collisions between methods a menu entry is only generated for the method that belongs to one window: the one which is itself the main window or the one most closely related to the main window. A button is assigned to the other method(s). Each assignment goes along with the instanciation of a $c\_ood_{i,j} \in M_{ASO}$ with $c\_ood_{i,j}$:dialog_object = button. Dialog box methods are always mapped to one or several $c\_ood_i \in M_{ASO}$, representing buttons or button groups. For each method c_ooa_i:action$_v$, which was mapped to a menu action or button, a dynamic link back to the OOA object is generated, i.e. $c\_ood_{i,j}$:MessagLinksTo = c_ooa_i. At a later stage, the code generator exploits these links and creates code for calling the method, whenever the menu entry or button is selected.

During global design each *complex* OOA class (e.g. a class representing a data entry form for a multilingual dictionary) has to be expanded to an aggregation of many (typically dozens or hundreds) simple OOD-classes. Each *exploitable* class attribute c_ooa_i:actions$_v$ is mapped to one or more $c\_ood_{i,j} \in M_{ASO}$, representing one interaction object for some simple component of the entry form (e.g. a listbox for selecting the target language). Each generated $c\_ood_{i,j}$ is linked to its origin class by an aggregation relation. For grouping attributes special c_i:content_types settings are available for the designer during OOA. *Simple* OOA classes (with only one content value or attribute) are directly adopted as OOD classes. To be *exploitable* an attribute needs a data type, known to the system and some qualifying meta information. These data are used for mapping each attribute and its contents to an AIO. An OOA attribute c_ooa_i:contents$_v$ = *Language* of a class *Language Environment* with c_ooa_i:content_type$_v$ = *string:20,* for example, is mapped to two $c\_ood_{i,1}$ and $c\_ood_{i,2}$, which represent a label with the value *Language* and an edit field of *length 20.* The AIOs are represented as $c\_ood_{i,1}$:dialog_object = *Static* and $c\_ood_{i,2}$:dialog_object = *Edit.* Two aggregations from the c_ooa_i representing *Language Environment* to the new objects $c\_ood_{i,1}$ and $c\_ood_{i,2}$ are generated. To facilitate later layout generation, $c\_ood_{i,1}$ and $c\_ood_{i,2}$ are connected by associations. This is an example for a simple recurrent object pattern automatically generated by the object parser.

If the generated OOD-structure needs modifications for efficiency or usability reasons, the designer may improve the OOD model specification, using again ODE or other available object editors and browsers. Manual design makes sense, whenever domain-dependent decisions concerning the user interface structure or functionality, have to be taken. The result of an activity instance $A_{2,j}$ is an OOD model specification $S_{2,j} = \{c\_ooa_k, c\_ood_{k,l}, c\_menu_r \mid c\_ooa_k, c\_ood_{k,l}, c\_menu_r \in M_{ASO} \wedge 1 \leq k \leq m_j \wedge 1 \leq l \leq n_j \wedge 1 \leq r \leq p_j\}$.

4.5 Activity A_3: Detailed object design

During this life cycle activity the system selects AIOs $u_i \in M_{UI}$ for all $c_k \in M_{ASO}$ with c_k:dialog_object = NIL. AMEs AIO selection mechanism uses techniques similar to the systems discussed in (de Baar et al. 1992; Vanderdonckt, Bodart 1993). Attribute data types, cardinalities and some meta information are evaluated for AIO selection. Information about interaction media and data types, maintained in the class hierarchies M_{Media} and M_{Data} is also available to AME´s AIO selection mechanism.

Specific AIOs ($u_{Function}, u_{Code}, u_{Event} \in M_{UI}$) support the integration of domain functionality, code fragments or event based user interface dynamics. OOD class representation objects $c\_ood_{k,l}$, whose AIO was already specified during global design, are revisited during detailed design. In some cases abstract interaction objects are refined to more specific types (e.g. from a group of single *buttons* to a *button group).* The knowledge for selecting abstract interaction objects can be expressed in the form of rules. For efficiency reasons, these rules are coded as *if-then-else* cascades in a global AME resource method. The method *MakeDialogObject,* which is inherited by each $c_k \in M_{ASO}$, invokes this resource method for choosing the AIO. To make the selection process more flexible, a great number of data type synonyms is known to the system. A designer can easily modify the generated interaction object assignments interactively. The assignment of a list of AIOs to one $c_k \in M_{ASO}$ in order to represent multi-modal interaction is also supported. Resulting from this activity is a model specification $S_{3,j}$, with specified slots c_i:dialog_object for all $c_k \in S_{2,j}$.

4.6 Activity A_4:Automatic specification of dialog behavior and dynamics

The remaining construction level components map OOD object features to interaction object features *(behavior mapping)* and build the specification for the dynamic properties of the entire interactive system.

Each $c_k \in M_{ASO}$ owns the same common *Behavior* method. For each $u_i \in M_{UI}$ a *specific* Behavior method *(e.g. ComboBox.Behavior*, if u_i represents a combo box*)* is provided. After the assignment c_k:dialog_object = u_i, the method u_i.Behavior is activated by c_k.Behavior. It specifies the mapping of the relevant c_k-slot contents to the features of the specific interaction object type u_i. The specification information is written to reserved c_k -slots.

In the target environment, the target source code generator uses this information for creating concrete interaction object classes with correct interactive properties. The behavior mapping process also provides information needed to generate menu activations, external application calls and code for embedding domain objects, which encapsulate event handlers or application code fragments. For generating these control specifications c_k-action$_v$ with specific c_k:action_types$_v$ (e.g. Create, Delete, Activate) have to be evaluated together with the dynamic message passing structure of their c_k-objects.

Interactive and automatic AME tools support the specification of message channels between OOA classes and of messages exchanged via these links. OOA-message links are interactively specified already in activity A_1. During activity A_2 some additional dynamic links are generated between each c_ooa$_k$, representing a window, and all c_ood$_{k,l}$, and c_menu$_r$, whose interaction objects can dynamically be referenced from this window at runtime. During activity A_4 inter-class method calls may also be modeled as typed dynamic links between M_{ASO}-objects representing OOA and OOD classes. The dynamic links specified in the model guide the creation of source code for window activation and deactivation during activity A_6.

An additional A_4-tool allows the specification of some domain-dependent user interface dynamics for the target ISS (e.g. the time- and situation-dependent change of the appearance of AKI-objects or the availability of previously invisible menu-items, if a domain-condition evaluates to *True)*. The tool supports the modeling of state variables (represented by some c_ood$_{k,l}$:contents$_v$ and c_ood$_{k,l}$:content_types$_v$), state-dependent conditions (e.g. *below value, above value, changed, exact value* or more complex conditions involving multiple attributes) and the specification of method code for condition changes. The result of this activity is a more detailed ISS specification $S_{4,j}$, $S_{3,j} \subset S_{4,j}$, which includes all additional slot settings (e.g. message links) as well as any method parameter and code specifications.

4.7 Activity A_5: Layout and style generation

To provide a realistic simulation of the application before code generation, prototypical instances $w_{k|instance} \in U_{Window}$ of all aggregated $c_k \in M_{ASO}$ representing windows or dialog boxes (c_k:as_frame = $w_{k|instance}$) and concrete interaction objects (CIOs) $u_{k|instance} \in U_{Control}, U_{Menu}, U_{DomainUI}$ for atomic $c_k \in M_{ASO}$ (c_k:de_instance = $u_{k|instance}$) are created within the AME environment. Instances inherit the look-and-feel characteristics from their classes w_i, $u_i \in M_{UI}$ of AME's AIO class-hierarchy.

Any $c_k \in M_{ASO}$ inherits the method *MakeLayout* from M_{ASO}. When activated, this method first examines, whether c_k represents a dialog box or a window. In this case the layout type is selected by counting the number of each interaction object type in the dialog box or window. A set of common window and layout alternatives was implemented as AME standard resource methods, which may be invoked by c_k.MakeLayout. They were collected by evaluating and comparing the window or dialog box types of several widely used GUI-applications for the office environment. By parameterizing an amazingly small group of basic types, all relevant layout alternatives, especially for dialog boxes, can be created automatically. A simplified example for selecting a layout type for a dialog box w_k is shown in the following:

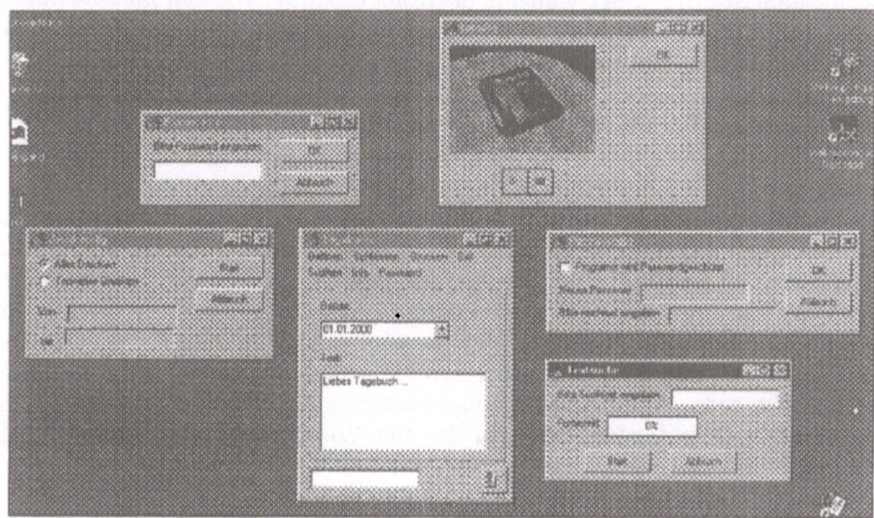

Fig. 7. Example ISS after Delphi code generation (Activity A_6)

If (w_0*:number(ComplexInteractionObjects) > 0) /*e.g. spreadsheets*/*
Then w_k*:layout_type := linear*
Else If (w_k*:number(Edit) > 4) Or* (w_k*:number(Editor) > 4)*
    Then w_k*:layout_type := entry_mask*
    Else If (w_k*:number(Edit) > 0)*
        Then w_k*:layout_type := entry_dialog*
        Else If (w_k*:number(Listbox) > 0)*
                    Or (w_k*:number(Combobox) > 0)*
            Then w_k*:layout_type := listbox_dialog*
            Else w_k*:layout_type := message_box.*

The layout generator also evaluates the neighborhood-associations specified between OOD classes to find semantically linked interaction objects. To facilitate layout generation, each window or dialog box is divided into rectangular areas. Each resource layout type (e.g. *entry_dialog)* defines in which rectangle instances of a certain interaction object type typically appear. The detailed geometrical design (spaces between elements, row and column ordering, width and height of the window) depends on the actual number of each element of a given type and is stored directly in the instanciated windows and interaction objects.

A preview of the layout of all windows is created by activating the layout methods. Presentational settings like colors, fonts or sizes are inherited from AME´s

AIO class hierarchy and can be changed by the designer. A designer can also add application- or user-specific presentation and layout rules or functions to the environment- and user-profile. Such rules are activated in a forward-chained mode. The designer may change the generated layout and presentation. Any changes will be stored and passed to the target code generator. The user interface specifications in the layout prototype are still independent of a specific GUI platform. The specification $S_{5,j}$, $S_{4,j} \subset S_{5,j}$, which is the result of A_5, includes all presentational and layout information in the form of the descriptor objects of the instanciated windows, dialog boxes and interaction objects.

4.8 Activity A_6: Target source code generation

Finally, the detailed design model $S_{5,j}$ that includes the specification of structure, dynamics, layout and presentation of the interactive system can be passed to one of the code generator tools. Two available parsers/generators exploit $S_{5,j}$ in order to create C++- or Delphi (i.e. object-oriented Pascal) source code. The source code, which could be seen as output specification $S_{6,j}$ can be translated by the appropriate compiler and automatically linked with domain method code. At the generator level, the designer still may modify the specification, before it is parsed and translated into target source code. Fig. 7 shows a simple example ISS that was generated with the *OBJECTWAND* Delphi code generator for the Windows 95 environment (Märtin, Humpl 1997).

5 Related work

When comparing AME to related model-based approaches, one must distinguish between ISS specification systems and ISS generators. Most related systems fall into one of these categories, whereas AME tries to combine the benefits of both.

Model-based ISS specification systems allow expert user interface developers to define the structural, functional and dynamic features of the user interface of applications in close coordination with the domain parts.

Systems like *UIDE, HUMANOID and MASTERMIND* (Szekely 1996) yield a high-quality and flexible detailed design-level model of the interactive system, which allows to represent application-independent and application-specific interactive requirements in great detail. Such systems demand the explicit specification of user interface functionality and dynamics. The systems provide mechanisms for representing runtime-dependent application dynamics. The resulting ISS design specification can be translated into a working prototype of the application. The systems support the generation of context-based help information as well as layout generation. In addition MASTERMIND incorporates techniques for automatically adapting the information presented by a user interface to varying screen or window sizes.

The *IDA* environment (Reiterer 1995) provides advanced tools for the construction of graphical user interfaces and uses an object-oriented approach for designing flexible, reusable interface templates. IDA includes an ergonomic design consulting system as well as a knowledge based quality assurance tool. *EXPOSE* is a consulting expert system for the design of highly-ergonomic user interfaces (Gorny 1995).

Model-based generators create user interface prototypes from domain data models. Generators are supposed to raise software productivity and to help application domain experts with the design of consistent user interfaces. However, the flexibility and the complexity of the generated user interfaces may be restricted.

Entity-Relationship (ER) models are exploited by the generators *TRIDENT* (Vanderdonckt, Bodart 1993) and *GENIUS* (Janssen et al. 1993). These systems need additional state-transition-specifications for dynamic modeling.

An object-oriented generator was described by (de Baar et al. 1992). It exploits attributes of domain object classes, action names and meta-data for generating application windows and their menus. In contrast to AME this system as well as the ER-based systems require explicit information about which of the domain data or object attributes will be grouped together in one target window. *JANUS* (Balzert et al. 1996) uses OOA-models for generating multi-window office applications. Each OOA class is mapped to exactly one window of the user interface. Ergonomic knowledge is used to translate class-attributes to AIOs and operations to menu-items of this window. Only static OOA-relationships are interpreted in order to decide how class attributes are presented and the generated windows are connected. No explicit dynamic modeling is supported. In contrast to AME specifications cannot be modfied by the designer at the global and detailed design levels. The *TADEUS* system (Elwert, Forbrig 1997) was developed as successor to EXPOSE and uses domain task models in addition to object-oriented models, dialog models and user models for generating multimedia applications. Other recent ISS design approaches are discussed in (Vanderdonckt 1996).

Conclusion

The AME environment provides an automated approach to ISS modeling and design. All activities of the development life cycle are supported by interactive and/or automatic tools for transforming intermediate input specifications into more detailed intermediate output specifications. Automatic design components provide standardized object-oriented and knowledge-based approaches for solving typical ISS design problems at all abstraction levels. Additional interactive tools rise the flexibility of the produced software systems. Cooperation between tools is easy, because all tools operate on the same object-oriented model specification. The model enters various refinement stages during its life cycle. In order to support reuse design results, model specifications may be stored after any life cycle activity. Iteration of the whole life cycle or parts of it are supported. This is an essential requirement for rapid application development environments.

Several application prototypes, including spreadsheets and multimedia office applications, were developed to compare the AME life cycle and the available modeling and generation tools to established conventional approaches. New application types or domains typically require sometimes many additional interaction object classes and additional construction knowledge based on software patterns. It is a large research topic in itself to find and represent new domain-dependent design knowledge categories. For the future we plan to establish a design mode that enables the system to learn previously unknown analysis patterns and their OOD mappings by watching the designer.

The system continuously develops in an evolutionary way. However, it is easy to integrate new resources and design knowledge into the AME prototype. AME does not offer solutions to all possible productivity problems. However, design time can be drastically reduced for design tasks, where the existing design knowledge can be used for generating the standard parts of the application. If available design resources fit the application domain, an OOA model with a dozen domain analysis classes can be automatically expanded into an OOD model with possibly several hundreds of M_{ASO}-objects and their interaction objects. Without programming, the resulting ISS includes correct interaction object mappings, a prototypical layout, prototypical presentation and style settings, links to all domain code methods, the menu hierarchy, complete application-independent interactive dynamics and part of the application-dependent dynamics.

References

de Baar, D.J.M.J., Foley, J., Mullet, K. (1992): Coupling Application Design and User Interface Design. In: Proc. of CHI '92, Bauersfeld, P. et al. (eds.), ACM Press, pp. 259-266

Balzert, H. et al. (1996): The JANUS Application Development Environment: Generating more than the User Interface. In: (Vanderdonckt 1996), pp. 183-206

Bernstein, P.A. (1996): Middleware: A Model for Distributed System Services. Commun. ACM. Vol. 39, No. 2: pp. 86-98

Boehm, B.W. (1981): Software Engineering Economics. Prentice-Hall, 1981

Coad, P., Yourdon, E. Object-Oriented Analysis. Prentice-Hall, 1991

Elwert, Forbrig (1997): Multimedia Data and Model-Based Development of Interactive Systems. In Proc. DSV-IS '97, pp.425-438

Fikes, R., Kehler, T. (1985): The Role of Frame-based Representation in Reasoning. Commun. ACM. Vol. 28, No. 9: pp. 44-52

Gamma, E. et al. (1995): Design Patterns. Elements of Reusable Object-Oriented Software. Addison-Wesley, Reading. Mass.

Gorny. P. (1995): EXPOSE - An HCI Counseling for User Interface Design. In: Proc. of INTERACT '95, Nordby, K. et al. (eds.), Elsevier, pp. 297-304

Janssen, C., Weisbecker, A., Ziegler, J. (1993): Generating User Interfaces from Data Models and Dialogue Net Specifications. In: Proc. INTERCHI '93, Ashlund, S. et al. (eds.), IOS Press, pp. 418-423

Märtin, C. (1990): A UIMS for Knowledge Based Interface Template Generation and Interaction. In: Proc. INTERACT '90, Diaper, D. et al. (eds.), Elsevier, pp. 651-657

Märtin, C. (1996a): Modellierung, Entwurf und automatische Konstruktion interaktiver Softwaresysteme. Entwurf der modellbasierten Entwicklungsumgebung Application Modeling Environment (AME), Ph.D. Thesis, University of Rostock

Märtin, C. (1996b): Software Life Cycle Automation for Interactive Applications: The AME Design Environment. In: (Vanderdonckt 1996), pp. 57-73

Märtin, C., Humpl, M. (1997): Generating Adaptable Multimedia Software from Dynamic Object-Oriented Models: The ObjectWand Design Environment.

In: Proc. HCI International 1997, Smith, M.J., Salvendy, G., Koubek, R.J. (eds.), Advances in Human Factors/Ergonomics, Vol. 21B, Elsevier, pp. 703-706

Meyer, B. (1995): Object-Success, Prentice-Hall, 1995

Reiterer, H. (1995): IDA - A Design Environment for Ergonomic User Interfaces. In: Proc. of INTERACT '95, Nordby, K. et al. (eds.), Elsevier, pp. 305-310

Rumbaugh, J. et al. Object-oriented Modeling and Design. Prentice-Hall, 1991

Salvendy, G. (ed.) (1997): Handbook of Human Factors and Ergonomics, 2nd Ed., John Wiley & Sons

Schmidt, D.C., Fayad, M., Johnson, R.E. (Guest Eds.) Softwre Patterns, Commun. ACM Vol. 39, No. 10, pp. 36-39

Shneiderman, B. (1997): Designing the User Interface, 3rd Ed., Addison-Wesley

Szekely, P. (1996): Retrospective and Challenges for Model-Based Interface Development. In: (Vanderdonckt 1996), pp. xxi-xliv

Vanderdonckt, J., Bodart, F. (1993): Encapsulating Knowledge for Intelligent Automatic Interaction Objects Selection. In: Proc. INTERCHI '93, Ashlund, S. et al. (eds.), IOS Press, pp. 424-429

Vanderdonckt, J. (ed.) (1996): Computer-Aided Design of User Interfaces. Proc. of CADUI '96. Press Universitaires de Namur.

Modeling fault-tolerant system behavior

Mario Dal Cin

1 Introduction

Aerospace and railroad control systems need to ensure the safety of passengers. To prevent financial losses, banking and telecommunication systems must offer high availability. Such safety- and mission-critical systems require high assurance. To this end, several formal methods for specifying and verifying non-functional system properties like timeliness, safety and liveness have been developed (Leveson et al. 1994, Leeb and Lynch 1996, Kirner and Davis 1996, Bruel et al.1996, Mok et al. 1996). These methods are intended to give system developers and customers greater confidence that the systems satisfy their requirements. A number of these verification methods are based on finite-state representations and have achieved considerable success in practical applications.

Safety stipulates that something bad does not happen; liveness guarantees that something useful happens (Lamport 1977). Other non-functional requirements like adaptability, smooth evolutionary structural changes or fault-tolerance have hitherto been investigated to a lesser extent. Among them, fault tolerance is important also for systems whose correct behavior has been verified. After all, an initially functionally correct system can fail in many ways due to wearout or inadequate handling and this loss of quality may not be detected in due time. As a consequence, the initially correct system may move through a series of undesirable, e.g. unsafe, states. Then it is possible that it moves to failure, hazard, and accident. Hence, fault tolerance is an important non-functional requirement for safety- and mission-critical systems and its assurance must consider all possible system behaviors, also unintended ones. Fault tolerance is generally addressed by incorporating redundancy in the design, such as component redundancy (soft- and hardware) or time redundancy.

The aim of this paper is to develop a framework which can be elaborated to allow the modeling and analysis of fault-tolerant system behavior to be undertaken. The framework is based on the notion of the finite state machine, since finite state machines provide intuitive and natural models for describing discrete-state and discrete-event systems which continuously receive inputs from and react to their environment. These reactive systems have wide application and vary greatly in scale. They include safety- and mission critical embedded systems.

We aim at a basic framework and, therefore, employ a simple finite state model. Our basic notions are that of a tolerance relation and an error relation over the state space of the system (Dal Cin 1997). Introducing tolerances and errors as binary relations gives us the possibility to exploit the relational calculus (Brink et al. 1997) and to deal explicitly with unintended state transitions. The intended benefits will be to simplify modeling erroneous behavior and tolerances.

We proceed as follows. In Section 2 the underlying fault model is discussed. Section 3 presents an example. Section 4 introduces tolerances and stable finite state machines. Section 5 introduces errors. Section 6 deals with fault-tolerant system behavior. In Section 7 we return to our example. Finally, in Section 8 we briefly discuss verifying fault-tolerant behavior by model checking. The Appendix summarizes our notation.

2 Unintended state transitions

According to Laprie (1992), an error is the consequence of a fault which can lead to a failure of the system. Within our abstract framework we are not interested in modeling faults, the physical causes of errors. We are rather interested in modeling errors. Nevertheless, studying the effects of errors on system behavior, one has to differentiate between errors caused by permanent faults and errors caused by temporary faults.

In research on fault tolerance it has long been recognized that an appropriate fault/error model plays a fundamental role, and that it is very hard to obtain one. Our fault/error model originates from the notion of erroneous state transitions. According to Lee and Anderson (1992): an erroneous transition of a system is any state transition to which a subsequent failure could be attributed. Thus, an erroneous state transition may lead to an erroneous state. An erroneous state of a system is any state which can lead to failure by a sequence of valid transitions. Specifically, there must exist a possible sequence of events which would, in the absence of external corrective actions and in the absence of erroneous transitions, lead from the erroneous state to a system failure. Hence, in order to investigate the consequences of erroneous state transitions we only need to consider temporary faults. A state transition due to a temporary fault occurs at most once in the considered period. Note, however, that a temporary fault may very well cause the system to take erroneous states permanently. Thus, though a temporarily faulty system component spontaneously recovers from a fault, the fault may have corrupted the system state and subsequent state transitions will perpetuate the contamination. Hence, a temporary fault can lead to failure.

On the other hand, the system may exhibit self-stabilizing recovery from temporary faults. Rushby (1996) discusses examples of self-stabilizing recovery. Hence, not all state transitions of a fault-tolerant system due to faults may lead to failure. We call these transitions as well as erroneous transitions unintended transitions. We will model unintended state transitions by binary relations over the state space of the system. (Such relations will be called errors). That is, we model errors explicitly as separate entities. This gives us the possibility of modeling uniformly errors caused by temporary faults and errors caused by permanent faults. Furthermore, this framework allows us to describe fault-tolerant behavior, such as fault masking or fail-stop, and to develop tests for fault tolerance in a concise way.

For a fault-tolerant system it makes little sense to try to verify, that upon an error it behaves as its specification. Ideally, a fault-tolerant system behaves in the presence of faults as the fault-free system. In reality, however, it exhibits most often a degraded but tolerable behavior due to recovery (performance degradation). Hence, we introduce the concept of tolerance as follows. A tolerance is a binary relation over the state space of a system such that two states are related (are in tolerance) if their difference can be tolerated. For example, the two states may give rise to different, but acceptable outputs. A tolerance is, clearly, reflexive and symmetric. A coarse tolerance , call it η, is given as follows: two states are η-related if and only if both are not erroneous. Roll-back recovery provides another example. If failures do not occur too frequently, a correct system state may be restored by roll-back recovery as long as the unintended state transition is due to a temporary or transient fault (temporary error) and does not affect the recovery mechanism. However, restoring the system state may only be feasible from certain (undesirable) states. We then may say that these states are in tolerance with the correct states. We wish to formalize these observations.

In this paper, systems modeled by sets of sequential finite state processes will be investigated. The processes are capable of reacting to events from a given finite set Ev. (A clock tick, denoted by t, is such an event). A process is sequential if it can perform at most one reaction at the same time. A sequential finite state process can be conveniently represented as a finite state machine (FSM). This is a finite set S of states over which a binary relation r_a is defined for each event $a \in$ Ev; r_a is called a transition relation or a reaction. The triple sr_aq means that the process reacts in state s to the event a by changing to state q. We denote a FSM by T = (S, Ev, R = $\{r_a \subseteq S \times S \mid a \in \text{Ev}\}$) and require that [R] $= \bigcup$ R is total (see Appendix). We call reaction r_a deterministic, if r_a is a univalent relation, i.e $r_a^c \cdot r_a \subseteq \iota S$. T is deterministic, if all reactions of T are deterministic; T is complete, if all reactions of T are total.

Thus, the system model is given by a set of processes each represented as a FSM. The state space of the model is the Cartesian product of the state spaces of these processes. We denote the AND-composition (Harel 1987) of two FSMs T1 and T2 by T1|T2.
We will begin by considering a small example of fault-tolerant systems modeled as a set of sequential processes.

3 Railroad crossing example

To illustrate our approach we consider a railroad crossing system similar to that introduced by Heitmeyer and Mandrioli (1996). It consists of a railroad track, a traffic signal (controlling light) for trains, a gate, and trains, see Figure 1. To keep the example simple we consider a one-way railroad and assume that trains arrive at the crossing at a sufficiently low rate, and leave it quickly enough, so that no train attempts to enter the crossing while another train is already there. When the train approaches region P it observes the traffic light. When off (or red), the train has to come to a halt within region P (state H). Region I is the crossing through which cars can travel, of course, only if the gate is up. After the train left the crossing, the gate is reopened. It may then be possible that a train approaches region I while the gate is opening. In this case, the gate must reverse its movement and start closing. Figure 1 illustrates the crossing by an activity chart (Harel 1987).

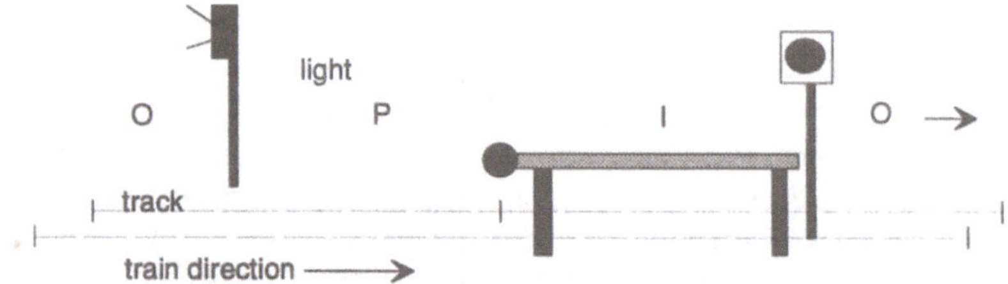

Regions of the railroad crossing system

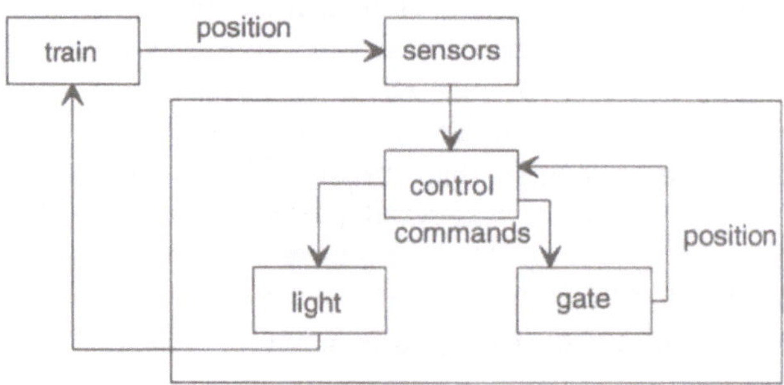

Activity chart of the railroad crossing system

Fig. 1. The railroad crossing example

The statechart (Harel 1987, Erikson and Penker 1998) is given by Figure 2. This so-called AND-composition specifies the behavior of the train, the light, the gate and the system controller. The statechart contains four finite state machines (finite automata or transition tables).

For instance, the gate model is defined by the FSM $G = (S^G, Ev^G, R^G)$

where S^G = {Open, Opening, Closing, Closed}
Ev^G = {lower, down, rise, up} and
R^G = {r_{lower} = {(Open, Closing), (Opening, Closing)},
r_{down} = {(Closing, Closed)},
r_{rise} = {(Closing, Opening), (Closed, Opening)},
r_{up} = {(Opening, Open)}}

State s_0 := Open is the default (initial) state of G. (Default states are indicated by arrows).

The AND-chart of Fig. 2 specifies a composed finite-state machine (composed automaton) with state space $S = S^{Train} \times S^{Light} \times S^{Gate} \times S^{Controller}$. The event set is the union of the individual event sets. Thus, the railroad crossing system has 96 states. In the following we consider only the crossing system (called GRC) and not its controller. The controller is assumed to be fault-free. The initial state is O1 = (O, Off, Open). If the system undergoes intended state transitions only, its (fault-free) transition relations are given by the state transition graph of Figure 3. Notice that the state space contains many more states which can not be reached from the initial state.

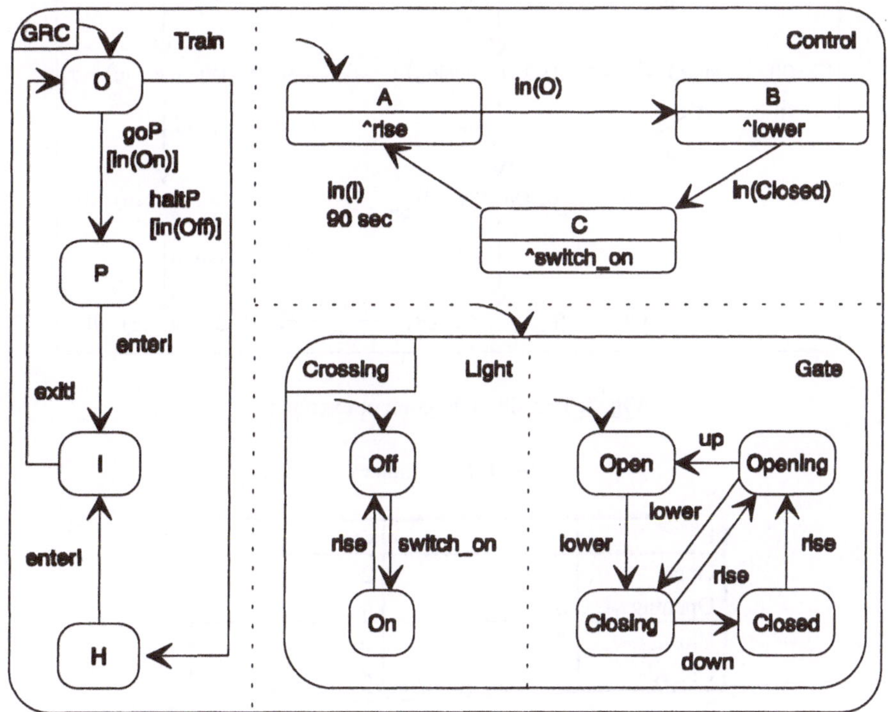

Fig. 2. Statechart showing the AND-composition of the railroad crossing system
States: O: train approaching the crossing, P: train passes the light without stopping, I: train is crossing, H: train halts

Some of the transitions between states are guarded by conditions. For example, goP[in(On)] (Whenever the train passes region P without stopping the green light must be on). The guarding condition must be true before the transition can take place.

In reality, however, it may become difficult to guarantee that the guards are observed by the system. The system may wear out, its environment may become adverse or its operator may introduce erroneous states. Hence, there may be a non-zero probability (depending on time) that an (unintended) transition takes place even if its guarding condition evaluates to false. For example, a guard for the light could be „switch_on [in (Closed)]“, i.e., the light becomes green, only if the gate is closed. This guard will not be observed when the light bulb is broken or, if there is a short-circuit. We will, however, not mention guards explicitly, since they are not needed for the following. Not observing a guard is a fault not an error; it may result in an error.

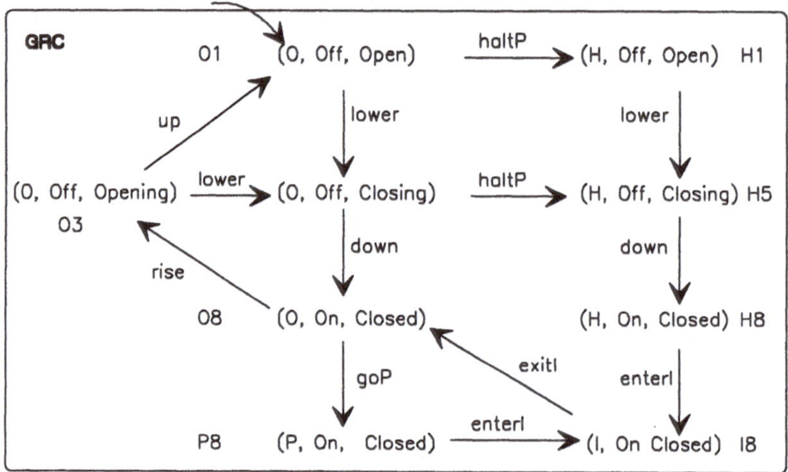

Fig. 3. Transition diagram of example

Table 1. State indices

i	Off	On
Open	1	2
Opening	3	4
Closing	5	6
Closed	7	8

For the following we chose indices for states as shown in Table 1. State I3 = (I, Off, Opening) can be considered in tolerance with state I8 = (I, On, Closed). Before the gate is fully open and the cars start moving the train may have left the crossing. If we consider safety, all states of Figure 3 are in tolerance with each other. They are, however, not in tolerance with the unsafe states P1 = (P, Off, Open) – the train passed the light without permission- or I1 = (I, Off, Open) – the train is crossing despite the open gate. The states I3 and I4 = (I, On, Opening) may be considered in tolerance with I8 = (I, On, Closed) but undesirable (see above). The 'non-productive' state O7 =(O, Off, Closed) – the train is not allowed to approach to the crossing despite the crossing is closed - is undesirable too and with no other state in tolerance. The remaining states can be classified similarly.

In Section 7 we will consider unintended state transitions of this example. If the overall design of the system is considered to be correct then unintended transitions, i.e. errors, can only occur because of a failure of one of the components of the system. Introducing tolerances and, likewise errors, as binary relations over state spaces gives us the possibility to exploit the calculus of relations and to deal explicitly with unintended state transitions.

4 Tolerances

Central to our approach is the concept of a tolerance space (Arbib 1967, Dal Cin 1975), introduced first by H. Poincare (1958). For finite domains, the notions of distance, nearness, neighborhood, etc. can be formalized by tolerance relations. A tolerance τ over a set X is a binary, reflexive, and symmetric relation over X and (X, τ) is a tolerance space. If $(x, x') \in \tau$, we say x is within tolerance of x'.
For any $x,y \in X$ let distance $d_\tau(x,y)$ be defined as follows: $d_\tau(x,y)$ is the least n for which there are $x_0, \dots, x_n \in X$ such that $x = x_0$, $y = x_n$ and $x_i \tau x_{i+1}$ for $0 \le i < n$; and $d_\tau(x,y) = \infty$, if there is no such n. The distance d_τ is a metric on X. Let $A \subseteq X$, the set $cl_\tau(A) = \bigcup_{x \in A} x\tau$ is called the τ-closure of A. The interior of A is $int_\tau = \{x \mid x\tau \subseteq A\}$, and $bnd_\tau(A) = cl_\tau(A) - int_\tau(A)$ is the τ-boundary of A. Obviously, $bnd_\tau(A) = cl_\tau(A) \cap cl_\tau(A^{\neg})$, hence, $bnd_\tau(A) = bnd_\tau(A^{\neg})$.

4.1 Stable finite state machines

We call a FSM stable if a (non-tivial) tolerance over its state space is given and this tolerance is 'stable' with respect to all state transitions of the FSM. For the following we stipulate that all FSMs are complete.

Definition (stability): Let T= (S, Ev, R) be a FSM and τ be a tolerance over its state set S. The tuple (T, τ) is called a FSM with tolerance. A complete FSM T with tolerance τ is called stable, if τ is stable under all state transitions of R, i.e. if
$r_a \bullet \tau \subseteq \tau$, $a \in Ev$.

The reactions of a stable FSM are metric-decreasing, i.e. $d_\tau(q,s) \le d_\tau(q',s')$ for $q' \in qr_a$ and $s' \in sr_a$, $r_a \in R$. This may be viewed as a kind of stability property, and, hence, the naming. Intuitively, the successors of states within tolerance of a stable

FSM all stay within tolerance under the reaction of any sequence of events. For example, the tolerance η of Section 2 is stable according to the definition of erroneous states. With τ we can express certain properties of states formally, e.g. $q((\tau\cdot\tau^{\neg})^{\neg}\cap(\tau^{\neg}\cdot\tau)^{\neg})s$ expresses that q is in tolerance with all and only those states that are in tolerance with state s.

Let F(T) be the set of all stable tolerances of the complete FSM T. F(T) is a complete, distributive lattice (with respect to set inclusion) which is a sublattice of the lattice of all tolerances of T (Dal Cin 1975). Thus, $F(T, \tau) = \{\sigma \in F(T) \mid \sigma \subseteq \tau$ and τ a tolerance over S$\}$ has an unique maximal element, or is empty. This maximal stable element of $F(T,\tau)$ will be denoted by τ^*. To verify stability of (T,τ) we can compute τ^* and then check whether $\tau = \tau^*$. The following straightforward procedure determines the maximal symmetric binary relation ρ_τ over S with $r_a^c\cdot\rho_\tau\cdot r_a \subseteq \rho_\tau \subseteq \tau$ for all $a \in$ Ev. Hence, $\tau^* = \iota \cup \rho_\tau$, if $F(T) \neq \emptyset$.

Stability check: Compute binary relations $\rho(k)$ as follows: (1) $\rho(1) = \tau$; (2) for all $k \geq 1$, as long as $\rho(k+1) \neq \rho(k)$, set $s\rho(k+1)s^*$ if and only if $s\rho(k)s^*$, $sr_a\bar{s}$ and $s^*r_a\bar{s}^*$ imply $\bar{s}\rho(k)\bar{s}^*$ for all $a \in$ Ev. Now, $\rho_\tau = \rho(k_0)$ where k_0 is the smallest index k with $\rho(k) = \rho(k+1)$; (3) set $\tau^* = \iota \cup \rho_\tau$ and check $\tau = \tau^*$?
Obviously, ρ_τ is symmetric since τ is symmetric. Moreover, ρ_τ is the maximal relation with the properties stated above, since $\sigma \subseteq \tau$ implies $\rho_\sigma \subseteq \rho_\tau$ and, hence, $\sigma = \rho_\sigma \subseteq \rho_\tau$ if $r_a^c\cdot\sigma\cdot r_a \subseteq \sigma$ for all $a \in$ Ev.

To check whether a deterministic FSM is stable we do not have to compute τ^*. The following proposition suggests a more efficient procedure: Choose $a \in$ Ev and compute $r_a \lrcorner \tau$ (see Appendix). If $\tau \not\subset r_a \lrcorner \tau$, then (T,τ) is not stable and stop; else choose an new event $a \in$ Ev.

Proposition 4.1: Let FSM T be deterministic and complete. The FSM with tolerance (T, τ) is not stable if and only if there exists an event a such that $\tau \not\subset r_a \lrcorner \tau$.

Proof: (a) From $\tau \subseteq r_a \lrcorner \tau$ for all $a \in$ Ev follows $r_a^c\cdot\tau\cdot r_a \subseteq r_a^c\cdot r_a\cdot\tau\cdot r_a^c\cdot r_a \subseteq \tau$ for all $a \in$ Ev, since T is deterministic. Hence, $r_{a*}\tau \subseteq \tau$ for all $a \in$ Ev and T is stable. (b) Let (T,τ) be stable, then $\tau \subseteq r_a\cdot r_a^c\cdot\tau\cdot r_a\cdot r_a^c \subseteq r_a\cdot\tau\cdot r_a^c = r_a \lrcorner \tau$ for all events, since T is complete. ◁

Broadly speaking a stable tolerance is a bisimulation. A bisimulation is a kind of invariant between the states of two FSMs expressing that they are behaviorally equivalent. On the other hand, not every bisimulation is stable and, hence not every reflexive and symmetric bisimulation qualifies as a tolerance. Janowski (1995) investigates several concepts of bisimulation in the context of fault tolerance.

Definition (bisimulation): Binary relation σ over the state space of T is a (strong) bisimulation of T, if $p\sigma q$ implies, for all $a \in$ Ev

(1) whenever $p\, r_a\, p'$ then, for some q', $q\, r_a\, q'$ and $p'\sigma\, q'$,

(2) whenever $q\, r_a\, q'$ then, for some p', $p\, r_a\, p'$ and $p'\sigma\, q'$.

This is equivalent to (1) $r_a^c \cdot \sigma \subseteq \sigma \cdot r_a^c$ and (2) $\sigma \cdot r_a \subseteq r_a \cdot \sigma$. Note that (1) is identical with (2) if σ is symmetric.

In words, T can take the same transitions out of state q as out of state p (and vice versa) and the target states are again related. Thus, suppose that σ is a bisimulation of a FSM T expressing that two states of T are indistinguishable from outside of T. If two states s and q of T are bisimilar, then one can not distinguish from outside between the behavior of T starting in s and some behavior of T starting in q (and vice versa). Note that ι, ν and 0 are bisimulations of any complete FSM. Obviously, $r_{a*}\nu \subseteq \nu$.

Proposition 4.2: Let T = (S, Ev, R) be a complete FSM and $a \in$ Ev.
(a) Every stable tolerance τ over S is a bisimulation of T.
(b) If T is deterministic and σ a reflexive bisimulation of T, then σ is stable.

Proof: (a) Let τ be stable, then $r_{a*}\, \tau \subseteq \tau$ implies $r_a^c \cdot \tau \cdot r_a \subseteq \tau$. Thus, $r_a^c \cdot \tau \cdot r_a \cdot r_a^c \subseteq \tau \cdot r_a^c$. It follows that $r_a^c \cdot \tau \subseteq \tau \cdot r_a^c$, since r_a is total. (b) Let T be deterministic and σ a bisimulation over S. Then, $r_a^c \cdot \sigma \subseteq \sigma \cdot r_a^c$ and, hence,

$r_a^c \cdot \sigma \cdot r_a \subseteq \sigma \cdot r_a^c \cdot r_a \subseteq \sigma$. $\triangleleft$

4.2 Composite and incomplete finite state machines

When modeling reactive systems it is often convenient to model their behavior by a set of concurrent state machines (Harel 1987). The overall behavior of the reactive system is then given by the AND- or parallel composition of these FSMs.

Definition (induced tolerance): Let $(T1, \tau_1)$ and $(T2, \tau_2)$ be two FSMs with tolerances and the same event set Ev, i.e. T1 = (S_1, Ev, R_1) and T2 = (S_2, Ev, R_2). The induced tolerance over $S_1 \times S_2$ of the parallel composition T1|T2 is defined as $\tau_1 \mid \tau_2 = pr_1 \lrcorner \tau_1 \cap pr_2 \lrcorner \tau_2$ where pr_i: $S_1 \times S_2 \to S_i$, i = 1,2 is a projection.

Proposition 4.3 If $(T1, \tau_1)$ and $(T2, \tau_2)$ are two stable FSMs then also $(T1|T2,\ \tau_1 \mid \tau_2)$ is stable.

Proof: $\tau_1 \mid \tau_2 = pr_1 \lrcorner \tau_1 \cap pr_2 \lrcorner \tau_2 = \iota_{S_1 \times S_2} \cup (pr_1 \cdot \tau_1 \cdot pr_1^c \cap pr_2 \cdot \tau_2 \cdot pr_2^c) =$

$pr_1 \cdot \tau_1 \cdot pr_1^c \cap pr_2 \cdot \tau_2 \cdot pr_2^c$, since τ_i is reflexive.

Let $s, s' \in S_1 \times S_2$ where $s = (q_1, q_2)$ and $s' = (q'_1, q'_2)$. Now, $s\,\tau_1 \mid \tau_2\, s'$ if and only if $s\,(pr_1 \cdot \tau_1 \cdot pr_1^c)\,s'$ and $s\,(pr_2 \cdot \tau_2 \cdot pr_2^c)\,s'$ if and only if $q_1 \tau_1 q'_1$ and $q_2 \tau_2 q'_2$. Consider event $a \in$ Ev. If $s\,\tau_1 \mid \tau_2\, s'$, then for stable FSMs $q_1(r_a^{1c} \cdot \tau_1 \cdot r_a^1)q'_1$ and $q_2(r_a^{2c} \cdot \tau_1 \cdot r_a^2)q'_2$. Hence, the successors of s and s' are again within tolerance $\tau_1 \mid \tau_2$. ◁

We stipulated that all FSMs are complete. However, modeling, reactive systems one often specifies incomplete FSMs, since not all events are relevant for all states.

Definition (stability of incomplete FSM): An incomplete FSM is stable if (i) $r_a * \tau \subseteq \tau$ and (ii) $(\tau \cdot r_a) \cdot v \subseteq r_a \cdot v$ for all $a \in$ Ev, i.e. $q \tau s$ only if q and s have successor states or both have no successors for every event.

Note that for any binary relation ρ over X, $\rho \cdot v_X$ is a relation (sometimes called domain of ρ) describing 'where ρ is defined'.

The notion of a tolerance gets its special bearing when we consider unintended state transitions, called errors.

5 Errors

As stated in Section 2, our fault/error-model is based on the notion of unintended state transitions. The unintended transition from s to q may be due to a permanent or temporary fault and q may be an erroneous state. An unintended state transition due to a temporary fault occurs at most once in the considered period. An unintended state transition caused by a permanent fault can occur whenever the system is in the state that gives rise to the erroneous transition.

Definition (error): Let T = (S, Ev, R) be a FSM, and τ a tolerance over S. Any binary and reflexive relation $\phi \neq \iota$ S over S is called an error of (T,τ). Error ϕ is called temporary, if ϕ is "effective only once". If error ϕ is due to a permanent fault, it changes the transition r_a of T to $r_a^\phi = r_a \cdot \phi$, $a \in$ Ev. Then also error ϕ is called permanent.

Definition (basic properties of errors): Error ϕ is compatible, if $\phi \subseteq \tau$. A temporary error $\phi \subseteq \tau^*$ is insignificant. It is small, if (i) $r_a * \phi \subseteq \tau$ and (ii) $\phi \cdot r_a \cdot v \subseteq r_a \cdot v$ for all $a \in$ Ev. A permanent error ϕ is small, if (i) $r_a^c \cdot \phi \cdot r_a^\phi \subseteq \tau$ and (ii) $\phi \cdot r_a \cdot v \subseteq r_a \cdot v$ for all $a \in$ Ev.

The reflexive relation $r_a * \phi$ is called a successor error of temporary error ϕ with respect to event a and ϕ is small if all its successor errors are compatible. In the following, we interpret a proper error $(s, s') \in \phi$ with $s' \neq s$ as an undesirable (or unintended) state transition of T. That is, T changes (with nonzero probability) into state s' when it should change into state s. This transition may be due to a permanent or temporary fault and s' may be an erroneous state.

As an example consider the scenario of Figure 4. Due to error $(q, s) \in \phi$ the error-prone system may enter state s instead of q. From state s it enters state u; from state q it enters state v or state w. Error (q, s) is small, if and only if $(u, v) \in \tau$ and $(u, w) \in \tau$, i.e. state u can be tolerated regardless whether the error-free system would have entered state v or w. Observe further that ϕ being small (or τ being stable) implies that $r_a^c \cdot \iota \cdot r_a = r_a^c \cdot r_a \subseteq \tau$. Hence, also $(v, w) \in \tau$. This, obviously, holds for deterministic reactions, but not necessarily for non-deterministic ones. However, if we relate r_a with intended state transitions only, then $r_a^c \cdot r_a \subseteq \tau$ should be required of any meaningful tolerance. On the other hand, if this requirement is met by a FSM with tolerance for all $a \in$ Ev, then the FSM can be seen as "deterministic up to tolerance".

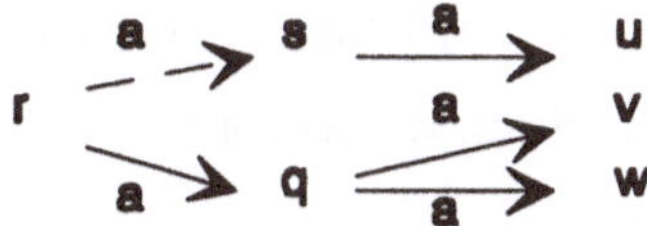

Fig. 4. Error

As with tolerances, we can now express properties of errors directly, e.g. $q\ (\phi \cdot \phi^c)^{\neg} s$ expresses that there is no unintended state transition from q and s to the same state of T.

6 Temporary errors

Systems fail, broadly speaking, for one of five reasons: inadequate specification, design or implementation errors, random component failures or inadequate handling. Throughout the remainder of this chapter, only temporary errors will be considered. Hence, only consequences of component failures and inadequate handling will be considered. Dal Cin (1998) focuses on permanent errors due to design and implementation faults. We investigate temporary errors by relational means. For instance, the next proposition gives us a relational formula to compute the largest small temporary error of any deterministic FSM. For the following we again stipulate that the FSMs are complete.

Proposition 6.1: Any small temporary error of a complete FSM with tolerance (T, τ) is a subset of $\chi = \bigcap_a (r_a \cdot (\tau \cdot r_a^c)^\neg)^\neg$. If T is deterministic, then χ is the largest small temporary error of (T, τ).

Proof: $r_a * \phi \subseteq \tau$ implies $r_a^c \cdot \phi \cdot r_a \subseteq \tau$ and, hence, $r_a^c \cdot \phi \cdot r_a \cdot r_a^c \subseteq \tau \cdot r_a^c$ implies (*) $r_a^c \cdot \phi \subseteq \tau \cdot r_a^c$ since $\iota \subseteq r_a \cdot r_a^c$. Inclusion (*) is equivalent to the Schröder equivalences. Hence, $\phi \subseteq (r_a \cdot (\tau \cdot r_a^c)^\neg)^\neg$ (Brink et al. 1997). Thus, if ϕ is small, then $\phi \subseteq \chi$.

On the other hand, (**) $r_a^c \cdot \phi_a \subseteq \tau \cdot r_a^c$, for a binary relation ϕ_a, implies $r_a^c \cdot \phi_a \cdot r_a \subseteq \tau \cdot r_a^c \cdot r_a \subseteq \tau$, if T is deterministic. Thus, every transition maps $\bigcap_a \phi_a$ into τ, i.e. $\bigcap_a \phi_a$ is small. Now $(r_a \cdot (\tau \cdot r_a^c)^\neg)^\neg$ is the largest set for which (**) holds. Hence, χ is the largest binary relation which is small. Furthermore, if T is deterministic, then $r_a^c \cdot \iota \cdot \subseteq r_a^c \cdot r_a \cdot r_a^c \subseteq \iota \cdot r_a^c \subseteq \tau \cdot r_a^c$, and hence, $\iota \subseteq \chi$, i.e. χ is reflexive. ◁

Likewise, all successor errors of ϕ are insignificant if and only if $\phi \subseteq \chi^* = \bigcap_a (r_a \cdot (\tau^* \cdot r_a^c)^\neg)^\neg$. Thus, in order to see whether all consequences of unintended state transitions specified by temporary error ϕ can be tolerated we, first, have to compute τ^* and then to check whether $\phi \subseteq \tau^*$ or $\phi \subseteq \chi^*$ holds. Recall, that $\tau = \tau^*$ if and only if the FSM is stable.

Masking of temporary faults can be modeled by compatible, small temporary errors of a stable FSM. Let '*stop*' be the set of stop-states of T. A stable FSM may be called a fail-stop FSM if $\phi \subseteq S \times cl_\tau(stop)$ are the only possible errors; i.e. all unintended state transitions lead to states that are in tolerance with the stop-states of T.

Example (roll-back recovery): Consider a simple recovery scheme that can be used with fail-stop systems. The system checkpoints its state after receiving an input. Upon an error the system is reset to the checkpoint allowing to relive the computation. Of course, this scheme works well for temporary faults only. If recovery fails, the system stops. A simplified model of this scheme is given as follows.

The model:
$T = (S, Ev, R)$ with
$S = \{ s_0, s_1, \ldots, s_n, s_f \mid n \geq 2\}$, $Ev = \{ in, t, rb, rst \}$ and $R = \{ r_{in}, r_t, r_{rb}, r_{rst} \}$
where s_0 is the default state, s_n the output-state, s_1 the checkpoint and s_f the faulty state of the system. A computation is modeled by the transition sequence $s_1\ s_2\ \ldots\ s_n$. Hence, the state transition relations are:

$$\begin{aligned}
&\textit{input } r_{in} = \{s_0 s_1\}\\
&\textit{computatinal step } r_t = \{s_n s_n, s_f s_f, s_i s_{i+1}; i = 1,\ldots,n-1\}\\
&\textit{rollback } r_{rb} = S_{0,n} \times s_1, \textit{where } S_{i,j} = S - \{s_i, s_j\}\\
&\textit{reset } r_{rst} = S \times s_0
\end{aligned}$$

We consider as tolerance over S the relation: $\tau = {}^{rs}\{s_i s_f; i = 1,2,\ldots,n-1\}$. It is not stable, since $r_t * \tau \not\subset \tau$. If we have a fail-stop system then error $\phi_0 = S \times cl_\tau(s_0) = S \times s_0$ is its only error. Temporary error ϕ_0 is not small, since $\phi_0 \cdot r_{in} \cdot v = S \not\subset r_{in} \cdot v = \{s_0\}$. Let us now consider error $\phi_f = S \times s_f$. This error is also not small, since $r_t * \phi_f \not\subset \tau$. Now assume that unintended transitions are detected in time. Then, $\phi' = \phi_f - s_n s_f$ is the possible error, i.e. if the system leaves state s_{n-1} the computation was successful. ϕ' is also not small. Hence, let us compute the relations $\phi_a = (r_a \cdot (\tau \cdot r_a^c)^\neg)^\neg$ for n = 3: ($S_k = S - s_k$):

$$\begin{aligned}
&\phi_{in} = \iota \cup {}^s(S_0 \times S)\\
&\phi_t = \iota \cup (s_0 \times S) \cup {}^s(s_1 s_f \cup s_2 s_3)\\
&\phi_{rb} = \iota \cup (s_0 \times S) \cup {}^s(s_1 s_f \cup s_2 s_f \cup s_1 s_2)\\
&\phi_{rst} = v.
\end{aligned}$$

Thus, $\phi = \bigcap \phi_a = \iota \cup s_1 s_f$ and $\phi \cdot r_a \cdot v \subseteq r_a \cdot v$; ϕ is not compatible but small; ϕ is not masked, since τ is not stable. ◁

Proposition 6.2: Compatible temporary errors of a stable FSM are insignificant and small, their consequences (induced errors) remain insignificant under the reaction of any sequence of events. The maximal compatible temporary error of (T,τ) is τ and τ* is the maximal insignificant and compatible temporary error of T. (The proof is obvious).

Up to now it is possible to check that something will happen immediately after an unintended state transition. It may, however, also be of interest to check that something will happen after at most *l* time units when an unintended state transition occurred, where time passes by one unit at each state transition. For instance, we may wish to check that roll-back recovery can bring the system back to a tolerable state after at most *l* time units. (Non-unit transitions can be constructed from a sequence of unit transitions; non deterministic transition times can also be implemented in the same way, by using stuttering (Campos et al. 1996)).

Definition (boundness of errors): Temporary error ϕ of (T,τ) is (*p*, *l*)-bounded, if for all finite sequences w of events with length equal or greater than *p*: $r_w * \phi \subseteq \tau^l$ with

$\phi \cdot r_w \cdot v \subseteq r_w \cdot v$ holds, and these inclusions do not hold for (*p*-1, *l*) or (*p*, *l*-1). Here, $r_w = r_{x_1} \cdot r_{x_2} \cdot ... \cdot r_{x_n}$ for $w = x_1 x_2 x_3 ... x_n$; $x_i \in$ Ev.

Intuitively, a FSM overcomes the influence of error ϕ, if after some time it enters only states which are, and from then on stay, in tolerance with correct states, i.e. if ϕ is (p,1)-bounded.

Definition (error correction): The FSM with tolerance (T,τ) is said to correct temporary error ϕ, if there is a *p* such that ϕ is (*p*,1)-bounded.

For instance, temporary error ϕ of a stable FSM is (1,1)-bounded, if it is small (and not compatible), since
$r_w^c \cdot \phi \cdot r_w = r_v^c \cdot r_{x_1}^c \cdot \phi \cdot r_{x_1} \cdot r_v \subseteq r_v^c \cdot \tau \cdot r_v \subseteq \tau$, $w = x_1 v$. (Note, that a FSM with bounded errors need not be stable and (0, 1)-bounded errors do not describe erroneous state transitions, all other errors may). On the other hand, gracefully degrading fault-tolerant systems for which a certain time elapses before the recovery becomes effective can be modeled by FSMs with tolerance and (*p*, *l*)-bounded errors with $l > 1$ and $p > 1$. A *t*-fault-tolerant FSM can mask at most t faults (Perraju et al. 1996). Let ϕ model its errors, then *t*-fault tolerance can be expressed as: all errors ϕ^i , i = 1,2, ...,t, are (0,1)-bounded. The FSM is *t*-fail-stop if, in addition, $\phi^{t+1} \subseteq S \times cl_\tau(stop)$.

Example (error detection and roll-back): Consider again the simple roll-back scheme. We now add a reliable error detection process TD and consider the model T|TD. TD has just two states: d_0 the default state and d_1 the state entered when an error is detected. Its tolerance is ι_D. Thus,
$TD = (D, Ev, R_D)$ with $D = \{d_0, d_1\}$ and $R_D = \{r_{in}^D, r_t^D, r_{rb}^D, r_{rst}^D\}$
where
$r_{in}^D = r_t^D = \{d_0 d_0[\text{not in}(s_f)], d_0 d_1[\text{in}(s_f)], d_1 d_0[not\,in(s_f)], d_1 d_1[in(s_f)]\}$
and $r_{rb}^D = r_{rst}^D = D \times d_0$.
If we can always tolerate the faulty state of T as long as the error is detected by TD, we can consider as appropriate tolerance on T|TD the relation $\tau' = (\tau | \iota_D) \cup {}^s((S_0 \times d_0) \times s_f d_1)$. Observe that

$$\tau' = {}^{rs}\big((s_i d_j, s_f d_j), (s_k d_0, s_f d_1)\ ;\ i = 1,...,n-1; k = 1,...,n; j = 0,1\big)$$

Temporary error $\phi_1 | \iota_D$ is still not compatible; it is, however, (1,1)-bounded, i.e. it is corrected ◁.

Boundness check: To see whether temporary error ϕ of a stable FSM is (*p*, *l*)-bounded, we check for increasingly greater *p* and *l* < |S| whether $\phi \subseteq \chi(\tau, p, l) = \bigcap_{z, |z| = p} (r_z \cdot (\tau^l \cdot r_z^c)^\neg)^\neg$ holds. If ϕ is (*p*,1)-bounded, $0 \le p \le 2\,|S| - 2$.

Table 2. State transitions of T^+

R^+	a	b	c	d	present states
r_x	a	a	b	b	next states
r_y	c	c	a	a	of T^+

Example (boundness): Consider FSM (T^+, τ_1) of Table 2 with $\tau_1 = {}^{rs}(ab \cup ac)$; T^+ is a stable FSM. Temporary errors may be $\phi_1 = {}^{rs}(ab)$ and $\phi_2 = {}^{rs}(ac \cup ad \cup bd)$. Error ϕ_1 is (0,1)- and (1,0)-bounded since ϕ_1 is compatible. Error ϕ_2 is (1,1)-bounded. Hence, ϕ_1 and ϕ_2 are corrected by T^+. Consider now tolerance $\tau_2 = {}^{rs}(ac \cup cd \cup db)$. ($T^+, \tau_2$) is not stable, since ac is mapped to ab by r_x. Furthermore, ac $\notin r_x \lrcorner \tau$. Now $\rho(2) = \rho(3) = {}^{rs}cd$ and, hence, $\tau_2{}^* = {}^{rs}cd$. Temporary error ${}^{r}cd$ is insignificant; temporary error ${}^{r}ac$ is compatible but not small, it is, however (1,0)-bounded. The largest small error is ${}^{rs}(ab \cup cd)$, since $(r_y \cdot (\tau \cdot r_y^c)^{\neg})^{\neg} = \nu$ and ϕ_3 = $(r_x \cdot (\tau \cdot r_x^c)^{\neg})^{\neg} = {}^{rs}(ab \cup cd)$. $\triangleleft$

In order to check boundness and other fault tolerance properties model checking can be employed. If it is fully automated, one can check properties of large systems (Burch et al. 1992). Model checking is based on searching the state space and can produce counterexamples when the search fails. In Section 8 we will discuss model checking of fault tolerance within our framework.

7. Errors of the railroad crossing system

Let us now return to our general railroad crossing example (in short, GRC) of Section 3 and consider its fault-tolerant behavior. The relevant part of the FSM with set of states Q $\subseteq$ S is given in Figure 3. We assume, that GRC is complete by stipulating, that reactions of events not relevant in a state do not change this state.

The light bulb may break with some (small) probability just when the train passes it. This defines error $\phi_L = {}^{r}\{(Pi, Pi-1), (Hi, Hi-1), i = 2,\ldots,8\}$ specifying On/Off-transition of the light when the train approaches the crossing or comes to an halt. Suppose ϕ_L models consequences of temporary faults. The statechart with error ϕ_L is shown in Figure 5. (Dashed transitions are taken only once). For example, in addition to O8 = (O, On, Closed) $\rightarrow^{goP}$ P8 = (P, On, Closed) we have O8 = (O, On, Closed) $\rightarrow^{goP}$ P8 (P8, P7) = P7 = (P, Off, Closed). Similarly, H7 $\in$ H5($r^{GRC}{}_{down} \cdot \phi_L$)
Aggregation of states are depicted as superstates; e.g. Out is the superstate when the train is outside the railroad crossing, and Phi is the superstate entered by an unintended state transition due to error ϕ_L.

When the train passed the light, the state of the light is not safety relevant, provided the light can be switched on again before the next train arrives. Hence, we may say that H7 is within tolerance of H8 and P7 is within tolerance of P8, i.e. $\tau = {}^{rs}(\nu_Q \cup$ H7H8 $\cup$P7P8 $\cup$ I7I8 $\cup$I3I7 $\cup$I3I8), (see Section 3). The FSM with tolerance (GRC,τ) is stable. Moreover, ϕ_L is (2,0)-bounded. Now consider error $\phi_{LP} = \phi_L \cup$ O8O7. Its effect on superstate Out is shown in Figure 6. The other superstates, particularly Phi, do not change. Error ϕ_{LP} is small; but GRC becomes unstable, since I8 τ I7 and $r_{exitI}{}^c \cdot$ I8I7 $\cdot\, r_{exitI}$ = O8O7 $\notin \tau$. Considering safety, the system changes into the undesirable state O7 with non-zero probability but not into erroneous states.
Now, consider error $\phi_G = {}^r$(H7H3 $\cup$ I7I3 $\cup$ P7P3), thus, $\phi_G \not\subset \tau$. This error models the possibility that the gate opens whenever the traffic light goes off. This effect may be due to a short circuit. Normally, this is a safe state transition except when the train is in state P7. More interesting is, therefor, the error $\phi_{LG} = \phi_{LP} \cdot \phi_G$, *i.e.*, $\phi_{LG} = \iota \cup \phi_{LP} \cup \phi_G \cup$H8H3$\cup$P8P3. The consequences of ϕ_{LG} are shown in Figure 7. Temporary error ϕ_{LG} is (2,0)-bounded. Moreover, $\phi_{LG} \not\subset \tau$ and the state transitions may lead to the erroneous state I1 and, hence, are erroneous.

Remark: Statecharts give us a convenient way also to model more complex FSMs. We simply denote $s_i\ \phi\ s_j$ by s_i >— s_j. For example, suppose the railway crossing system can enter a superstate which describes a degraded mode of performance. This state is entered when the controller is out of order and given as follows: The traffic light for the train is flashing, the gate is open but a red light is flashing to warn cars. An approaching train blows a horn. Figure 1 illustrates the railway crossing system. The degradable railway crossing system can be modeled as shown by Figure 8, where $\phi = {}^r\phi^+$ is the error. (Some of the conventional notations for statecharts of Harel (1987) have been used). Superstate Phi models the system being in a degraded sate. When the train leaves this state the system enters again the default state O1.

8 Model checking fault tolerance

Many interesting properties of system behavior can be expressed as formulas of propositional temporal logic, since temporal logic permits us to make statements about changes in time, e.g. that a logic formula may be true at some point in time. Computational tree logic (CTL) is such a temporal logic. It is a branching time logic whose operators permit explicit quantification over all possible futures (Burch et al. 1992). It provides the path quantifiers A (always) and E (exists) and temporal operators F (eventually), G (always), U (until) and X (next).

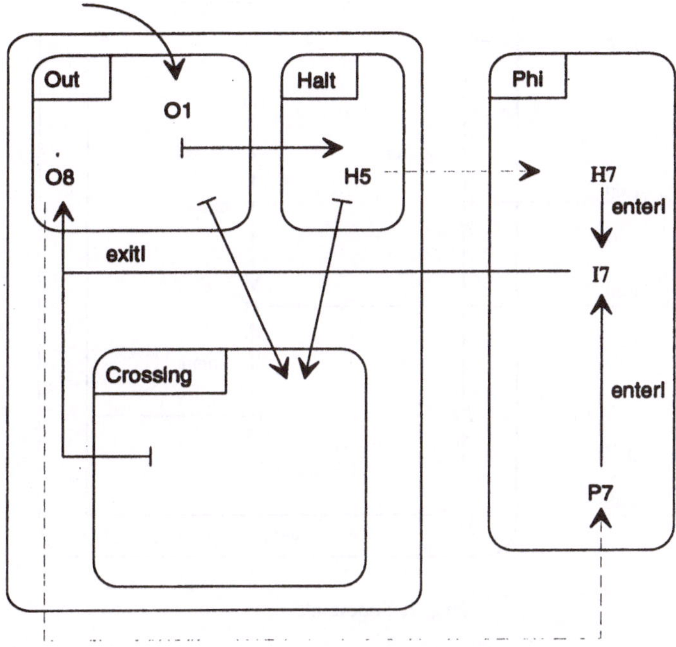

Fig. 5. GRC with temporary state transition error

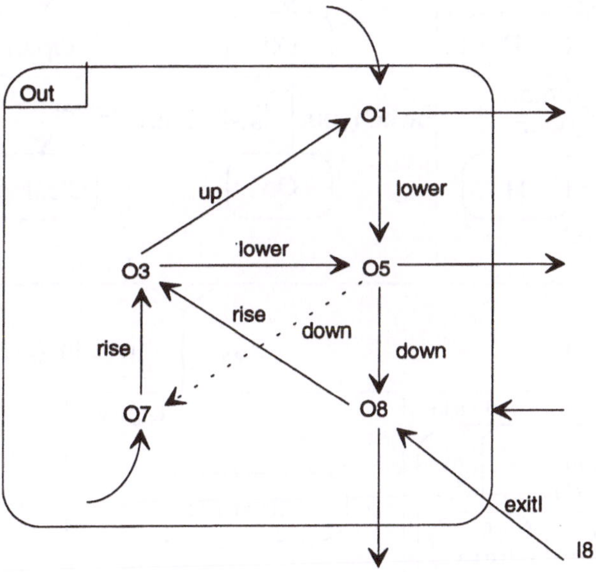

Fig.6. Superstate

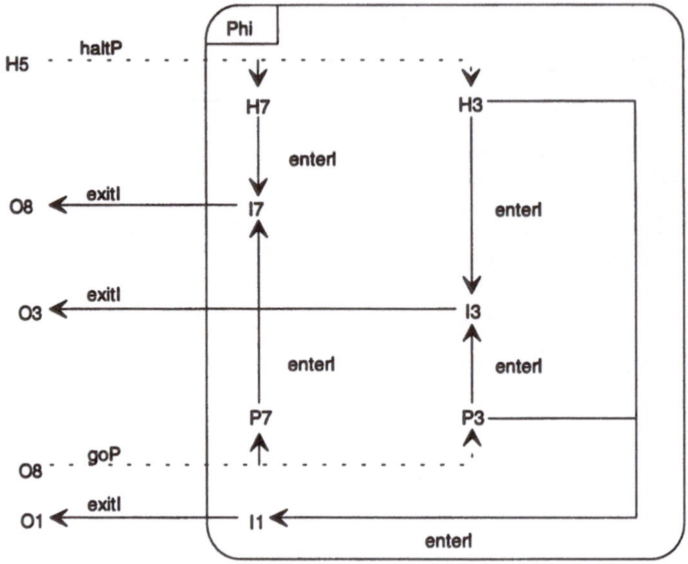

Fig. 7. Superstate

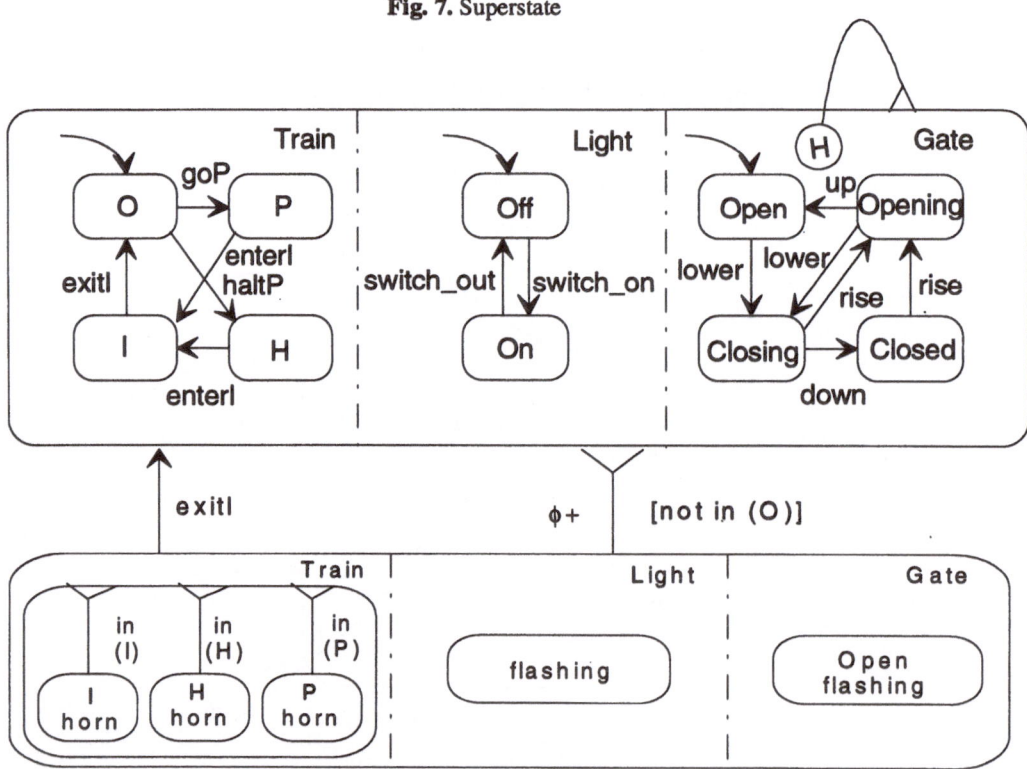

Fig. 8. A degradable railroad crossing system

For instance, Fg is true of a sequence of state transitions, if there exists a state in this sequence that satisfies the logic formula g or EXg is true for state s if there exists a sequence $\pi = s_0 s_1 \ldots$starting at $s = s_0$, such that g is true for state s_1.The syntax of CTL is defined by:

(a) Every atomic proposition is a CTL-formula.

(b) If f and g are CTL-formulas, then so are $\neg f$, $f \wedge g$, EXf, EGf and E[f U g]. (f U g : f holds until g holds).

The following abbreviations are used:

$$AXf \equiv \neg EX \neg f$$
$$EFf \equiv E[\, true\; U f]$$
$$AFf \equiv \neg EG \neg f$$
$$AGf \equiv \neg EF \neg f$$
$$A[f\, U\, g] \equiv \neg E[\neg g\; U \neg f \wedge \neg g] \wedge \neg EG \neg g$$

Model checking CTL-formulas can be fully automated, and it has been shown that properties of large systems can be checked (Burch et al., 1992). Model checking can, obviously, also be employed to verify dependability requirements like reliability and availability.

Given a FSM T = (S, Ev, R), an initial state s_0, a set V of propositional variables and a labeling function L specifying which propositions are true in each state, i.e. L: V ⟶ 2^S assigns to each proposition the set of states at which it is true. We define the *labeled* transition system T_L as T_L = (S, s_0, [R], V, L) where [R] = $\bigcup_{Ev}$ R. If formula f is true in state s_0 of T_L, we write, as usual, $T_L, s_0 \models f$.

It is now straightforward to specify dependability properties of a FSM T as logical formulas and to employ symbolic model checking (McMillan 1993). For example, let $a \in$ V represent the atomic proposition which is true when the system is available. A typical formula is AGa which is true in state s, if a is always true along all paths emanating from state s, i.e. if the system is always available when, and after, it entered states s. On the other hand, formula $AF(\neg a)$ is true, if failure is inevitable when state s is entered. Fault coverage is incomplete, if formula $EF(\neg a)$ is true in a system state. Let us identify a predicate with the set of states which makes it true. Then another example of a CTL-formula expressing dependable behavior is AG EF{s_0}: from any state it is possible to roll-back to the initial state. We can employ CTL and symbolic model checking also to express and check fault tolerance requirements of a FSM with tolerance. As example consider temporary error ϕ of a FSM with tolerance (T, τ) and let FSM TT be defined as the product FSM:

$$TT = (S \times S, Ev, RR) \text{ with } RR = \{\, t_a \mid (s, s')t_a = sr_a \times s'r_a;\ a \in Ev\},$$

Now, TT_L = (S×S, q_0, [RR], {τ, ϕ }) where $q_0 \in S \times S$ is a labeled transition system. Again we identified a predicate with the set of states which makes it true. E.g., the property 'to be in tolerance' is specified as the subset τ of the state set of TT. TT exhibits fault-tolerant behavior if it stays within τ. Hence, the requirement, that T masks temporary error ϕ is: TT_L, $q_0 \models \phi \Rightarrow AG\tau$ and T corrects ϕ, if TT_L, $q_0 \models \phi \Rightarrow A(\text{true U } AG\tau)$.

The tolerance of a stable FSM is a bisimulation and, hence, an informative question is whether a corresponding labeled transition system of a stable FSM when starting in state s or starting in state q which is within tolerance of s will satisfy the same CTL-formulas (Browne et al. 1989). To check fault-tolerant behavior of several models we employed SMV (McMillan 1993), a tool for checking finite state systems against specifications in CTL. The language allows the specification of synchronous or asynchronous, deterministic or abstract non-deterministic finite state machines. Hence, SMV allows non-deterministic state assignments, needed to model unintended state transitions.

Appendix

We use capital letters to denote sets, lower case to denote elements of sets, and Greek letters to denote relations; $X^{\neg}$ is the complement and $|X|$ the cardinality of X. Special relations over X are: $\iota_X = \{(x, x), x \in X\}$the identity relation, $\nu_X = X \times X$ the universal relation, and $0_X = \varnothing$ the empty relation over X.

Definitions: Given binary relations ρ, $\sigma \subseteq X \times Y$, binary relations τ and α over X, χ a binary relation over Y and $S \subseteq X$, $R \subseteq Y$. Then:

1. converse: $\rho^c = \{(y, x) \mid (x, y) \in \rho\} \subseteq Y \times X$,
2. image: $S\rho = \{y \mid \exists x \in S, x\rho y\}$, Peirce product: $\rho R = \{x \mid \exists y \in R, x\rho y\}$,
3. composition: $\rho \cdot \sigma = \{(x, y) \mid \exists x' : x \rho x', x' \sigma y\}$, obviously $\rho \cdot \sigma = \varnothing$ if $X \cap Y = \varnothing$,
4. image of τ under ρ: $\rho^c \cdot \tau \cdot \rho$;
5. coimage of χ under ρ: $\rho \cdot \chi \cdot \rho^c$
6. $\rho * \tau = \iota Y \cup \rho^c \cdot \tau \cdot \rho \subseteq Y \times Y$
7. $\rho \dashv \chi = \iota X \cup \rho \cdot \chi \cdot \rho^c \subseteq X \times X$
8. exponentiation: $\tau^0 = \iota X$ and $\tau^n = \tau^{n-1} \cdot \tau$
9. reflexive extension: ${}^r\alpha = \iota X \cup \alpha$,
10. symmetric extension ${}^s\alpha = \alpha \cup \alpha^c$,
11. pr_{jk} is the projection on the j-th and the k-th component of a relation; $dom(\alpha) = pr_1 \alpha$, $ran(\alpha) = pr_2 \alpha$,

Abbreviations: Instead of $(x, y) \in \rho$ and $\{x\}\rho$ we usually write $x\rho y$ and $x\rho$, respectively. Instead of (x, y) and {(x, y)} we usually write xy. We write ι, ν, and 0 for ι_X, ν_X and 0_X, respectively, if no confusion arises.

Relation τ is stable under α if $\alpha^c \cdot \tau \cdot \alpha \subseteq \tau$, i.e. if it contains its image. If τ is reflexive and stable then $\alpha * \tau \subseteq \tau$. Relation α is univalent (or deterministic) if $\alpha^c \cdot \alpha \subseteq \iota_X$, it is total (or complete) if $\iota_X \subseteq \alpha \cdot \alpha^c$, it is surjective if $\iota_X \subseteq \alpha^c \cdot \alpha$ and injective if $\alpha \cdot \alpha^c \subseteq \iota_X$.

The definition of $\alpha | \tau$, clτ, bndτ, intτ, and dτ is given in Section 4.

References

Arbib, M. (1967): Tolerance automata. Kybernetika Cislo 3: 223-233

Brink, C., Kahl, W., Schmidt, G. (eds.) (1997): Relational Methods in Computer Science (Advances in Computing Science). Springer, WienNewYork

Browne, M.C., Clarke, E.M. and Grumberg O. (1989): Reasoning about networks with many identical finite state processes. Inf. Comput. 81; 13-31

Bruel, J.M., France, R.B., Benezekri, A., Raynaud, Y. (1996): A real-time specification environment based on Z and graphical object-oriented modeling techniques. In Proceedings of the First IEEE High-Assurance Systems Engineering Workshop HASE 96, Niagara on the Lake. Canada. IEEE

Burch, J.R., Clarke, E.M., McMillan, K.L., Dill, K.L., Hwang, L.J. (1992): Symbolic model checking: 10^{20} states and beyond. Inf. Comput. 98, 2: 142 - 170

Campos, S., Clarke E., Minea, M. (1996): Analysis of real-time systems using symbolic techniques. In: Hartmeyer and Mandrioli (eds.) (1996): Formal Methods For Real-Time Computing. J. Wiley, New York, pp. 216 - 235

Dal Cin, M. (1975): Fuzzy-state automata: their stability and fault tolerance. Int. Journ. of Comp. a. Inform. Sciences Vol. 4: 63 - 80, and: Modification tolerance of fuzzy-state automata. Int. Journ of Comp. a. Inform. Sciences Vol. 4: 81 - 93

Dal Cin, M. (1997): Verifying fault-tolerant behavior of state machines. In Proceedings of the Second IEEE High-Assurance Systems Engineering Workshop HASE 97, Bethesda, Maryland. IEEE

Dal Cin, M. (1998): Checking modification tolerance. Internal Report 98-2. IMMD3 University of Erlangen-Nürnberg

Eriksson, H.-E., Penker M. (1998): UML Toolkit. J. Wiley, New York

Harel, D. (1987): Statecharts: a visual formalism for complex systems. Science of Computer Programming 8: 231 - 274

Heitmeyer, C, Mandrioli, D. (eds.) (1996): Formal Methods For Real-Time Computing. J. Wiley, New York

Janowski, T. (1995): Bisimulation and Fault-Tolerance. PhD thesis, Department of Computer Science, University of Warwick

Kirner, T.G., Davis, A.M. (1996): Nonfunctional requirements of real-time systems. In: Advances in Computers Vol. 42, pp. 1 - 37. Academic Press, New York

Lamport, L. (1977): Proving the correctness of multiprocess programs. IEEE Trans. Software Engineering SE-3,2: 125-143

Laprie, J.-C. (1992): Dependability: Basic Concepts and Terminology. Dependable Computing and Fault-Tolerant Systems Vol. 5, Springer Verlag, Wien NewYork

Lee, P.A., Anderson T. (1990): Fault Tolerance, Principles and Practice. Springer Verlag, Wien NewYork

Leeb, G., Lynch, N. (1996): Proving safety properties of the steam-boiler problem. In: Formal Methods for Industrial Applications. Springer Lecture Notes in Computer Science 1165, pp. 318 - 338. Springer Verlag, Berlin Heidelberg NewYork

Leveson, N.G., Heimdahl, M., Hildreth H., Rose, J.D. (1994): Requirements specification for process-control systems. IEEE Trans. Software Engineering. SE. 20,9: 684 - 706

McMillan, K.L., (1993): Symbolic Model Checking. Kluwer, Boston Dordrecht London

Mok, A.K., Stuart, D.A., Jahanian, F. (1996): Specification and analysis of real-time systems: modechart language and toolset. In: Heitmeyer and Mandrioli (eds.) (1996): Formal Methods For Real-Time Computing. J. Wiley, New York, pp. 33 - 53

Perraju, T.S., Rana, S.P., Sarkar, S.P. (1996): Specifying fault tolerance in mission critical systems. In Proceedings of the First IEEE High-Assurance Systems Engineering Workshop HASE 96, Niagara on the Lake, Canada. IEEE

Poincare, H. (1958): The Value of Science. Reprinted by Dover

Rushby, J. (1996): Reconfiguration and transient recovery in state machine architectures. Proceedings of the IEEE International Symposium on Fault-Tolerant Computing FTCS-26, Sendai, Japan. IEEE, pp. 6 - 15

Applications of artificial neural networks

David William Pearson and Gérard Dray

1 Introduction

In this article we consider some theoretical aspects of neural networks and some of their varied applications. The theoretical aspects are presented from the point of view of a system, basically input/state/output. For the applications, we consider large systems: from production systems, through biological and chemical systems and on to environmental systems.

Informally, the "state" part of the system is some sort of process that takes a set of input variables lumped together in a vector, performs some operations on them and outputs the results. In this way the state part of the system is like a black-box, ie it cannot be seen by the outside world. This is a very practical situation because very frequently all that we know of a process is what we input into it and what we observe at the ouput.

Very often we wish to find a mathematical model of a system in order to predict its behaviour in different circumstances and/or to develop control strategies in order to modify its behaviour. Some systems can be modelled by classical techniques involving differential or difference equations, for example spring-mass-damper systems, electrical circuits, heat diffusion. However, there are some systems for which there are no known classical models, for example certain ecological systems and biological systems including medical systems (the human body is a system). For these systems we can usually apply inputs to them and measure the resulting outputs, thus generating data samples. It is this type of system and situation, i.e. with relevant data available, that we are interested in.

Neural networks are useful tools for tackling this type of problem. When grouped together with other methods such as fuzzy-set theory, fuzzy logic, genetic algorithms and nonlinear methods they form a relatively new branch of mathematics/informatics called Soft-Computing.

This article is very general in nature and is addressed to a wide audience of people not specialised in artificial neural networks and/or mathematics. Hence we do not go into great detail. The reader interested in carrying out further study in the topics covered in this article should consult some introductory texts such as Haykin (1994) for artificial neural networks and Nguyen and Walker (1997) for fuzzy logic, or the other references cited for more specialised subjects.

The article is split into two main parts. The first part is devoted to a general overview and classification of artificial neural networks and the second is devoted to some applications.

2 Overview and classification of artificial neural networks

Originally, artificial neural networks were developed as a means of modelling and understanding biological systems, in particular animal brains and thought processes. A lot of important work has been done on brain function and thought processes and a lot more needs to be done before we will fully understand how and why brains function as they do: this has become to be known as the "mind-body" problem. In recent years however it has been noticed that artificial neural networks have interesting mathematical properties in themselves regardless of their biological meaning. In this article we are concerned more with the mathematical aspects and applications of artificial neural networks rather than the biological issues. Some interesting and thought provoking references for the mind-body problem are Lockwood (1989), Penrose (1995) and Stapp (1993).

As a very general classification, neural networks can be split into two different classes, feedforward networks and recursive networks. In this section we present some general aspects of these two classes of network. We also illustrate a type of hybrid network, called ANFIS, which combines a neural network structure with fuzzy logic.

2.1 Feedforward neural networks

Artificial neural networks are a special type of input/output system. A feedforward neural network is basically a nonlinear mapping from an input space to an output space. A network is built up of nodes, the nodes correspond to neurons and each node has a functional behaviour as follows, illustrated for the nth node

$$y_n = \sigma_n(g_n((x_{ni})_{i=1,\ldots,I_n}(p_{nj})_{j=1,\ldots,J_n}))$$

where y_n is the output of the node, $(x_{ni})_{i=1,\ldots,I_n}$ are the inputs into the node, $(p_{nj})_{j=1,\ldots,J_n}$ are controllable parameters and it is possible that $I_n \neq I_m$ and $J_n \neq J_m$ for other nodes m. The functions σ_n and g_n can be discrete, analytic, continuous or discontinuous. The connection of all or some nodes by input/output relations defines a directed graph of nodes. If there are N nodes then node m (predecessor) is connected to node n (successor) if there exists a unique m∈{1,...,N} with $x_{ni} = y_m$. The general situation is illustrated in Fig.1.

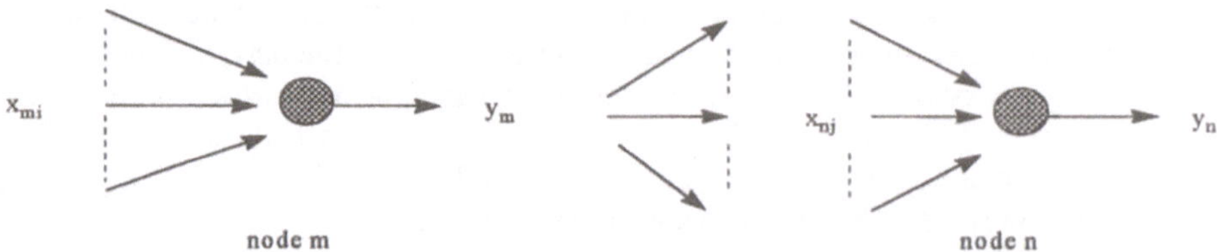

Fig.1. Directed graph of nodes

There are two cases to consider. If there are no feedback terms then the directed graph determines a partial ordering of the nodes, if there are feedback terms then there is no ordering.

A node may have free or external inputs, not coming from the outputs of predecessor nodes (sometimes referred to as bias inputs in the literature). The output of a node may also be free in that it is not fed into the input of another node. Each node is an input/output system. Nodes can also be grouped together into subsystems and these subsystems can be concatenated together, using free inputs and outputs, to form larger input/output systems. For example in Fig.2. the set of nodes encircled by the dotted line form a subsystem which can be joined to other subsystems via the free inputs and outputs.

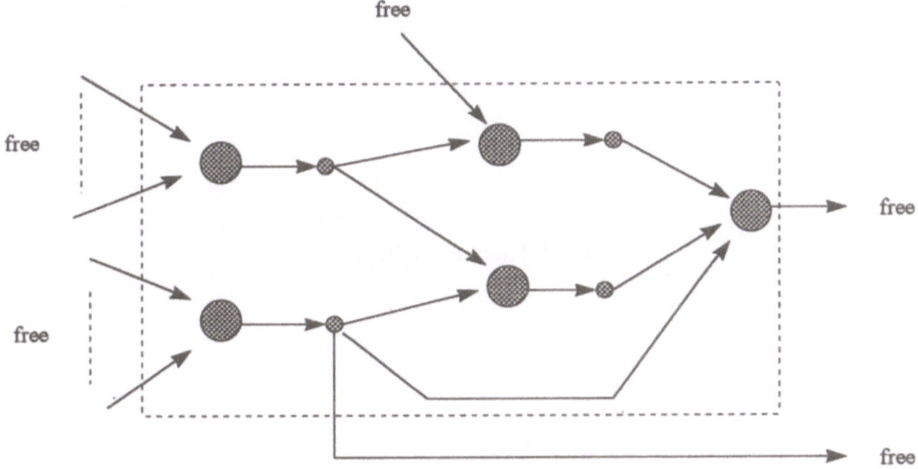

Fig.2. Concatenation of graphs

A hierarchy of "layers" is defined by the concatenation of these subsystems:

- 1st layer or input layer: nodes with free input only
- Lth layer: nodes depending on node(s) of layers $\leq$ L-1
- final or output layer: nodes with free output only

A typical hierarchy of layers is illustrated in Fig.3.

If there are feedback loops in the network then these can be parameterised by logical time to provide a partial ordering as illustrated in Fig.4.

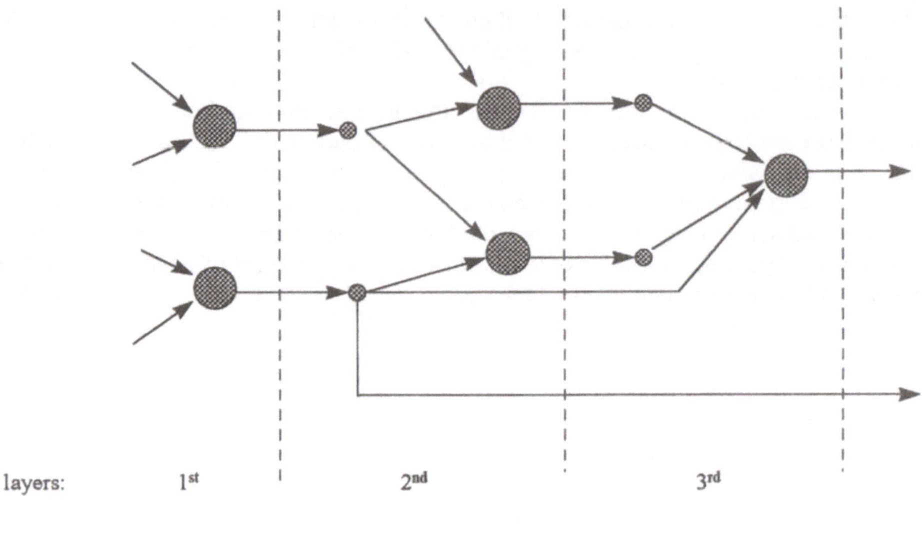

Fig.3. Hierarchy of layers

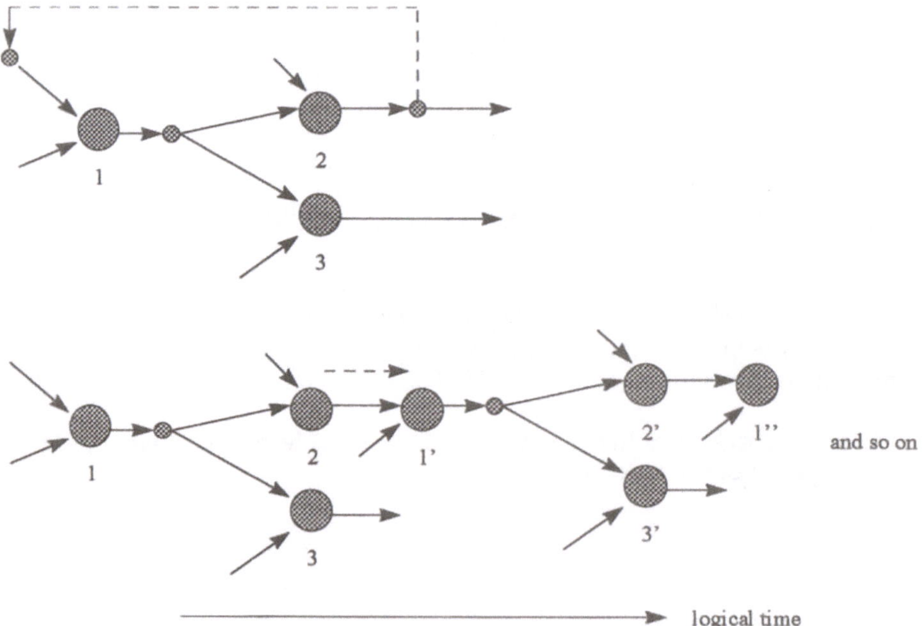

Fig.4. Parameterisation by logical time

The functions σ_n and g_n are usually chosen from the following classes:

a) σ_n and g_n discrete

b) σ_n and g_n analytic, for example $R^{I_n} \to R$

c) σ_n continuous or discontinuous with g_n continuous

For example we may have:

i) $g_n = \sum_{i=1}^{I_n} p_{ni}(x_{ni} - q_{ni}) = \sum_{i=1}^{I_n} p_{ni}x_{ni} + \tilde{p}$ and $y_n = \tanh(g_n)$

ii) $g_n = \exp(\gamma \sum_{i=1}^{I_n} p_{ni}(x_{ni} - q_{ni})^2)$ and $y_n = \tilde{r}_n g_n + \tilde{\tilde{r}}_n$

where p_{ni}, q_{ni}, r_n are parameters.

Fundamental to most types of neural networks are input/output data pairs (x_k, y_k), $k = 1, K, D$ where there are D data pairs available. The neural network is then "fitted" to the data by defining an error function, based on the difference between the neural network output values and the actual data values, which is minimised by adjusting the parameters p_{ni}, q_{ni}, r_n. A standard error function is simply

$$E = \sum_{k=1}^{D} \left(y_k - f(x_k)\right)^2$$

where $f(x_k)$ denotes the composition of all the functions σ_n and g_n acting on the kth input vector, i.e. the global neural network function. This error is usually minimised by gradient descent methods collectively known as backpropagation methods, the act of minimising the error function is known as training. There are several different types of network and error function structures, a good general introduction to the subject can be found in Haykin (1994).

One problem associated with this optimisation, or training, process is to avoid over training the network. By over training we mean that the network is optimised in order to model exactly the data set, but when presented with new data it is unable to generalise. One way of overcoming this problem is to split the data into three subsets, a training subset, a validation subset and a test subset. This is usually done at random and as a general rule of thumb 50% of the data are reserved for training, 25% for validation and 25% for testing. The three data subsets are used as follows. First of all a value for the error function E is chosen, the network parameters are adjusted so that the error value is reached for the training data only, then the validation data are presented to the network and the error function is calculated for these validation data. The target value for E is then reduced and the process repeated. At each stage the values for E based on training and validation data are stored. When the algorithm converges and the error cannot be reduced any further for the training data, the error values for training and validation are plotted, the resulting graph will look something like Fig.5.

We see from Fig. 5. that there is usually a value for the training error where the validation error is a minimum, the neural network trained to this value of training error is in general the optimal neural network in terms of precision for the training data and generalisation capabilities based on the validation data.

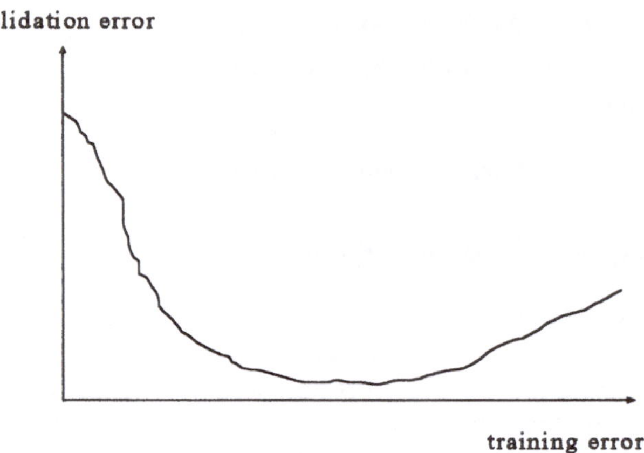

Fig.5. Validation error versus training error

Finally the neural network is presented with the test data. These data have not been used for training or for optimisation of generalisation capabilities and so they represent a real test for the network.

2.2 Recursive neural networks

A feedback, or recursive, neural network can be treated and analysed as a dynamical system. For example Funahashi and Nakamura (1993), Pearlmutter (1989) and Pearson (1996) have developed various methods of fitting a given network to a data sequence by adjusting the synaptic weights.

There are different types of recursive network, a very well known type of recursive network is the Hopfield network (Hopfield 1984). A lot of interest has been generated in this type of network due to its possible application to combinatorial optimisation problems such as the travelling salesman problem. However the results remain inconclusive and a lot more research needs to be carried out before these techniques can be reliably applied to real problems.

For this article, as an example of a recursive network, we take some time to show how a network structure can be built up in logical steps for one particular type of application.

First of all, as an illustrative example, assume that we are given certain quantities, call them {A,B,C,D,E}, and we know that there exist dynamical relations between these quantities such as "an increase in quantity A results in an increase in quantity B" etc. The objective is to begin with a coarse qualitative model of the relations and then refine this to a quantitative model by tuning the parameters on data coming from the physical system to be modelled.

To begin with we denote the five quantities by labelled circles, drawn freely on a sheet of paper for example. We then ask an expert, i.e. someone who has a knowledge of the dynamical behaviour of the system, to note the general relations between the quantities in a coarse fashion. We ask the expert to consider each pair of quantities in turn and decide whether or not there is a relation between them, if he/she decides that there is a relation then the expert traces a directed arc between the circles and marks a "+" if an increase in the first quantity causes an increase in the second and a "-" if a decrease is caused. The absence of an arc indicates that the expert considers that the quantitites are not directly related. Notice that the arcs are directed and that a quantity can be related to itself. The result of this procedure will be a diagram something like Fig.6. which is sometimes referred to as an entity/relationship diagram.

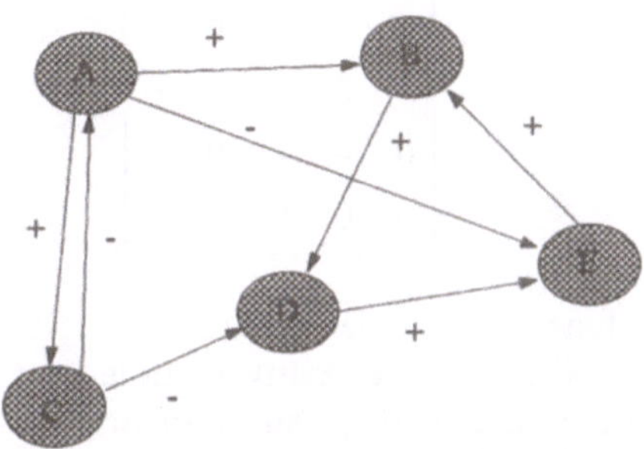

Fig.6. Entity/relationship diagram

These relations are noted in matrix form by marking five rows and five columns labelled {A,B,C,D,E}, then considering each column in turn and leaving blank, placing a "+" or a "-" in the corresponding row according to the absence of an arc, an arc with a "+" or an arc with a "-". The resulting matrix for Fig.6. is then illustrated in Fig.7.

	A	**B**	**C**	**D**	**E**
A			–		
B	+				+
C	+				
D		+	–		
E	–			+	

Fig.7. Matrix for Fig.6.

Now, assuming that the quantities {A,B,C,D,E} vary as a function of time, we define a state vector at time instant t as follows

$$x_t = \begin{bmatrix} x^1(t) \\ x^2(t) \\ x^3(t) \\ x^4(t) \\ x^5(t) \end{bmatrix} = \begin{bmatrix} A(t) \\ B(t) \\ C(t) \\ D(t) \\ E(t) \end{bmatrix}$$

Then we define a "symbolic" weight matrix, W, corresponding to Fig.7

$$W = \begin{bmatrix} 0 & 0 & - & 0 & 0 \\ + & 0 & 0 & 0 & + \\ + & 0 & 0 & 0 & 0 \\ 0 & + & - & 0 & 0 \\ - & 0 & 0 & + & 0 \end{bmatrix}$$

where the empty spaces have simply been replaced by zeros. Now, we imagine that the total increase (or decrease) in a quantity at time step $t+1$ is some nonlinear function of the other quantities having an influence on it at time step t (the influences being determined by the weight matrix W of course) and so we can write

$$x^k(t+1) = \sigma\left(\sum_{j=1}^{5} w_{kj} x^j(t)\right), \; k = 1, \mathrm{K}, 5$$

where the $w_{ij} \in \{0, -, +\}$ (for the moment) are the elements of the matrix W and σ is a nonlinear function. We assume that the relations can reach saturation points, where an increase in the inputs does not increase the outputs and so for simplicity we usually choose the function tanh for σ.

We then assume that data sequences are available for the particular physical system that we are trying to model. For example

$$\left\{ \begin{bmatrix} x^1(0) \\ \mathrm{M} \\ x^5(0) \end{bmatrix}, \begin{bmatrix} x^1(1) \\ \mathrm{M} \\ x^5(1) \end{bmatrix}, \mathrm{K}, \begin{bmatrix} x^1(T) \\ \mathrm{M} \\ x^5(T) \end{bmatrix} \right\}$$

where we have $T+1$ sequential values for each of the five variables and in general $T>>5$ (or the corresponding dimension for a system with more or less variables). The objective is then to calculate values for the nonzero elements of the matrix W, knowing that for each nonzero element we have extra expert information because we know whether the element is positive or negative. Also, based on the characteristics of the tanh function, we can reasonably limit each nonzero element to lie in the interval [–2, 2].

There are certain choices for the error function that we want to minimise by tuning the elements of the matrix W. For example we might want to be able to predict the behaviour of the system for several iterates in the future in which case we would try to minimise an error function of the form

$$E=\sum_{t=1}^{T}\sum_{k=1}^{5}\left(\sigma^{t}(\sum_{j=1}^{5}w_{kj}x^{j}(0))-x^{k}(t)\right)^{2}$$

where $\sigma^{t}(\sum_{j=1}^{5}w_{kj}x^{j}(0))$ denotes the function σ iterated t times. Alternatively one is sometimes confronted with the problem of predicting simply one step ahead and the error function can be modified to the following

$$E=\sum_{t=1}^{T}\sum_{k=1}^{5}\left(\sigma(\sum_{j=1}^{5}w_{kj}x^{j}(t-1))-x^{k}(t)\right)^{2}$$

Methods have been developed to minimise these functions under the constraints imposed by the positions and signs of the elements in the matrix W (Gerault and Pearson 1996, Gerault and Pearson 1997).

2.3 ANFIS a neural network/fuzzy logic hybrid approach

ANFIS is a method that allows us to tune the parameters of a Fuzzy Inference System (FIS) using a gradient error backpropagation technique (Jang 1993). An introduction to fuzzy logic goes beyond the objectives of this article and so we refer the interested reader to Nguyen and Walker (1997) for a comprehensive introduction to this very important discipline.

A Fuzzy Inference System (FIS) represents a type of model using fuzzy reasoning in order to map an input space to an output space. A FIS is composed of a set of if-then fuzzy rules and a fuzzy reasoning process.

There exist several types of FIS, depending on the type of rules used. In this paper we will introduce one of the most commonly used : the Takagi-Sugeno type rules. In this kind of FIS the output of each rule is a linear combination of input variables plus a constant term, and the final output is the weighted average of each rule's output. A classical representation of a two inputs (x,y) and one output (z) Takagi-Sugeno FIS is given in Fig.8.
The FIS represented has two Takagi-Sugeno rules of the form

Rule 1 : if x is A1 and y is B1 then z1 $=$ p1 $\times$ x $+$ q1 $\times$ y $+$ r1
Rule 2 : if x is A2 and y is B2 then z2 $=$ p2 $\times$ x $+$ q2 $\times$ y $+$ r2

where : Aj and Bj are membership functions; and pj, qj, and rj are the linear parameters.

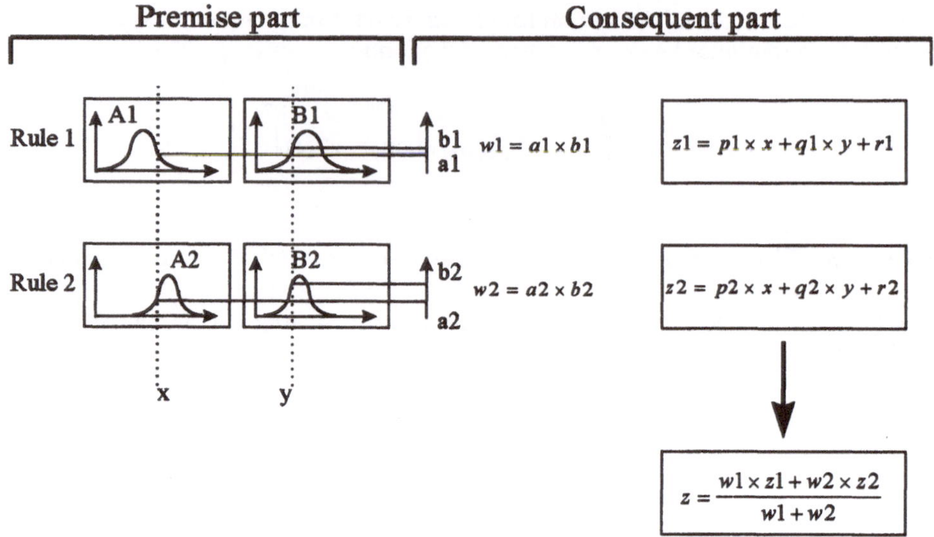

Fig.8. Takagi-Sugeno FIS

The reasoning process can be decomposed into 4 steps :

1. Compute the membership values for each of the rules and each input (a1,a2,b1,b2 in the example of Fig.8.)
2. Combine the membership values with a fuzzy AND operator (multiplication in the example of Fig.8.) in the premise part to get the firing strength of each rule
3. Compute the consequence of each rule (linear combination of inputs plus a constant for the example in Fig.8.)
4. Combine all the outputs of all the rules (weighted average for the example in Fig.8.)

The ANFIS method consists in representing a FIS by an adaptive network and performs a learning procedure to tune the parameters according to a set of inputs and outputs.

An adaptive network is a multi-layer feedforward network with two kinds of nodes : adaptive nodes and fixed nodes. The adaptive nodes realise functions between inputs and one output, and the parameters of the function can be optimised by the learning procedure. The fixed nodes realise functions between inputs and one output that cannot be changed by the learning procedure.

The ANFIS representation of the FIS described in Fig.8. is given in Fig.9. Squares represent adaptive nodes and circles fixed nodes.

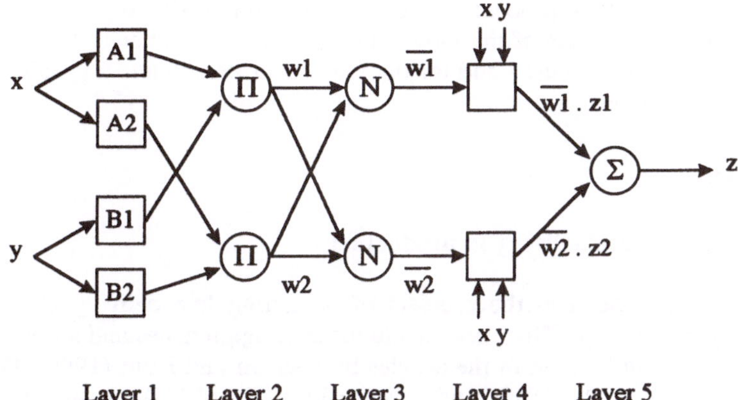

Fig.9. ANFIS representation of the FIS described in Fig.8.

Layer 1 performs the computation of the membership values. Commonly, the membership functions are Gaussians.

$$A_j(x) = \exp\left[-\left(\frac{x - c_j}{a_j}\right)^2\right]$$

Layer 2 performs the fuzzy AND operation between the premise elements of a rule using multiplication.

$$w_j = A_j(x) \times B_j(y)$$

Layer 3 performs the computation of the relative firing strengths of the rules

$$\overline{w}_j = \frac{w_j}{\sum_{i=1}^{\text{number of rules}} w_i}$$

Layer 4 performs the computation of the weighted output of a rule

$$\overline{w}_j \times z_j = \overline{w}_j \times (p_i x + q_i y + r_i)$$

Layer 5 performs the computation of the FIS output

$$z = \sum_{j=1}^{\text{number of rules}} \overline{w}_j \times z_j$$

The author of the method proposes a hybrid procedure in order to perform the tuning of the different parameters of the model. For the linear parameters a classical least square error technique is used and for the non-linear parameters a gradient descent technique is performed.

3 Applications

3.1 Prediction of late delivery in production

This application concerned the problem of predicting late delivery of products 5 working days in advance. The work lead to different approaches and an international collaboration as can be seen in the articles by Pearson and Hunt (1996), Pearson et al. (1996), Peton et al. (1997) and Swiniarski et al. (1995). A summary of the different approaches and achieved results is presented by Gerault et al. (1997).

One of the main difficulties associated with the problem was the paucity of data. It was fairly obvious in looking at the raw data that there were cycles present. However we were not able to confirm the existence of cycles because data were only available for a period of approximately one year. A second difficulty was that there were no obvious choices for input parameters, the data were in fact time series of late deliveries, predicted orders and real orders. Experiments showed that the best results were obtained when 13 parameters were calculated from the data and used as inputs. The input parameters were:

- the difference between real orders and predicted orders for the day and for previous 4 days
- late delivery state for the day and for previous 4 days
- the day of the week
- the day of the month
- the month

the single output parameter was, of course, the predicted state of late delivery 5 days in the future.

Various methods were tested on the available data, including classical linear regression a feedforward neural network and a recursive neural network. The structure of the feedforward neural network is presented in Fig.10. This structure was chosen after several trials on the data.

For our initial trials we used 75% of the data to train the neural network and optimise the parameters (validation of model) and then tested the network using the remaining 25%. For these trials we kept the time series nature of the data and selected the first 75% of the data for training and then the final 25% for testing. The training versus validation error curve is shown in Fig.11, this corresponds well with the theoretical curve illustrated in Fig.5. above.

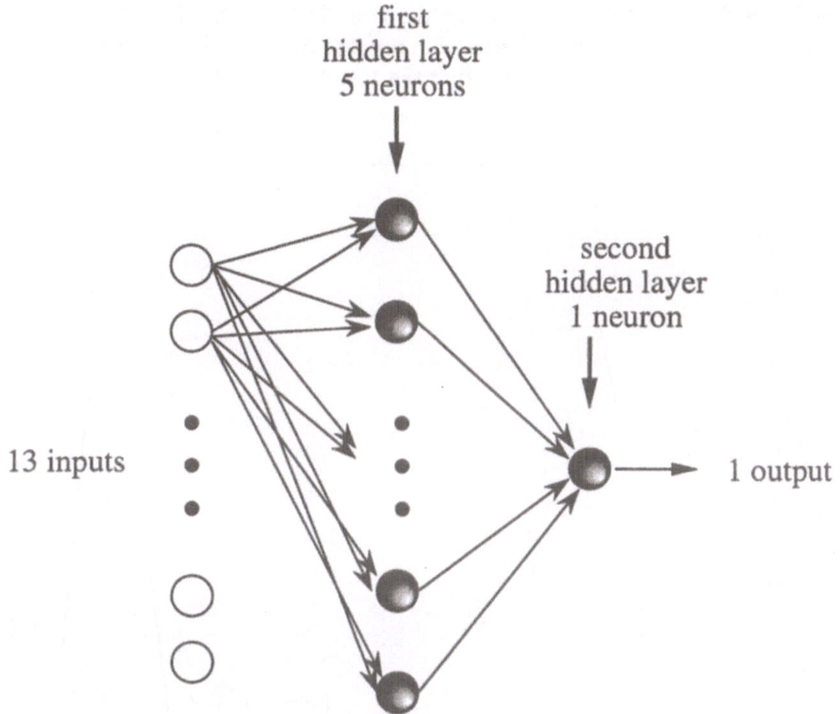

Fig.10. Neural network structure for tardy delivery prediction

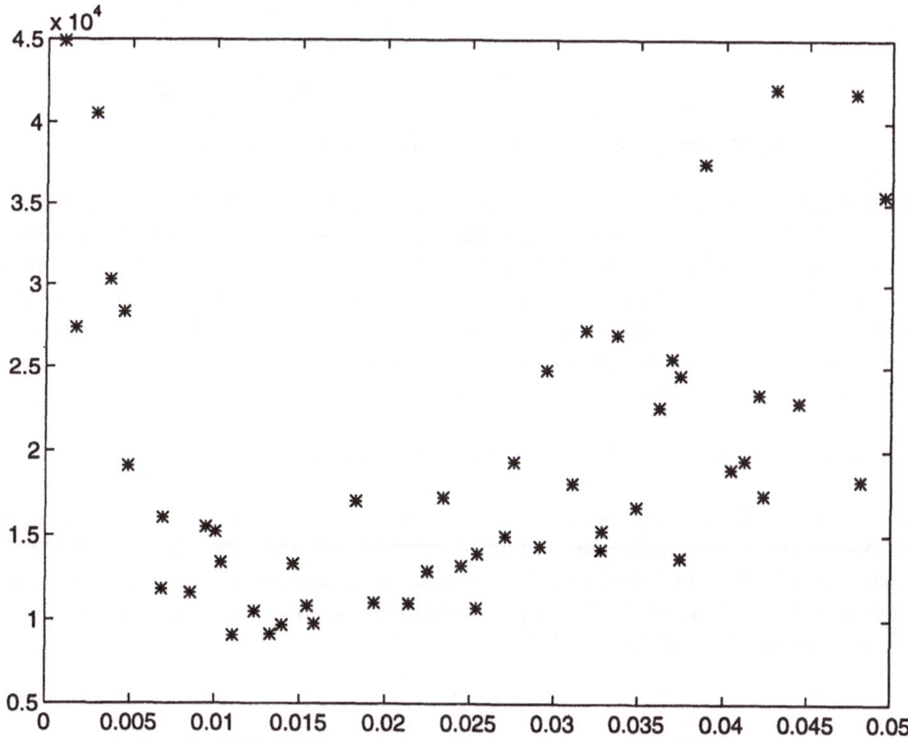

Fig.11. Example of training error versus validation error

In actual fact it was an error to choose the first 75% of the data for training and validation because certain phenomena were present in the final 25% which were not present in the initial part, this meant that the results were not as good as they could have been as can be seen in Fig. 12.

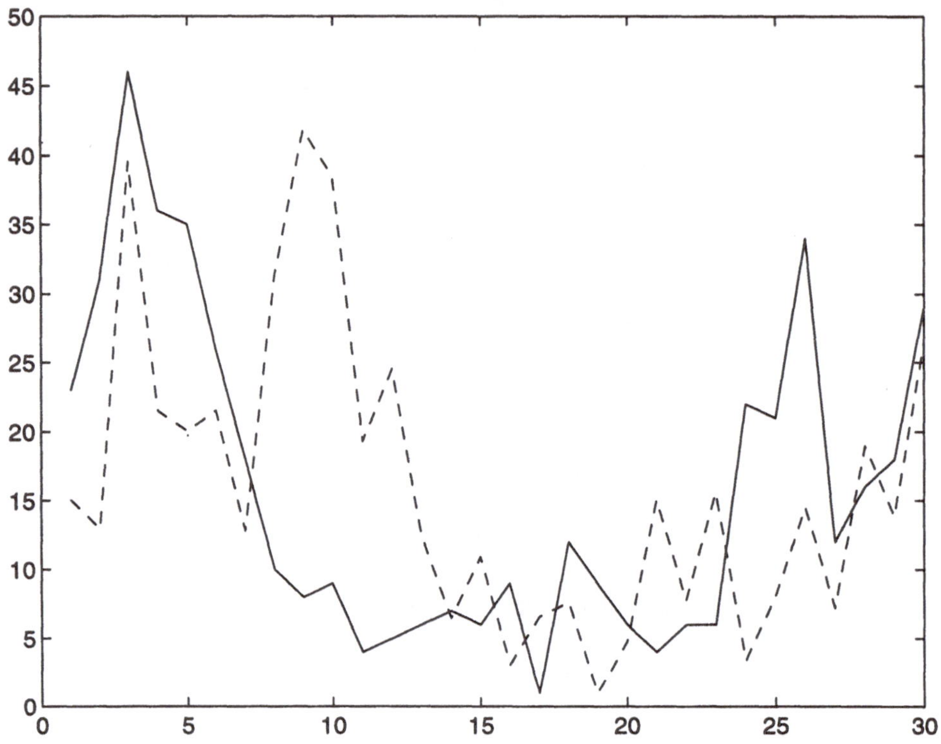

Fig.12. Real (solid line) and predicted (dotted line) tardy delivery

This problem was cured in later work when we applied a fuzzy logic based method to the data. It also needs to be mentioned that the end-user of this application does not seek to predict the behaviour of his system to a high precision point-wise. He is looking more for the general trend of his system in order to take action to avoid a possible catastrophic situation, i.e. to avoid the situation where he is unable to meet his clients orders. A prototype prediction system is in the testing phase at present, installed in a highly automated factory in the South of France.

3.2 Prediction of biodegradability of molecules in soils

The objective of this application was to use certain structural parameters of molecules, such as molecular weight, number of chlorine atoms etc, plus environmental characteristics, such as soil type, temperature, humidity etc, as input parameters and to predict the biodegradability of the molecules in different environments in terms of their half-lives.

Preliminary investigations were based upon a neural network approach. The results were promising but not excellent. In the previous application precise point-wise prediction was not necessary, however in this application it was an important factor.

Later research showed that a hybrid approach such as ANFIS provided the best results, most of the approaches and results are summarised in the dissertation by Battaglia (1996). As a comparison we illustrate in Fig.13. the results obtained by a straightforward linear regression approach, a neural network and the hybrid ANFIS. The results are presented in terms of the correlations between the predicted values and the true values, in each case we present the correlations based on the test data.

Method	*Correlation*
Linear regression	0.62
Neural network	0.68
ANFIS	0.84

Fig.13. Comparison of methods

3.3 Predicting the rate of spread of wildland fires

The types of models presented in this article are very applicable in the domain of environmental systems. Frequently these systems are very difficult to model using more classical techniques such as differential equations, in many cases there are no known mathematical models of the underlying physical phenomena and there can be great uncertainty and imprecision in parameters. For these reasons a soft-computing approach can sometimes be the only way forward in building a mathematical model of a system.

In the case of the wildland fire application we were confronted with the above mentioned difficulties, along with the added problem of data paucity. Due to the success obtained with the neural-fuzzy ANFIS approach for the two applications described above our preliminary investigations were based on this technique and are reported by Sauvagnargues et al. (1997).

Data were obtained during controlled field burning experiments in the Lozère region of Southern France. The data concerned meteorogical conditions (wind speed, wind, temperature, relative humidity, ...) and environmental parameters (altitude, slope, fuel moisture, biomass, ...) and the parameter to be predicted by the model was the propagation speed of the fire.

Some typical results are presented in Fig.14.

Real Value m/h	Predicted Value m/h
90	196
150	121
235	267
240	253
218	109
50	73

360	247
40	118

Fig.14. Comparison of real and predicted speeds

Quite obviously the results presented in Fig.14. are not very precise. However it is worth noting that they are better than those predicted by most of the classical methods actually in use. Work is in progress to improve on these results and to widen the range of the model to include data from other regions in the South of France.

References

Battaglia, D. (1996): Détermination de la biodégradabilité d'une molécule, DEA dissertation, University of Montpellier.

Funahashi, K-I. and Nakamura, Y. (1993): Approximation of Dynamical Systems by Continuous Time Recurrent Neural Networks, Neural Networks, Vol 6, No 6, pp 801-806.

Gerault, R. and Pearson, D.W. (1996): A dynamical model for production systems, Eleventh International Conference on Systems Engineering, Las Vegas, United States.

Gerault, R. and Pearson, D.W. (1997): A Dynamical Model for Production Systems Prediction, International Symposium on Intelligent Industrial Automation, Nîmes, France.

Gerault, R., Pearson, D.W., Peton, N., Hunt, F. and Dray, G. (1997): Several Methods Usable in Production Systems Prediction, to appear in Mathematical and Computer Modelling.

Haykin, S. (1994): Neural Networks: A Comprehensive Foundation, Macmillan.

Hopfield, J.J. (1984): Neurons With Graded Response Have Collective Computational Properties Like Those of Two-state Neurons, Proc. Nat. Acad. Sci. USA Biophysics, Vol 81, pp 3088-3092.

Jang, J-S.R. (1993): ANFIS : Adaptive-Network-Based Fuzzy Inference System, IEEE Transactions on systems, Man, and Cybernetics Vol. 23, NO. 3, pp 665- 685

Lockwood, M. (1989): Mind, Brain & the Quantum, Blackwell.

Nguyen, H.T. and Walker, E.A. (1997): A First Course in Fuzzy Logic, CRC Press.

Pearlmutter, B.A. (1989): Learning State Space Trajectories in Recurrent Neural Networks, Neural Computation, pp 263-269.

Pearson, D.W. (1996): Local Approximation of Control Systems using Artificial Neural Networks, Neurocomputing , Vol 11, pp 43-54.

Pearson, D.W. and Hunt, F. (1996): Tardiness Prediction in Production Systems Using Neural Network Generalisation Capabilities, International Symposia on Intelligent Industrial Automation and Soft Computing, Reading.

Pearson, D., Dobnikar, A., Petelin, B. and Dray, G. (1996): Estimating neural network based predictors for production processes, Computational Engineering in Systems Applications, Lille, France.

Penrose, R. (1995): Shadows of the Mind, Vintage.

Peton,N., Dray,G. and Pearson,D.W. (1997): Rupture Prediction in Higly Automated Production Process Using Fuzzy Logic Based Method, International Symposium on Fuzzy Logic, Zurich, Switzerland.

Stapp, H.P. (1993): Mind, Matter, and Quantum Mechanics, Springer-Verlag.

Sauvagnargues, S., Dusserre,G., Dray,G., Peton,N. and Pearson,D.W. (1997): Determination of a Wildland Fire Rate of Spread Model using Fuzzy Logic, International Symposium on Soft Computing, Nîmes, France.

Swiniarski, R., Hunt,F., Chalvet,D. and Pearson,D.W. (1995): Feature Selection using Rough Sets and Hidden Layer Expansion for Rupture Prediction in a Highly Automated Production Process, International Conference on Systems Science, Wroclaw, Poland.

The method of equivalence in robotics

Krzysztof Tchoń

1 Robotic prerequisites

A manipulation robot is a technical system capable of affecting its environment purposefully, in a way resembling the human manipulation. The function of manipulation is executed by a mechanical device called a manipulator. The robotic manipulator consists of a certain number of rigid bodies called links, connected to each other by joints. Links form a chain that begins at a fixed base of the manipulator, and terminate at its end. Relative motions of consecutive links, accomplished at the joints, have usually one degree of freedom and can be described either as rotations or as translations. Accordingly, the joints are referred to as revolute or prismatic. In typical manipulator designs the joints are driven by independent actuators. Forces or torques exerted by actuators play the role of control inputs. From anthropomorphic perspective the manipulator acts as a substitute of the human arm, while the end-effector, often topped with a gripper, replaces the human hand.

In order to be able to manifest a purposeful behaviour, the manipulator should be equipped with a control system. Given a task to be accomplished by the manipulator, the control system decides how to actuate the joints. Designing control algorithms for manipulators belongs to the most fundamental tasks of contemporary robotics. These algorithms can be divided into kinematic or dynamic. The latter control algorithms exploit a model of manipulator's kinematic and dynamics, the former content ourselves with the kinematics alone. Several approaches have been elaborated in robotics to the synthesis of control algorithms for manipulators. This study will be concerned with a very systematic or even paradigmatic approach that is based on equivalence between the models of manipulator's kinematics and dynamics and simple mathematical models called normal forms. Furthermore, applicability of the equivalence method will be extended to manipulators with flexible joints and to wheeled platforms of mobile robots.

Before proceeding further, we need to introduce a formal description of the kinematics and dynamics of a robotic manipulator. To this aim, consider a manipulator with n joints (degrees of freedom) with two fixed co-ordinate frames: a base co-ordinate frame at manipulator's base, and a moving co-ordinate frame at manipulator's end-effector. Assume that rotations and translations at manipulator's joints are unlimited. Then position of a revolute joint can be modelled by a circle S^1, whereas positions of a prismatic joint are described by points of a real line R^1. In consequence, positions of n joints including r revolute and $p = n - r$ prismatic can be characterised by the Cartesian product $Q = T^r \times R^{n-r}$ called the joint manifold. Elements $q \in Q$ of the joint manifold will be referred to as manipulator's configurations. Obviously, Q is an analytic manifold of dimension n, and simultaneously an Abelian Lie group. Given the base and end-effector frames, we define manipulator's kinematics as a map associating with every manipulator's configuration a position and an orientation of the end-effector with respect to the base frame. Positions and orientations of a rigid body compose the special Euclidean group SE(3). Mathematically, the kinematics are identified with an analytic map

$$k: Q \to SE(3) \tag{1}$$

between the joint manifold Q and the task manifold SE(3). Notice that the special Euclidean group is a Lie group, and a semi-direct product of R^3 by the special orthogonal group SO(3), therefore $\dim SE(3) = 6$. After introducing co-ordinates into Q and SE(3) the kinematics (1) are rendered a co-ordinate representation

$$k: R^n \to R^6, y = k(q). \tag{2}$$

Now suppose that the end-effector of a manipulator has to travel along a prescribed trajectory $y_d(t), t \in I \subseteq R$ in the task manifold. Given the co-ordinate representation of kinematics (2), we want to find a joint trajectory $q_d(t)$ that realises $y_d(t)$, i.e. for every $t \in I$

$$y_d(t) = k(q_d(t)). \tag{3}$$

The problem of determining $q_d(t)$ is referred to as the *inverse kinematic problem* (Nakamura 1991), (Murray et al. 1994). A solution to this problem serves simultaneously as a kinematic control algorithm of the manipulator. As can be seen, solving the inverse kinematic problem amounts to solving a system of non-linear equations (3) with respect to $q_d(t)$, that can be underdetermined (if $n > 6$) or overdetermined (if $n < 6$). In the case of nonredundant manipulators ($n = 6$), the number of indeterminates is equal to the number of equations. The existence, uniqueness, and smoothness of solutions to the inverse kinematic problem depends critically on whether kinematic equations (3) are regular (independent) or singular (dependent). In this context recall that a configuration $q \in Q$ is regarded as singular whenever

$$\operatorname{rank} \frac{\partial k}{\partial q}(q) < \min(n,6), \tag{4}$$

otherwise q is regular. As long as k(q) stays regular, several methods of solving the inverse kinematic problem are available (Nakamura 1991). However, the problem gets complicated, if the resulting joint trajectory is to pass through or nearby singular configurations. In section 3 of this study we shall show how the singular inverse kinematic problem can benefit from an application of the equivalence method.

Let $q, \dot{q} \in R^n$ stand for position and velocity co-ordinates of a manipulator with n degrees of freedom, and let $L(q,\dot{q}) = T(q,\dot{q}) - V(q)$ denote the Lagrangian defined as the difference between kinetic and potential energy of the manipulator. It is known that for a manipulator the kinetic energy defines a Riemannian metric $T(q,\dot{q}) = 1/2\dot{q}^T A(q)\dot{q}^T$, $A(q)$ being a symmetric, positive definite, $n \times n$ inertia matrix. Assuming that each manipulator's joint is has its own actuator, the dynamics of the manipulator can be characterised by standard Euler-Lagrange equations

$$\frac{d}{dt}\frac{\partial L(q,\dot{q})}{\partial \dot{q}} - \frac{\partial L(q,\dot{q})}{\partial q} = u, \tag{5}$$

where $u \in R^n$ denotes control torques and forces exerted at the joints. Exploiting the form of the kinetic energy, we derive from (5) the equations of manipulator's dynamics (Canudas de Wit et al. 1996)

$$A(q)\ddot{q} + B(q,\dot{q}) = u. \qquad (6)$$

Vector $B(q,\dot{q})$ in (6) comprises Coriolis and centripetal terms expressed by Christoffel symbols of the first kind associated with the Riemannian metric $T(q,\dot{q})$ as well as gravity terms defined by the gradient of the potential energy. After a substitution $q = x^1, \dot{q} = x^2$ into (6) and an addition of kinematic equations (2) we obtain an affine control system with outputs as the mathematical model of kinematics and dynamics of a robotic manipulator

$$\begin{cases} \dot{x}^1 = x^2 \\ \dot{x}^2 = F(x^1,x^2) + G(x^1)u \end{cases} \qquad (7)$$
$$y = k(x^1).$$

Hereabove $G(x^1) = A(x^1)^{-1}$ is a symmetric, positive definite matrix, $F(x^1,x^2) = -\ G(x^1)B(x^1,x^2)$, x^1, x^2, $u \in R^n$, $y \in R^6$. Assume, as previously, that manipulator's end-effector should move along a prescribed task trajectory $y_d(t)$, $t \in I \subset R$. Then the *output tracking problem* for the manipulator described by (7) amounts to finding out a control $u_d(t)$ such that system (7) driven by $u_d(t)$ produces an end-effector trajectory $y(t)$ convergent to $y_d(t)$, i.e.

$$\lim_{t\to\infty} y(t) = y_d(t). \qquad (8)$$

A more realistic mathematical model takes into account the fact that in real manipulators the drive transmission from actuators to joints is not ideally rigid, but suffers from flexibility of transmission devices like gears, etc. A Lagrangian model of dynamics of a flexible joint manipulators employs separate position and velocity co-ordinates q^1, $\dot{q}^1 \in R^n$ for the joints and q^2, $\dot{q}^2 \in R^n$ for the actuators, describes elastic linkages between actuators and joints by an extra potential energy term $1/2(q^2 - q^1)^T K(q^2 - q^1)$ containing a diagonal, positive definite matrix K of joint stiffness, and distinguishes the kinetic energy of actuators $1/2\dot{q}^{2T}J\dot{q}^2$ (J - a diagonal, positive definite matrix of actuator inertias) from the kinetic energy of the joints $1/2\dot{q}^{1T}A(q^1)\dot{q}^1$. Therefore, the Euler-Lagrange dynamics equations for a flexible joint manipulator become (Canudas de Wit et al. 1996)

$$\begin{cases} A(q^1)\ddot{q}^1 + B(q^1,\dot{q}^1) - K(q^2 - q^1) = 0 \\ J\ddot{q}^2 + K(q^2 - q^1) = u. \end{cases} \qquad (9)$$

In dynamic equations (9), $u \in R^n$ denotes control forces or torques produced by the actuators. Having introduced new state variables $x^1 = q^1$, $x^2 = \dot{q}^1$, $x^3 = q^2$, $x^4 = \dot{q}^2$, and after including the kinematics (2) as an output map, the mathematical model of kinematics and dynamics of a flexible joint manipulator takes again the form of an affine control system with outputs. Equations of this control system are defined in a 4n-dimensional state space, and with notations $F(x^1,x^2) = -\ A(x^1)^{-1}(B(x^1,x^2)\ -\ Kx^1)$, $G(x^1,x^3) = \ -J^{-1}K(x^3 - x^1)$ they can be represented concisely in the form of the following affine control system

$$\begin{cases} \dot{x}^1 = x^2 \\ \dot{x}^2 = F(x^1, x^2) + Kx^2 \\ \dot{x}^3 = x^4 \\ \dot{x}^4 = G(x^1, x^3) + J^{-1}u \end{cases} \tag{10}$$
$$y = k(x^1),$$

The output tracking problem for a flexible joint manipulator consists in determining, for a prescribed end-effector trajectory $y_d(t)$, $t \in I \subset R$, a control $u_d(t)$ such that the resulting end-effector trajectory y(t) of system (10) converges to the prescribed one,

$$\lim_{t \to \infty} y(t) = y_d(t). \tag{11}$$

Eventually, let us assume that $q, \dot{q} \in R^n$ denote position and velocity co-ordinates of a mobile platform of a wheeled mobile robot consisting of a rigid robot's body and a number of rigid wheels moving on the ground without slipping. The non-slipping requirement imposes on robot's motion a collection of m velocity constraints in the Pfaffian form

$$R(q)\dot{q} = 0, \tag{12}$$

where R(q) is an $m \times n$ full rank constraint matrix depending analytically on q. Constraints (12) are usually non-holonomic (Murray et al. 1994), (Canudas de Wit et al. 1996). They can be incorporated into Euler-Lagrange equations following the d'Alembert principle that states that the work of forces and torques extorting the satisfaction of velocity constraints along feasible infinitesimal displacements is equal to zero. This being so, the dynamic equations of a wheeled mobile platform assume the form

$$\begin{cases} A(q)\ddot{q} + B(q, \dot{q}) = R^T(q)\lambda + C(q)u \\ R(q)\dot{q} = 0, \end{cases} \tag{13}$$

where $\lambda \in R^m$ denotes a vector of indeterminate Lagrange multipliers, $u \in R^{n-m}$ is a control, and C(q) is an $n \times (n\text{-}m)$ full rank control matrix. The meaning of A(q) and $B(q, \dot{q})$ is the same as in the case of the rigid manipulator. In order to eliminate λ from equations (13), it suffices to introduce an $n \times (n\text{-}m)$ matrix S(q) whose columns span the kernel of R(q) (Canudas de Wit et al. 1996), (Campion et al. 1996), so

$$R(q)S(q) = 0. \tag{14}$$

Clearly, the velocity constraints are equivalent to the existence of a vector $\eta \in R^{n-m}$ such that

$$\dot{q} = S(q)\eta. \tag{15}$$

A substitution of (15) into the first group of equations in (13), after suitable manipulations, yields

$$\eta = F(q, \eta) + G(q)u, \tag{16}$$

where

$$\begin{aligned} F(q, \eta) &= -(S^T(q)A(q)S(q))^{-1}(S^T(q)A(q)\frac{\partial(S(q)\eta)}{\partial q}S(q)\eta + S^T(q)B(q, S(q)\eta)), \\ G(q) &= (S^T(q)A(q)S(q))^{-1}S(q)^T C(q). \end{aligned} \tag{17}$$

Expressions (15), (16) will be referred to as the mathematical model of kinematics and dynamics of a wheeled mobile platform. This model will be exploited in section 3 in order to solve a position tracking problem for an exemplary mobile robot.

2 Equivalence

The concept of system equivalence belongs to the most pivotal tools of system theory. Its usefulness in solving diverse system problems, including the problems of system synthesis, results from the following reasoning. Given a synthesis problem formulated in the framework of a system σ, we first establish an equivalence between σ and another system σ', transform the problem to new system σ' then solve the problem in σ', and eventually transform the solution back to original system σ. The way of solving synthesis problems outlined above will be referred to as the *method of equivalence.* It should be clear that since the essence of the method of equivalence lies in replacing an original synthesis problem for σ by an equivalent problem addressed in equivalent system σ', it is desirable that system σ' be as simple as possible. Such a system is often called a normal form of σ. The relation of system equivalence can be perceived as a specific instant of the modelling relation. However, what distinguishes the equivalence is the reversibility of equivalence transformations, guaranteeing that a solution derived for the equivalent system transforms strictly and uniquely to the original system.

The method of equivalence possesses a long tradition in the study of stability and bifurcations of dynamical systems (Arnold 1983), (Guckenheimer and Holmes 1983). The concepts of topological and differentiable equivalence of dynamical systems serve as prototype equivalence relations in system and control theory. A systematic investigation of equivalence of diverse mathematical objects including vector fields, differential forms, Riemannian metrics and variational problems has been pioneered by Cartan (Olver 1995). A mixture of dynamical system, homotopy theory, and Cartan's ideas has been employed in order to establish equivalence of vector field distributions, differential 1-forms, and non-linear control systems (Jakubczyk and Przytycki 1984), (Jakubczyk 1990), (Gardner 1992), (Zhitomirski 1992), (Respondek and Zhitomirski 1995). In the latter case, the most powerful equivalence relation proved to be the feedback equivalence, accomplished either by static or by dynamic feedback. Among celebrated recent results of control theory there are linearization theorems referring to the feedback equivalence between non-linear and linear control systems (Jakubczyk and Respondek 1980), (Fliess 1990), (Charlet et al. 1991), (Gardner 1992), (Jakubczyk 1993), (Fliess et al. 1995). Last but not least, we want to mention RL (left-right) equivalence of smooth maps used in singularity theory (Golubitsky and Guillemin 1973), (Martinet 1982), (Arnold et al. 1983). Some of the aforesaid equivalence concepts will contribute in the sequel to providing solutions to the inverse kinematic problem and to the output tracking problem for robotic manipulators.

Let us begin with manipulator's kinematics $y=k(q)$ defined by expression (2). Given a configuration q of the manipulator, we say that the kinematics k are locally

RL-equivalent to the normal form kinematics k_0, if there exist open neighbourhoods U of q and $V \supseteq k(U)$ as well as local diffeomorphisms (analytic invertible maps with analytic inverses) $\varphi: U \to \varphi(U), \psi: V \to \psi(V)$ such that

$$k_0 = \psi \circ k \circ \varphi^{-1}. \tag{18}$$

Expression (18) means that in suitable co-ordinate systems in the source and target spaces the kinematics k can be given the normal form k_0.

Now we shall introduce the concept of feedback equivalence of control systems. To this aim, consider two affine control systems in R^n, with m controls, defined by the following equations

$$\dot{x} = f(x) + g(x)u = f(x) + \sum_{i=1}^{m} g_i(x)u_i \tag{19}$$

and

$$\dot{\xi} = F(\xi) + G(\xi)v = F(\xi) + \sum_{i=1}^{m} G_i(\xi)v_i, \tag{20}$$

where f, F, g_i, G_i, i=1, ..., m are smooth vector fields. Suppose additionally that the point 0 is an equilibrium point in both systems, so $f(0) = F(0) = 0$, and let $U \subset R^n$ be an open neighbourhood of 0. Control system (19) will be subject to the following smooth transformations defined for $x \in U$:

1. A local change of co-ordinates $\xi = \varphi(x)$ in the state space. (21)
2. A static state feedback $u = \eta(x) + \psi(x)v$.

In order to ensure invertibility of feedback transformations, it is assumed that φ is a local diffeomorphism and that matrix $\psi(x)$ has full rank for every $x \in U$. The action of feedback transformations on control system (19) results in

$$\dot{\xi} = \frac{\partial \varphi}{\partial x}(f + g\eta) \circ \varphi^{-1}(\xi) + \frac{\partial \varphi}{\partial x} g\psi \circ \varphi^{-1}(\xi)v. \tag{22}$$

Control systems (19), (20) are called locally static feedback equivalent when (20) coincides with (22). It turns out that system (19) is locally static feedback libearizable, if it is feedback equivalent to a linear system (20), i.e. to the control system

$$\dot{\xi} = A\xi + Bv. \tag{23}$$

Necessary and sufficient conditions for static feedback linearizations have been discovered by Jakubczyk and Respondek (1980). A comprehensive theory of feedback linearization has been developed by Nijmeijer and van der Schaft (1990), Isidori (1995) as well as Marino and Tomei (1995).

As can be seen, the static state feedback is based on a modification of control system (19) using a current value of the state variable x(t). Recently, it has been proved that applicability of linearization technique extends remarkably, if a dynamic feedback is employed, defined by the following transformations:

1. A dynamic compensation accomplished by a compensator of the form

$$\begin{cases} \dot{z} = K(x,z) + L(x,z)v \\ u = M(x,z) + N(x,z)v. \end{cases} \tag{24}$$

2. A local co-ordinate change $\xi = \varphi(x,z)$.

The dynamic compensator described above is an affine control system with a state variable $z \in R^p$, parameterized by the state variable $x \in R^n$ of system (19). A coupling of this system with compensator (24) produces a new affine control system with extended state space containing two variables (x,z), equipped with new input v and output u, i.e.

$$\begin{pmatrix} \dot{x} \\ \dot{z} \end{pmatrix} = \begin{pmatrix} f(x) + g(x)M(x,z) \\ K(x,z) \end{pmatrix} + \begin{pmatrix} g(x)N(x,z) \\ L(x,z) \end{pmatrix} v \tag{25}$$
$$u = M(x,z) + N(x,z)v.$$

Extended system (26) is required to be input-output invertible (Nijmeijer 1990), (Isidori 1995). Subject to a co-ordinate change φ included in (24), system (25) becomes an affine control system in R^{n+p}

$$\dot{\xi} = \frac{\partial\varphi}{\partial x}(f + gM) \circ \varphi^{-1}(\xi) + \frac{\partial\varphi}{\partial z} K \circ \varphi^{-1}(\xi) + (\frac{\partial\varphi}{\partial x} gN + \frac{\partial\varphi}{\partial z} L) \circ \varphi^{-1}(\xi)v. \tag{26}$$

Analogously to the terminology adopted for static state feedback linearization, we shall call control system (19) locally dynamically feedback linearizable, if system (26) is linear, i.e. has the same form as (23). It turns out that in order to linearize a system by dynamic feedback, we first need to design a dynamic compensator, then define extended system (25), and finally make this system linear by a diffeomorphic change of extended state co-ordinates. Obviously, if p=0, no dynamic compensator is designed and dynamic feedback is equivalent to static feedback. Necessary and sufficient conditions for dynamic feedback linearization have been obtained by Fliess (1990), Fliess et al. (1995), and Jakubczyk (1993) by differential algebraic methods, and independently by Gardner (1992) who used Cartan's approach. For a specific dynamic compensator, referred to as the Brunovsky compensator, checkable sufficient conditions for dynamic feedback linearization have been derived by Charlet et al. (1991). In particular, it has been proved that, in order to be effective, dynamic linearization requires that system (19) have sufficiently many inputs. For single input systems static and dynamic linearizations are equivalent.

3 Application to robotics

This section will be devoted to presenting a sample of results concerned with a design of control algorithms for manipulation and mobile robots, assisted by the method of equivalence.

3.1 Inverse kinematic problem

Let $k: R^n \to R^n, y = k(q)$, denote the kinematics of a non-redundant manipulator. Given a desirable task trajectory $y_d(t)$, $t \in I \subset R$, we want to find a joint trajectory $q_d(t)$ solving for every $t \in I$ the kinematic equation $y_d(t) = k(q_d(t))$. Assume temporarily that $q_0 \in R^n$, $k(q_0) = y_d(0)$, is a regular configuration of the manipulator, i.e. rank of the Jacobian matrix of k at q_0 is equal to n. Therefore, it follows from the inverse function theorem that there exist open neighbourhoods U of q_0 and V of $k(q_0)$, and diffeomorphisms $\varphi: U \to \varphi(U)$, $\psi: V \to \psi(V)$,

$x = \varphi(q)$, $\eta = \psi(y)$, such that kinematics k can be transformed locally to the normal form

$$k_0(x) = \psi \circ k \circ \varphi^{-1}(x) = x. \tag{27}$$

Furthermore, it is easily checked that $\varphi(q) = k(q) - k(q_0), \psi(y) = y - k(q_0)$. Having defined normal form (27), we express the desirable trajectory in new co-ordinates as $\eta_d(t) = y_d(t) - y_d(0)$ and solve immediately the inverse kinematic problem obtaining $x_d(t) = \eta_d(t)$. Eventually, this trajectory is transformed back to original co-ordinates, yielding a solution

$$q_d(t) = \varphi^{-1}(y_d(t)) \tag{28}$$

valid in the neighbourhood U.

The case of regular kinematics can be regarded as just a simple exemplification of RL-equivalence method in the context of robotics, that in fact is a prelude to more advanced applications coming into play when the manipulator is forced to assume singular configurations. To analyse this case in some detail, suppose again that the manipulator is non-redundant. Let $q_0 \in R^n$ stand for a singular configuration. Then the following procedure of solving the inverse kinematic problem in the presence of singular configurations has been set forth by Tchoń and Muszyński (1997b).

1. Given the kinematics $y=k(q)$, determine the locus of singular configurations $S = \{q \in R^n \mid \operatorname{rank}\frac{\partial k}{\partial q}(q) < n\}$, and compute the set of singular values of kinematics $k(S)$.
2. Determine the set of singular time instants T_s containing time instants $t_s \in I$ at which desirable trajectory $y_d(t)$ takes on its values in $k(S)$.
3. If current time instant $t \in T_s$, solves the regular inverse kinematic problem using the method presented above to obtain a regular piece of solution $q_d(t)$.
4. If $t \in T_s$, the joint trajectory enters a singular configuration q_s. In this case, by the employment of suitable equivalence transformations $\varphi: U \to \varphi(U)$ and $\psi: V \to \psi(V)$, defined on open neighbourhoods U of q_s and V of $k(q_s)$, find a normal form $k_0(x) = \psi \circ k \circ \varphi^{-1}$ of kinematics $k(q)$.
5. Transform the desirable trajectory to new co-ordinates, setting $\eta_d(t) = \psi(y_d(t))$.
6. Solve the inverse kinematic problem for k_0 and $\eta_d(t)$, i.e. find a piece of trajectory $x_d(t) \in \varphi(U)$ such that $\eta_d(t) = k_0(x_d(t))$.
   Transform $x_d(t)$ back to original co-ordinates according to the formula $q_d(t) = \varphi^{-1}(x_d(t))$, where $q_d(t) \in U$, producing a singular piece of the joint trajectory.
7. Repeat, if necessary, steps 3 ÷ 8 of this procedure.

Compute $q_d(t)$, $t \in I$, by gluing together regular and singular pieces of the joint trajectory.

In order to apply the above procedure efficiently, several specific points need a further clarification. First, we want to mention that a solution to the regular inverse kinematic problem is equivalent to solving the following set of non-autonomous differential equations

$$\dot{q} = \left(\frac{\partial k}{\partial q}(q)\right)^{-1} \dot{y}_d(t) \tag{29}$$

initialised at $q(0)$ that in turn appears to be the limit at $t \to \infty$ of the solution $z(t)$ of the system

$$\dot{z} = \alpha \left(\frac{\partial k}{\partial q}(z) \right)^{-1} (y_d(0) - k(z)), \tag{30}$$

started at any $z(0) \in R^n$, and converging with a rate $\alpha > 0$. Second, it should be expected that the set of singular time instants T_s will be discrete or even finite, most of $t \in I$ being regular. Therefore, while following a regular piece of trajectory computed according to (29), (30), a decision of switching the procedure from regular to singular mode can be made on the basis of a dexterity measure of the manipulator (Nakamura 1991). Undoubtedly, a crucial step of the procedure (no 4) consists in defining a normal form k_0 and equivalence transformations φ, ψ establishing RL-equivalence of the kinematics to the normal form. Clearly, this form should be simple enough to allow for solving the inverse kinematic problem at step 6. The role of equivalence transformations resolves itself to accomplishing a passage from the original desirable trajectory to the normal form trajectory at step 5, and then a reverse passage from the normal form solution to the original co-ordinate solution at step 7. As far as the normal forms are concerned, a fairly complete collection of them has been obtained by Tchoń (1991), and Tchoń and Muszyński (1997a). In the last reference, it has been demonstrated that a normal form of non-redundant kinematics at singular configurations of corank 1 (i.e. such that rank of the Jacobian matrix is equal to $n-1$) depends on a single integer invariant called the differential degree of kinematics. In particular, given the kinematics $k: R^n \to R^n$ with a singular configuration q_0 of corank 1, it has been proved that around q_0 the kinematics are RL-equivalent to the quadratic normal form

$$k_0(x) = (x_1^2, x_2, \ldots, x_n),$$

provided that for a vector $v \in \mathrm{Ker}\frac{\partial k}{\partial q}(q_0)$

$$d\left(\det \frac{\partial k}{\partial q}\right)(q_0)v \neq 0.$$

A wide range of applicability of the quadratic normal form has been discovered by (Muszyński 1996). A computation of equivalence transformations is closely related to the determination of normal forms. If a normal form has been established then these transformation can always be computed with arbitrary accuracy by the Taylor series expansion technique. This method has been applied successfully in (Tchoń 1997b). On the other hand, in some cases equivalence transformations can be found analytically and even globally, after a careful analysis of the inverse kinematic problem in its specific context (Muszyński and Tchoń 1997).

3.2 Tracking in rigid manipulators

Consider model (7) of kinematics and dynamics of a rigid robotic manipulator with n degrees of freedom. Given a desirable task trajectory $y_d(t)$, $t \in I \subset R$, we need to find a control $u_d(t)$ that applied to the manipulator produces a joint trajectory $x^1{}_d(t)$

converted by the kinematic map $k(x^1)$ into $y_d(t)$. It turns out that the output tracking problem in (7) can be decomposed into two subproblems: the inverse kinematic problem yielding a joint trajectory $x^1_d(t)$ that realises $y_d(t)$, and a *state tracking problem* formulated as follows. Given the model of manipulator's dynamics

$$\begin{cases} \dot{x}^1 = x^2 \\ \dot{x}^2 = F(x^1, x^2) + G(x^1)u \end{cases} \tag{31}$$

and a desirable state trajectory $x^1_d(t)$, $t \in I \subset R$, find a control $u_d(t)$ such that the corresponding state trajectory of system (31) steered by $u_d(t)$ tracks $x^1_d(t)$, i.e. $x^1(t) \to x^1_d(t)$ along with $t \to \infty$. Equivalently, having solved the inverse kinematic problem with respect to $x^1_d(t)$, our task is to design a control algorithm able to track this trajectory in affine control system (31).

A synthesis of the tracking control algorithm will be based on the static state feedback linearization of system (31). To this aim, let us observe that feedback transformations (21) consisting of the trivial co-ordinate change $\xi^1 = x^1, \xi^2 = x^2$ accompanied by feedback

$$u = -G(x^1)^{-1}F(x^1, x^2) + G(x^1)^{-1}v \tag{32}$$

make system (31) linear of the form

$$\begin{cases} \dot{\xi}^1 = \xi^2 \\ \dot{\xi}^2 = v. \end{cases} \tag{33}$$

The state tracking problem in (33) amounts to defining a control $v_d(t)$ driving the state of system (33) in such a way that $\xi^1(t)$ tracks $x^1_d(t)$. But the tracking problem in (33) has a straightforward solution

$$v_d(t) = \ddot{x}^1_d - K_1(\xi^2 - \dot{x}^1_d) - K_0(\xi^1 - x^1_d), \tag{34}$$

where K_1, K_0 are diagonal gain matrices with positive entries. Returning to original system (31), we conclude by (32) and (34) that the state tracking algorithm becomes

$$u_d(t) = -G(x^1)^{-1}F(x^1, x^2) + G(x^1)^{-1}(\ddot{x}^1_d - K_1(x^2 - \dot{x}^1_d) - K_0(x^1 - x^1_d)). \tag{35}$$

It easily checked that in fact control (35) ensures the exponential convergence of the state trajectory $x^1(t)$ to $x^1_d(t)$. The control law (35) is known in robotics as the *computed torque control*. It plays the role of a reference control algorithm of rigid manipulators (Murray et al. 1994), (Canudas de Wit et al. 1996).

3.3 Tracking in flexible joint manipulators

In this subsection we shall study the output tracking problem in model (10) of a flexible joint manipulator with n degrees of freedom. Thus, given a task trajectory $y_d(t)$ defined on a time interval $I \subset R$, we need to find a control $u_d(t)$ of the affine control system with outputs (10), producing a joint trajectory $x^1(t)$ that converges to $x^1_d(t)$ solving the kinematic equations $y_d(t) = k(x^1_d(t))$, $t \in I \subset R$. As in the previous subsection, the output tracking problem will be separated into the inverse kinematic problem and a *joint tracking problem*. The former problem can be solved according to the procedure introduced in subsection 3.1, providing a desirable joint trajectory $x^1_d(t), t \in I \subset R$. The latter problem consists in determining a control signal $u_d(t)$ that makes the joint trajectory ($x^1(t)$) of affine control system

$$\begin{cases} \dot{x}^1 = x^2 \\ \dot{x}^2 = F(x^1, x^2) + Kx^2 \\ \dot{x}^3 = x^4 \\ \dot{x}^4 = G(x^1, x^3) + J^{-1}u \end{cases} \tag{36}$$

to track prescribed joint trajectory $x^1_d(t)$. Analogously to the case of tracking in rigid manipulators, after exclusion of the inverse kinematic problem, solving the output tracking problem in flexible joint manipulators becomes equivalent to designing a control algorithm enabling the joint trajectory of (36) to track desirable trajectory $x^1_d(t)$. This last problem, however, can be solved by employing the static state feedback linearization of affine control system (36). To accomplish the linearization, consider a change $\xi = \varphi(x)$ of state co-ordinates in (36), defined as follows

$$\begin{gathered} \xi^1 = x^1, \xi^2 = x^2, \xi^3 = F(x^1, x^2) + Kx^3, \\ \xi^4 = \frac{\partial F}{\partial x^1} x^2 + \frac{\partial F}{\partial x^2}(F(x^1, x^2) + Kx^3) + Kx^4. \end{gathered} \tag{37}$$

Due to invertibility of stiffness matrix K, it is easily deduced that map φ defined above is a global diffeomorphism. Furthermore, in new co-ordinates system (36) can be written down as

$$\begin{cases} \dot{\xi}^1 = \xi^2, \dot{\xi}^2 = \xi^3, \dot{\xi}^3 = \xi^4, \\ \dot{\xi}^4 = H(x^1, x^2, x^3, x^4) + KJ^{-1}u, \end{cases} \tag{38}$$

where

$$H(x^1, x^2, x^3, x^4) = \frac{d}{dt}\left(\frac{\partial F}{\partial x^1} x^2 + \frac{\partial F}{\partial x^2}(F(x^1, x^2) + Kx^3) + KG(x^1, x^3)\right),$$

computed along a trajectory of (36), collects all components of ξ^4 that do not depend on control explicitly. By definition, matrix KJ^{-1} premultiplying control in (38) is invertible. This being so, a feedback

$$u = -(KJ^{-1})^{-1} H(x^1, x^2, x^3, x^4) + (KJ^{-1})^{-1} v \tag{39}$$

will eventually make system (38) linear of the form

$$\dot{\xi}^1 = \xi^2, \dot{\xi}^2 = \xi^3, \dot{\xi}^3 = \xi^4, \dot{\xi}^4 = v. \tag{40}$$

Clearly, the joint trajectory tracking problem in linear control system (40) amounts to finding a control $v_d(t)$ ensuring the tracking of $\xi^1_d(t) = x^1_d(t)$. A solution is found immediately as

$$v_d(t) = \xi^{1\,(4)}_d - K_3(\xi^4 - \dddot{\xi}^1_d) - K_2(\xi^3 - \ddot{\xi}^1_d) - K_1(\xi^2 - \dot{\xi}^1_d) - K_0(\xi^1 - \xi^1_d), \tag{41}$$

where gain matrices K_i, $i=0\div3$, should be chosen in such a way that the tracking error in (40) + (41) converge to zero exponentially.

Control algorithm (41) may be regarded as a solution of the tracking problem addressed in normal form (40) of affine control system (36). In order to transform this algorithm to the original system, we need to use invertibility of feedback transformations (38), (39). A final result is the following joint tracking algorithm in the flexible joint manipulator (36)

$$u_d(t) = -(KJ^{-1})^{-1}H(x^1,x^2,x^3,x^4) +$$
$$(KJ^{-1})^{-1}(x^{1}{}_d{}^{(4)} - K_3(\frac{\partial F}{\partial x^1}x^2 + \frac{\partial F}{\partial x^2}(F(x^1,x^2)+Kx^3)+Kx^4 - \dddot{x}^1_d) - \tag{42}$$
$$K_2(F(x^1,x^2)+Kx^3 - \ddot{x}^1_d) - K_1(x^2 - \dot{x}^1_d) - K_0(x^1 - x^1_d)).$$

Similarly as in the case of rigid manipulators, control law (42) plays the role of a paradigmatic tracking algorithm for flexible joint manipulators (Canudas de Wit et al. 1996).

3.4 Position tracking in mobile robots

Eventually, consider a wheeled mobile platform (a tricycle) equipped with a pair of fixed rear wheels, driven by actuators, and one front wheel adjusted by a driver. It will be assumed that the platform moves in a flat terrain and that its motion can be characterised by co-ordinates $q = (x,y,\theta,\varphi) \in R^2 \times S^1 \times (-\pi/2,+\pi/2)$, where x,y,θ denote position and orientation of the platform, φ is the orientation of the steering wheel. A non-slipping condition of the wheels translates into velocity constraints $R(q)\dot{q} = 0$ defined by a matrix

$$R(q) = \begin{bmatrix} \sin\theta & -\cos\theta & 0 & 0 \\ \cos\theta\sin\varphi & \sin\theta\cos\varphi & -\cos\varphi & 0 \end{bmatrix}. \tag{43}$$

It is easily deduced from (43) that the velocity constraints generate the following kinematic equations (15) of the platform (Murray et al. 1994)

$$\begin{cases} \dot{x} = \eta_1 \cos\theta \\ \dot{y} = \eta_1 \sin\theta \\ \dot{\theta} = \eta_1 \tan\varphi \\ \dot{\varphi} = \eta_2, \end{cases} \tag{44}$$

where η_1, η_2 may be interpreted as velocities: the linear velocity of the platform and the angle velocity of reorienting the steering wheel, respectively. Suppose for simplicity that corresponding accelerations are treated as controls. In this way dynamic equations (16) of the platform become just

$$\dot{\eta}_1 = u_1, \dot{\eta}_2 = u_2. \tag{45}$$

Given equations (44), (45) of the mobile platform, we shall examine the following *position tracking problem.* For a desirable trajectory $\zeta_d(t) = (x_d(t), y_d(t))$ of the platform, find a control $u_d(t)$ such that the actual position trajectory $\zeta(t) = (x(t), y(t))$ produced in (44) by this control track $\zeta_d(t)$, i.e. $\zeta(t) \to \zeta_d(t)$ as $t \to \infty$. A solution to this tracking problem will constitute an output tracking control algorithm for affine control system (44), (45) endowed with an output function

$$\zeta_1 = x, \ \zeta_2 = y. \tag{46}$$

Our approach at designing the tracking algorithm will adopt the equivalence method, namely the dynamic feedback linearization technique (d'Andrea-Novel et al. 1995), (Canudas de Wit et al. 1996). (It is easily checked that system (44) is not static feedback linearizable). To this aim we choose outputs (46) as so-called linearizing outputs (Charlet et al. 1991). These outputs will be differentiated along trajectories of system (44) as long as two independent controls appear on the right hand sides. More specifically, relying on (44) ÷ (46) we compute

$$\begin{cases} \dot{\zeta}_1 = \dot{x} = \eta_1 \cos\theta \\ \ddot{\zeta}_1 = \ddot{x} = u_1 \cos\theta - \eta^2{}_1 \tan\varphi \sin\theta \\ \dddot{\zeta}_1 = \dddot{x} = \dot{u}_1 \cos\theta - 3u_1\eta_1 \tan\varphi \sin\theta - \eta_1{}^2\eta_2 \dfrac{\sin\theta}{\cos^2\varphi} - \eta_1{}^3 \tan^2\varphi \cos\theta \end{cases} \tag{47}$$

and

$$\begin{cases} \dot{\zeta}_2 = \dot{y} = \eta_1 \sin\theta \\ \ddot{\zeta}_2 = \ddot{y} = u_1 \sin\theta + \eta^2{}_1 \tan\varphi \cos\theta \\ \dddot{\zeta}_2 = \dddot{y} = \dot{u}_1 \sin\theta + 3u_1\eta_1 \tan\varphi \cos\theta + \eta_1{}^2\eta_2 \dfrac{\cos\theta}{\cos^2\varphi} - \eta_1{}^3 \tan^2\varphi \sin\theta. \end{cases} \tag{48}$$

Expressions (47), (48) will serve us as a point of departure toward defining a dynamic compensator (24). Regarding u_1 as a new state variable driven by new control v_1,

$$u_1 = z_1, \; \dot{z}_1 = v_1, \tag{49}$$

we re-write the 3rd order derivatives of ζ_1, ζ_2 in the following form

$$\begin{cases} \dddot{\zeta}_1 = \dddot{x} = v_1 \cos\theta - 3u_1\eta_1 \tan\varphi \sin\theta - \eta_1{}^2\eta_2 \dfrac{\sin\theta}{\cos^2\varphi} - \eta_1{}^3 \tan^2\varphi \cos\theta \\ \dddot{\zeta}_2 = \dddot{y} = v_1 \sin\theta + 3u_1\eta_1 \tan\varphi \cos\theta + \eta_1{}^2\eta_2 \dfrac{\cos\theta}{\cos^2\varphi} - \eta_1{}^3 \tan^2\varphi \sin\theta. \end{cases} \tag{50}$$

Subject to the next time differentiation, equations (50) transform to

$$\begin{cases} \zeta_1{}^{(4)} = x^{(4)} = \dot{v}_1 \cos\theta - \eta_1{}^2\dot{\eta}_2 \dfrac{\sin\theta}{\cos^2\varphi} + f_1(\theta, \varphi, \eta_1, \eta_2, z_1, v_1) \\ \zeta_2{}^{(4)} = y^{(4)} = \dot{v}_1 \sin\theta + \eta_1{}^2\dot{\eta}_2 \dfrac{\cos\theta}{\cos^2\varphi} + f_2(\theta, \varphi, \eta_1, \eta_2, z_1, v_1), \end{cases} \tag{51}$$

where functions f_1, f_2 collect terms that are not essential for further mathematical developments. Now we extend the compensator by former control variable v_1 and add to (49) the equation

$$v_1 = z_2, \; \dot{z}_2 = w_1, \tag{52}$$

with w_1 denoting a new control. For the sake of definiteness, we substitute $u_2 = w_2$. Eventually, equations (51) will be given the following form

$$\begin{pmatrix} \zeta_1{}^{(4)} \\ \zeta_2{}^{(4)} \end{pmatrix} = \begin{pmatrix} x^{(4)} \\ y^{(4)} \end{pmatrix} = \begin{pmatrix} f_1 \\ f_2 \end{pmatrix} + \begin{bmatrix} \cos\theta & -\eta_1{}^2 \dfrac{\sin\theta}{\cos^2\varphi} \\ \sin\theta & \eta_1{}^2 \dfrac{\cos\theta}{\cos^2\varphi} \end{bmatrix} w = f + Dw. \tag{53}$$

It is easily observed that matrix D premultiplying new control vector w is invertible provided that $\eta_1 \neq 0$, i.e. when the platform is moving. Under this last assumption we are in a position to introduce new control r via an identity

$$w = -D^{-1}f + D^{-1}r. \tag{54}$$

In this way we have designed the following dynamic compensator for system (44), (45)

$$\begin{cases} \dot{z}_1 = z_2, \ \dot{z}_2 = w_1 \\ u_1 = z_1, \ u_2 = w_2 \\ w = -D^{-1}f + D^{-1}r. \end{cases} \tag{55}$$

Let us include the dynamic compensator into system (44), (45), and define an extended system

$$\begin{cases} \dot{x} = \eta_1 \cos\theta, \ \dot{y} = \eta_1 \sin\theta, \\ \dot{\theta} = \eta_1 \tan\varphi, \ \dot{\varphi} = \eta_2, \\ \dot{\eta}_1 = z_1, \ \dot{\eta}_2 = w_2, \\ \dot{z}_1 = z_2, \ \dot{z}_2 = w_1, \\ w = -D^{-1}f + D^{-1}r \end{cases} \tag{56}$$

with 8-dimensional state space and 2-dimensional control r. We shall apply to (56) a co-ordinate change $\xi = \phi(x, y, \theta, \varphi, \eta_1, \eta_2, z_1, z_2)$ defined as

$$\begin{cases} \dot{\xi}_1 = x, \quad \dot{\xi}_2 = \eta_1 \cos\theta, \quad \dot{\xi}_3 = z_1 \cos\theta - \eta_1^2 \tan\varphi \sin\theta \\ \dot{\xi}_4 = z_2 \cos\theta - 3z_1\eta_1 \tan\varphi \sin\theta - \eta_1^2\eta_2 \dfrac{\sin\theta}{\cos^2\varphi} - \eta_1^3 \tan^2\varphi \cos\theta, \\ \dot{\xi}_5 = y, \quad \dot{\xi}_6 = \eta_1 \sin\theta, \quad \dot{\xi}_7 = z_1 \sin\theta + \eta_1^2 \tan\varphi \cos\theta, \\ \dot{\xi}_8 = z_2 \sin\theta + 3z_1\eta_1 \tan\varphi \cos\theta + \eta_1^2\eta_2 \dfrac{\cos\theta}{\cos^2\varphi} - \eta_1^3 \tan^2\varphi \sin\theta \end{cases} \tag{57}$$

that is a diffeomorphism under condition $\eta_1 \neq 0$. Eventually, we conclude that in new co-ordinates extended system (56) gets linearized and assumes the following form

$$\begin{cases} \dot{\xi}_1 = \xi_2, \quad \dot{\xi}_2 = \xi_3, \quad \dot{\xi}_3 = \xi_4, \quad \dot{\xi}_4 = r_1, \\ \dot{\xi}_5 = \xi_6, \quad \dot{\xi}_6 = \xi_7, \quad \dot{\xi}_7 = \xi_8, \quad \dot{\xi}_8 = r_2. \end{cases} \tag{58}$$

Linear control system (58) may be regarded as a normal form of original system (44), (45) created by the dynamic state feedback. The position tracking problem in the normal form resolves itself to finding a control r such that $\zeta_d(t) = (\xi_{1d}(t), \xi_{5d}(t))$ will be tracked. But this last problem has the following straightforward solution thanks to linearity of system (58)

$$\begin{cases} r_1 = \xi_{1d}^{(4)} - k_3(\xi_4 - \dddot{\xi}_{1d}) - k_2(\xi_3 - \ddot{\xi}_{1d}) - k_1(\xi_2 - \dot{\xi}_{1d}) - k_0(\xi_1 - \xi_{1d}) \\ r_2 = \xi_{5d}^{(4)} - l_3(\xi_8 - \dddot{\xi}_{5d}) - l_2(\xi_7 - \ddot{\xi}_{5d}) - l_1(\xi_6 - \dot{\xi}_{5d}) - l_0(\xi_5 - \xi_{5d}). \end{cases} \tag{59}$$

Gain coefficients k_i, l_i in the above control algorithm should ensure an exponential decay of tracking errors to zero. In order to reconstruct a tracking control for the original system we observe that an inclusion of co-ordinates change formula (57) into (59) expresses control r as a function of original co-ordinates. Therefore, such a control can be plugged into dynamic compensator (55), culminating in a determination of a position tracking algorithm for the mobile platform (44), (45).

4 Conclusion

The content of this study situates at the crossroad of system theory and robotics. It has been demonstrated that the method of equivalence, having its roots in system theory, can provide solutions to the fundamental robotic problems, including the inverse kinematic problem and the tracking problems for manipulation as well as mobile robots. Furthermore, the solution provided by the method of equivalence is always in a sense canonical, serving as a reference point for other approaches developed in robotics.

References

Arnold, V.I., Varchenko, A.N., Gussein-Zade, S.M. (1985): Singularities of Differentiable Maps. Birkhäuser, Boston

Arnold, V.I. (1983): Geometrical Methods in the Theory of Ordinary Differential Equations. Springer-Verlag, New York

Campion, G., Bastin, G., d'Andrea-Novel, B. (1996): Structural properties and classification of kinematic and dynamic models of wheeled mobile robots. IEEE Trans. Robotics & Automat. 12: 47-62

Canudas de Wit, C., Siciliano, B., Bastin, G., (1996): Theory of Robot Control. Springer-Verlag, London

Charlet, B., Levine, J., Marino, R. (1991): Sufficient conditions for dynamic state feedback linearization. SIAM J. Control & Optimiz. 29: 38-57

d'Andrea-Novel, B., Campion, G., Bastin, G. (1995): Control of nonholonomic wheeled mobile robots by state feedback linearization. Int. J. Robotics Research 14: 543-559

Fliess, M. (1990): Generalized controller canonical form for linear and nonlinear dynamics. IEEE Trans. Autom. Control 35: 994-1001

Fliess, M., Levine, J., Martin, Ph., Rouchon, P. (1995): Flatness and defect in nonlinear systems: introductory theory and applications. Int. J. Control 61: 1327-1361

Gardner, R.B. (1989): The Method of Equivalence and Its Applications. SIAM, Philadelphia

Golubitsky, M., Guillemin, V. (1973): Stable Mappings and Their Singularities. Springer-Verlag, Berlin

Guckenheimer, J., Holmes, P.J. (1983): Nonlinear Oscillations, Dynamical Systems, and Bifurcations of Vector Fields. Springer-Verlag, New York

Isidori, A. (1995): Nonlinear Control Systems. Springer-Verlag, London

Jakubczyk, B., Respondek, W. (1980): On linearization of control systems. Bull. Acad. Polon. Sci. Ser. Sci. Math. 28: 517-522

Jakubczyk, B., Przytycki, F. (1984): Singularities of k-tuples of vector fields. Dissertationes Mathematicae, 213: 1-64

Jakubczyk, B. (1990): Equivalence and invariants of nonlinear control systems. In: Sussmann, H.J. (ed.): Nonlinear Controllability and Optimal Control. M. Dekker, New York, pp. 177-218

Jakubczyk, B., (1993): Invariants of dynamic feedback and free systems. In: Proceedings of the European Control Conference, ECC'93, Groningen, pp. 1510-1513

Marino, R., Tomei, P. (1995): Nonlinear Control Design. Prentice Hall, London

Martinet, J. (1982): Singularities of Smooth Functions and Maps. Cambridge University Press, Cambridge

Murray, R.M., Li, Z., Sastry, S.S. (1994): A Mathematical Introduction to Robotic Manipulation. CRC Press, Boca Raton

Muszyński, R., Tchoń, K. (1996): Normal forms of non-redundant singular robot kinematics: Three DOF worked examples. J. Robotic Systems 13:765-791

Muszyński, R., Tchoń, K. (1997): A solution to the singular inverse kinematic problem for the ASEA IRb-6 manipulator mounted on a track. Pre-print, submitted for publication

Nakamura, Y. (1991): Advanced Robotics: Redundancy and Optimization. Addison-Wesley, New York

Nijmeijer, H., van der Schaft, A. (1990): Nonlinear Dynamical Control Systems. Springer-Verlag, New York

Olver, P.J. (1995): Equivalence, Invariants, and Symmetry. Cambridge University Press, Cambridge

Respondek, W., Zhitomirski, M.Y. (1995): Feedback classification of nonlinear control systems on 3-manifolds. Math. Control, Signals, and Syst. 8: 299-333

Tchoń, K. (1991): Differential topology of the inverse kinematic problem for redundant robot manipulators. Int. J. Robotics Research 10: 492-504

Tchoń, K., Muszyński, R. (1997a): Singularities of non-redundant robot kinematics. Int. J. Robotics Research 16: 60-76

Tchoń, K., Muszyński, R. (1997b): Singular inverse kinematic problem for robotic manipulators: A normal form approach. IEEE Trans. Robotics & Automat., to appear

Zhitomirski, M.Y. (1992): Typical Singularities of Differential 1-forms and Pfaffian Equations. AMS, Providence

Manufacturing Algebra: a new mathematical tool for discrete-event modelling of manufacturing systems

Enrico Canuto, Francesco Donati, Maurizio Vallauri

1 The elements of the Manufacturing Algebra

The *Manufacturing Algebra* was developed during several years of research and more recently within the ESPRIT Basic Research HIMAC-8141 (Hierarchical Management and Control of Manufacturing Systems). For a comprehensive list of publications see the enclosed References. The Manufacturing Algebra is a methodology specifically conceived for investigating and modelling discrete manufacturing systems at various degrees of accuracy. Such an endeavour was motivated by the apparent limitations of the current approaches - first of all Queueing Theory and Petri Nets - whenever applied to the field of engineering here considered. In fact they were adapted to manufacturing problems, but not originally tailored to meet their requirements, whereas that is an essential feature of the Manufacturing Algebra.

The approach is axiomatic. Some properties are introduced as postulates and many other properties are deduced from them as logical consequences. The abstract concepts originated from the analysis of actual situations occurring in production plants. In order to facilitate their application, the mathematical elements introduced as postulates have been named like the specific physical elements of the manufacturing systems which inspired them. The fundamental elements are:

i) the *Objects:* material parts (raw materials, semi-finished products, finished products) and reusable parts (fixtures, tools and equipments),
ii) the *Manufacturing Operations:* MO,
iii) the *Storage Units:* SU,
iv) the *Production Units:* PU,
v) the *Resource Units:* RU,
vi) the *Control Units:* CU.

The abbreviations MO, SU, PU, RU and CU, used to denote the abstract mathematical elements, have been introduced to avoid possible confusions between the mathematical elements themselves and the physical elements having the same name.

The Manufacturing Algebra appears an easily accessible methodology which does not require from the user any sophisticated mathematical knowledge. Given a real manufacturing process and its production plant (the factory), the Manufacturing Algebra provides the tools for building up mathematical models capable of describing the same physical process at different levels of detail and accuracy, by using always and only the same six basic mathematical elements. Starting from the simplest models - deterministic (i.e. without uncertainty), linear, time-invariant - the accuracy of the description can be refined by introducing the uncertainty through stochastic models and removing the assumptions of linearity and time-invariance.

The description of the manufacturing and production processes can be brought up to the desired level of detail, by making use of the important property that a network of PUs (Production Units), SUs (Storage Units) and RUs (Resource Units), controlled by a Control Unit (CU), builds up again a single Production Unit to be regarded as an aggregated PU of higher level, but corresponding to a single mathematical element valid at any aggregation level. The possibility of grouping and disaggregating the manufacturing and production elements opens the way for designing hierarchical control systems, as it was made during the development of the case study at the conclusion of the HIMAC project.

The *Factory Dynamics* (to be intended as a part of the Manufacturing Algebra) concerns the dynamics of the production processes taking place in the factories. The notion of dynamic system, corresponding to the approach originally developed by Kalman and being nowadays the design foundation for all continuous- and discrete-time automatic control systems, has been extended to include discrete-event dynamic systems. What results is a unique definition of dynamic system (intended as a complex mathematical element) which includes both the classical continuous- and discrete-time systems as well as the discrete-event systems. That made possible the formulation of the production process dynamics in terms of state equations, with all the advantages inherent in such an approach.

The first applications seem to be very promising. One example is the design and simulation of a three-level hierarchical real-time control of a quite complex shop-floor production, derived as a test case from a real manufacturing system: the Machine Tool Division of a German industry. Among the most positive facts emerged during the test case development were the simplicity of the factory description and of the control design and the performance robustness of the hierarchical control in presence of internal and external irregularities.

This article is subdivided into three parts. The first part treats the *Manufacturing models* describing the sequence of manufacturing phases necessary to complete the planned finished products, out of raw materials and components. Their mathematics requires only two elements: *Objects* and *Manufacturing Operations*. The second part treats the *Factory Dynamics* in terms of discrete-event dynamic systems. Any production plant of a factory can be modelled as a network of three dynamic elements, PU, SU and RU, controlled by one or more CUs. The role of the CUs is to assign and schedule in real-time all the MOs of the manufacturing models of the planned finished products. The third part illustrates the manufacturing models, the factory network and the control hierarchy of the HIMAC case study. The Appendix summarizes the original event mathematics developed as part of the Manufacturing Algebra.

2 Manufacturing models

The first part of the Manufacturing Algebra is concerned with the so-called *manufacturing models*, i.e. the formal description of the different operations and objects needed to complete a product starting from its components and raw materials. Only two mathematical elements are necessary:

i) the *Objects* describing any material part used in manufacturing, like raw materials, semi-finished and finished products and also reusable parts like fixtures and tools;
ii) the *Manufacturing Operations*, MO, describing the transformations undergone by the material parts in order to produce other parts until the desired finished products are obtained.

2.1 The object set

The object types. An *object* is a mathematical element o describing any material part, which is involved in a manufacturing process as a single and different unit. The universe of all the objects is denoted by $\mathcal{O}$ and is assumed to be a numerable set. Successive parts having the same geometry, the same aspect and the same technological properties have to be considered as different objects, since they can be enumerated. Very often however it is necessary to group in the same subset objects which are similar and interchangeable from a technological point of view.

An *object type k* is a mathematical element describing such technological similarities. The set $\mathcal{K}$ of the elements k, called the object type set, is assumed to be finite and with cardinality n_k. An object type can also be defined as a value of the function $type{:}\mathcal{O}\rightarrow\mathcal{K}$, written as $k = type(o)$, assigning one and only one element $k\in\mathcal{K}$ to each object $o\in\mathcal{O}$.

Two objects $o_1\in\mathcal{O}$ and $o_2\in\mathcal{O}$ are said to be *equivalent* with respect to an object type k, if $type(o_1) = type(o_2) = k\in\mathcal{K}$. The type-equivalence partitions the universe $\mathcal{O}$ into a finite set of equivalence classes $\mathcal{O}_k$, each class being the numerable subset of all the elements $o\in\mathcal{O}$ sharing the same type k. It is a common practice in manufacturing systems (think of mass production) to use or produce several repetitions of the same object type and to be interested in the quantity of such repetitions as demanded for instance by a customer. The concept of numerical object quantities derives from the concept of type equivalence.

The quantity vector space. Consider a subset $M\subset\mathcal{O}$.

The object *quantity* $q(k)$ of the type k in M is defined as the cardinality of the subset $M\cap\mathcal{O}_k$.

The *quantity vector* $\boldsymbol{q}(M)$ is defined as the n_k-size vector of all the quantities $q(k)$ for $k\in\mathcal{K}$. The subsets $O\subset\mathcal{O}$ possessing the same quantity vector $\boldsymbol{q}(O)$ are said to be type-equivalent and make up a family $[O]$ of type-equivalent sets. The null quantity vector will be denoted by $\boldsymbol{0}$. Such concepts are illustrated in Figure 1.

A quantity vector $\boldsymbol{q}$ has been just defined as an n_k-tuple of non negative integer numbers, counting the objects of the same type k in a subset. It is useful, for the some reasons, to extend quantity vectors $\boldsymbol{q}$ to include the whole field of rational numbers.

i) In manufacturing processes the objects of a type k are counted in two different conditions: when they appear after having been produced from other objects or when they disappear having been employed to produce other objects. The former quantity will be assumed as positive, $q(k) > 0$, and referred to as produced or *delivered quantity*. The latter one will be assumed as negative, $q(k) < 0$, and referred to as consumed or *drawn quantity*.

ii) Very often objects quantities are not expressed as integer values, but as a ratio with respect to a reference quantity. For instance, when considering a box of N objects of the type k, a drawing of $n < N$ objects can be better expressed as the negative ratio $q(k) = -n/N$.

Therefore, the space $\mathcal{Q}$ of the quantity vectors will be defined as the space of the n_k-tuples of rational numbers, being a linear vector space over the rational field.

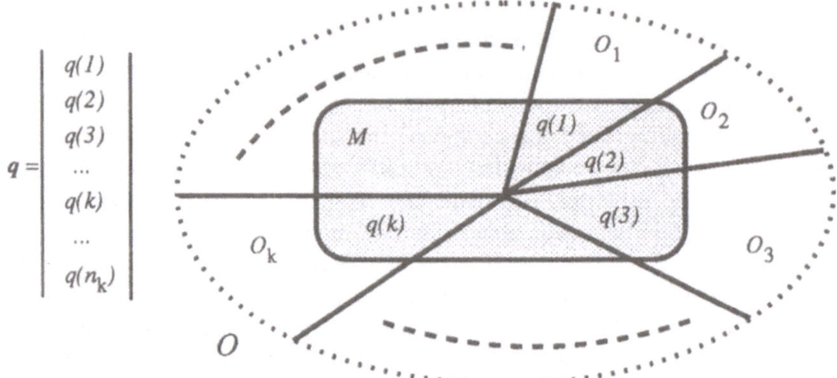

Fig. 1. A partition of the object set $\mathcal{O}$ into type-equivalence classes and the quantity vector q of a subset M.

Beside the classical vector operations of linear spaces, $\mathcal{Q}$ is endowed with a further pair of binary operations. Let x, y and z be quantity vectors $\in \mathcal{Q}$.

The *logic union* $z = x \cup y$ of two quantity vectors x and y is defined as the quantity vector z such that $z(k) = \max\{x(k), y(k)\}$ for $k \in \mathcal{K}$. The expression $x \geq y$ means $x = x \cup y$.

The *logic intersection* $z = x \cap y$ of two quantity vectors x and y is defined as the quantity vector z such that $z(k) = \min\{x(k), y(k)\}$ for $k \in \mathcal{K}$. The expression $x \leq y$ means $x = x \cap y$.

The above pair of operations, being idempotent, commutative, associative and satisfying the absorption law, imposes the lattice structure to $\mathcal{Q}$.

The distributive property. The vector addition in $\mathcal{Q}$ is distributive with respect to the logic intersection. Given x, y and $z \in \mathcal{Q}$ it holds

$$x+(y \cap z) = (x+y) \cap (x+z),\ (y \cap z)+x = (y+x) \cap (z+x) \tag{1}$$

and the equalities easily follow by applying to all components $k \in \mathcal{K}$ the equality $x(k)+min(y(k),z(k)) = min(x(k)+y(k),x(k)+z(k))$.

Event sequences of object quantities. In the course of the manufacturing processes taking place in a factory, quantities of objects appear and disappear in a discrete way (by jumps) rather than as a continuous flow. In other words, if one considers for instance a storing place, the quantity of the stored objects remains constant for finite time intervals and then jumps suddenly to another value after a drawing or delivery operation. Although such operations are not really instantaneous, it is convenient to refer them to a single time instant $t \in \mathcal{R}$, when there is no interest in their detailed evolution, but rather in their result. The sequence of jumps of the object quantities in a storage place or in other factory units can be represented by the concepts of event and event sequence introduced in the Appendix.

Events and event sequences of object quantities can be constructed using as the set of facts the quantity space $\mathcal{Q}$.

A *quantity event* will be denoted by $e_q = (t,q)$, $t \in \mathcal{T}$ being the occurrence time in the time set $\mathcal{T}$ and $q \in \mathcal{Q}$ a quantity vector describing the object quantities arising or disappearing at time t. The event e_q is called *delivery event* if $q \geq 0$ and *drawing event* if $q \leq 0$. The set of all possible quantity events is denoted by $\mathcal{E}_q = \mathcal{T} \times \mathcal{Q}$.

A *sequence of quantity events* will be denoted by σ_q and the event sequence operations defined in the Appendix - namely restriction, time-shifting and addition - apply. The set of all possible sequences of quantity events is indicated by $\Sigma(\mathcal{E}_q)$.

2.2 The manufacturing operations

A manufacturing operation A (shortly MO) being a mathematical operator transforming quantity vectors into other quantity vectors, is capable of modelling any phase of the real manufacturing processes, where object quantities of different types are transformed into other object quantities. To make available detail levls in progress, when modelling manufacturing processes, three different mathematical representations (or models) of the same MO are defined. They are in order of increasing complexity:

i) the *balance vector representation*, equal to a single quantity vector $b_a \in \mathcal{Q}$, called the balance vector,
ii) the *input-output representation*, equal to a pair of quantity vectors $(u_a, y_a) \in \mathcal{Q} \times \mathcal{Q}$, called the input and output vectors,
iii) the *event sequence representation*, equal to a finite quantity event sequence $\sigma_a \in \Sigma(\mathcal{E}_q)$.

Each of the above representations defines the following sets of manufacturing models:

i) the balance vector set $\mathcal{A} = \mathcal{Q}$, equal to the quantity space $\mathcal{Q}$,
ii) the input-output set $\mathcal{B} = \mathcal{Q} \times \mathcal{Q}$, equal to the cartesian product of two quantity spaces,
iii) the event sequence set $\mathcal{C} \subset \Sigma(\mathcal{E}_q)$, equal to the set of all the finite quantity event sequences.

By equipping each model set with suitable algebraic operations, it will be possible to create more complex MOs starting from finite subsets of elementary MOs. Two correspondence functions $\Lambda_{cb}: \mathcal{C} \to \mathcal{B}$ and $\Lambda_{ba}: \mathcal{B} \to \mathcal{A}$ can be defined, mapping a more complex model into a simpler one and inducing an equivalence class partition in the sets of more complex models, as it will be clarified later on.

2.3 The balance vector representation

The balance vectors and their operations. A MO A is represented by a quantity vector $\boldsymbol{b}_a \in \mathcal{Q}$, called the *balance vector* of A. The negative components of $\boldsymbol{b}_a$ denote the quantity of the objects which are transformed (or consumed) in the course of the manufacturing phase represented by A. The positive components denote the quantity of the objects which are produced at the end of the phase A. The balance vector cannot describe any sequence or precedence in the consumption or production of objects, but provides only the quantitative list of the different types of objects employed and produced by that MO. For such reasons, a balance vector will be referred to as the *MO bill-of-materials*.

The balance vector set $\mathcal{A}$, equal to the quantity space, is a linear vector space with the usual vector addition and scalar multiplication.

i) The addition of two MOs, A and B, defines a new MO, $C = A+B$, represented by the balance vector sum $\boldsymbol{b}_c = \boldsymbol{b}_a + \boldsymbol{b}_b$.
ii) The multiplication of a MO A by a scalar α defines a new MO B represented by the balance vector $\boldsymbol{b}_b = \alpha \boldsymbol{b}_a$.
iii) The null MO, O, is represented by the zero balance vector $\boldsymbol{0}$.

Independent and feasible MOs. Since $\mathcal{A}$ is a linear vector space, all the results of the linear algebra apply. A finite set $\mathcal{F} = \{A_1,...,A_h,...,A_n\}$ of MOs is said to be *linear independent*, whenever none of the balance vectors $\boldsymbol{b}_h$ can be expressed as a linear combination of the other ones. A necessary and sufficient condition for linear independence is that the balance matrix $B = [\boldsymbol{b}_1 ... \boldsymbol{b}_h ... \boldsymbol{b}_n]$, collecting all the balance vectors of the set $\mathcal{F}$, be full rank and $\text{rank}(B) = n$. The linear subspace of $\mathcal{A}$ spanned by $\mathcal{F}$ will be denoted by $\mathcal{A}_f$.

Given a linear independent set $\mathcal{F}$ spanning the subspace $\mathcal{A}_f \subseteq \mathcal{A}$, any MO $A \in \mathcal{A}_f$ represented by the balance vector $\boldsymbol{b}_a$ can be uniquely represented in $\mathcal{A}_f$ by a vector $\boldsymbol{m}_a$ which is the solution of the linear equation

$$\boldsymbol{b}_a = B\boldsymbol{m}_a \tag{2}$$

The vector $\boldsymbol{m}_a$ is a rational vector of dimension n, the same size of the subspace $\mathcal{A}_f$. The vector $\boldsymbol{m}_a$ is called the *bill-of-manufacturing-operations (BOMO)* of A.

Not all the balance vectors of $\mathcal{A}$ can represent MOs taking part to real manufacturing processes. Two main restrictions have to be introduced:

i) the bill-of-materials have to be integer-value vectors,
ii) the BOMOs have to be integer-value and non negative vectors.

A MO $A \in \mathcal{A}$ is said to be a *feasible MO* if all the components of its balance vector $\boldsymbol{b}_a$ are integers, i.e. $b_a(k) \in \mathcal{Z}$ whatever be $k \in \mathcal{K}$. The set $\mathcal{M} \subseteq \mathcal{A}$ of the feasible MOs is defined as the set of the MOs represented by integer n_k-tuples. Addition of feasible MOs and their multiplication by integer scalars still yield feasible MOs.

Consider now a basis $\mathcal{F}$ of n feasible and linear independent MOs, its elements A_h being called elementary MOs. The basis $\mathcal{F}$ spans a subspace $\mathcal{A}_f \subseteq \mathcal{A}$ containing both feasible and not feasible MOs. The set of all feasible MOs spanned by $\mathcal{F}$ is the countable set $\mathcal{M}_f \subseteq \mathcal{A}_f$ of the balance vectors **b** which are a solution of the equation

$$b = Bm \tag{3}$$

for all vectors $m \geq 0$ having integer components. In other words, the feasible MOs obtained as a linear combination of the elementary MOs of $\mathcal{F}$, are represented by integer and nonnegative BOMOs. The set $\mathcal{M}_f$ defined by Equation (3), is called the *feasible set spanned* by the basis $\mathcal{F}$.

2.4 The input-output representation: definition

A MO A is represented by an ordered pair of quantity vectors (u_a, y_a), that is by an element of the Cartesian product $\mathcal{B} = \mathcal{Q} \times \mathcal{Q}$. The components of the vector u_a, called the *input quantity vector*, represent the consumed object quantities whereas the vector y_a, called the *output quantity vector*, represents the produced object quantities. The balance vector b_a is related to the input-output vectors by the equation:

$$b_a = y_a - u_a \tag{4}$$

Some information is clearly lost in the balance vector computation, acting as a projection $\Lambda_{ba}: \mathcal{Q} \times \mathcal{Q} \to \mathcal{Q}$. Indeed Λ_{ba} partitions $\mathcal{B}$ into equivalence classes $M(b)$ which are in one-to-one correspondence with the balance vector space $\mathcal{Q}$. The loss of information occurs for all the object types k having non zero quantities both in the input and in the output vectors: $u_a(k) \neq 0$ and $y_a(k) \neq 0$. Objects of this kind, to be defined in Section 2.7, are called *semifinished objects*, because they are produced during manufacturing processes as intermediate objects to be later employed as input objects by subsequent manufacturing phases. A special case occurs when the input and output quantities of the same type k are equal: $u_a(k) = y_a(k)$. Objects of this kind correspond to a class of semifinished objects called *reusable objects*. They are employed to perform MOs, but at the end of the relevant manufacturing phase are given back unchanged among the products.

The graphical symbol of an input-output representation is shown in Figure 2.

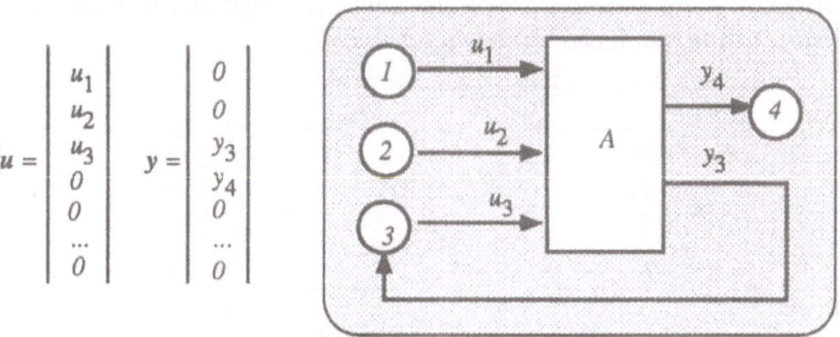

Fig. 2. Symbol of an input-output MO representation.

The manufacturing operation A is represented by a box and by arrows linked to circles, describing input and output object types. The arrows points from the input circles to the MO box and from the MO box to the output circles. The quantities of the input-output object types are written near the corresponding arrows. The object types which are not used by *A* are dropped from the graphical symbol. Figure 2 helps to understand how input-output representations can be employed for describing real manufacturing operations. The MO *A* has three types of input objects. Two of them, the object types 1 and 2, are assembled to produce the output type 4; the third one, the object type 3, appears at the same time as an input and output object. Several types of manufacturing objects, like fixtures, tools and setups, can be described like the object type 3.

Input-output representations make possible a more accurate description of the real manufacturing phases, not just because they allow to model the semifinished and reusable object flow, but because they permit to express input-output *causality*, completely absent in the balance vector model. In fact, by separating the input quantity vectors from the output and by ordering them, a *cause-effect* relation is formulated. Although time is not yet explicitly mentioned, the representation includes the order which exists between input object consumption and output object production. Causality between input and output objects allows to express precedence between different MOs. To this end, two kind of operations (or compositions) will be defined, providing a tool for building more complex MOs, with an explicit description of their precedence rules, although not constrained by a precise position in time. Precedence rules are necessary when the output objects of a MO have to be employed as input objects of another MO. In this case a *compensation of object quantities* is allowed between MOs and the pair of MOs is said to be interacting. Objects which can be compensated belong to a class of types called *semifinished objects*.

Different MOs may have the same object type, say *k*, among their input or output objects. Two MOs having common objects are called *interacting* and the opposite case is called *non interacting*. A pair of MOs $A_h = (\boldsymbol{u}_h, \boldsymbol{y}_h)$ and $A_j = (\boldsymbol{u}_j, \boldsymbol{y}_j)$ are non interacting iff $\boldsymbol{u}_h \cap \boldsymbol{y}_j = \boldsymbol{u}_j \cap \boldsymbol{y}_h = \boldsymbol{0}$, in other words, no output object exists which can be at the same time input object of the other MO. The definition can be extended to an arbitrary set of MOs and also to a single one. A single MO $A = (\boldsymbol{u}, \boldsymbol{y})$ is not self-interacting iff $\boldsymbol{u} \cap \boldsymbol{y} = \boldsymbol{0}$. A subset $\mathcal{N} \subset \mathcal{B}$ of non interacting operations is called a *non interacting subset*. Interaction can be shown graphically.

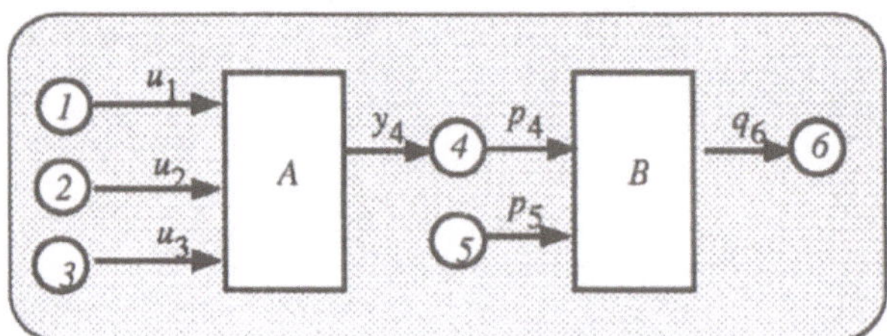

Fig. 3. Interaction between a pair of MOs.

In Figure 3 a pair of interacting MOs, describing a two-phase assembling, is illustrated: the MO $B = (\boldsymbol{p}, \boldsymbol{q})$ has an input object, the object type 4, which is also the output object of $A = (\boldsymbol{u}, \boldsymbol{y})$; it means that the quantity vectors $\boldsymbol{y}$ and $\boldsymbol{p}$ have the fourth component which is different from zero, i.e. $\boldsymbol{y} \cap \boldsymbol{p} \neq \boldsymbol{0}$.

2.5 Composition of input-output representations

Addition and multiplication. A pair of MOs can be composed in two ways:

i) in *parallel*: no precedence is established between the MOs, meaning that no object compensation is permitted,

ii) in *series*: either MO precedes the other one, meaning that the output objects of the preceding MO can be used as input objects of the following MO.

The series composition demonstrates that input-output representations cannot just be added as for the balance vectors. Instead the set $\mathcal{B} = \mathcal{Q}\times\mathcal{Q}$ must be equipped with addition and multiplication.

Addition or parallel composition. Given a pair of input-output representations $A = (\boldsymbol{u}_a, \boldsymbol{y}_a)$ and $B = (\boldsymbol{u}_b, \boldsymbol{y}_b) \in \mathcal{B}$, their sum $C = (\boldsymbol{u}_c, \boldsymbol{y}_c)$ is defined by (see also Figure 4)

$$\boldsymbol{u}_c = \boldsymbol{u}_a + \boldsymbol{u}_b,\ \boldsymbol{y}_c = \boldsymbol{y}_a + \boldsymbol{y}_b \tag{5}$$

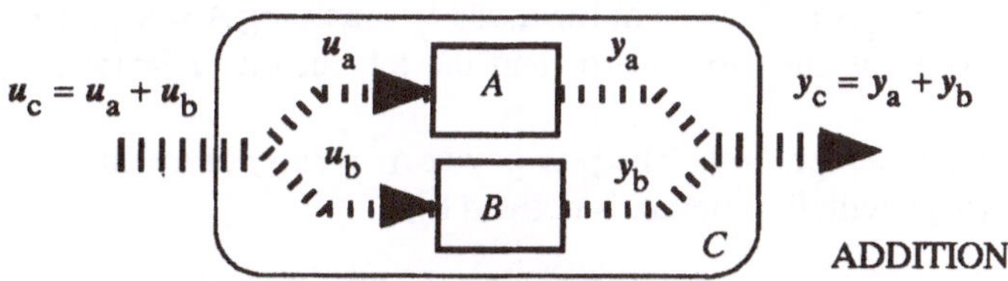

Fig. 4. Addition of two MOs.

Addition is commutative and associative. The null representation is defined by $O = (\boldsymbol{u}=0, \boldsymbol{y}=0)$. The additive inverse of $A = (\boldsymbol{u}_a, \boldsymbol{y}_a) \in \mathcal{B}$ is defined by $B = -A = (-\boldsymbol{u}_a, -\boldsymbol{y}_a)$ or by the equality $A+B = O$.

Scalar multiplication or parallel repetition. Given $A = (\boldsymbol{u}_a, \boldsymbol{y}_a) \in \mathcal{B}$, the scalar product of A times the rational scalar α is defined by $B = \alpha A = (\alpha\boldsymbol{u}_a, \alpha\boldsymbol{y}_a)$.

Multiplication or series composition. To define multiplication, it is necessary to recall a binary operation, namely the logic intersection, defined in the quantity vector space $\mathcal{Q}$. Given an ordered pair (A,B) of input-output representations, where $A = (\boldsymbol{u}_a, \boldsymbol{y}_a) \in \mathcal{B}$ is the preceding MO and $B = (\boldsymbol{u}_b, \boldsymbol{y}_b) \in \mathcal{B}$ is the following MO, their product $C = BA$, $C = (\boldsymbol{u}_c, \boldsymbol{y}_c) \in \mathcal{B}$, is defined by (see also Figure 5)

$$\boldsymbol{u}_c = \boldsymbol{u}_a + \boldsymbol{u}_b - \boldsymbol{y}_a \cap \boldsymbol{u}_b,\ \ \boldsymbol{y}_c = \boldsymbol{y}_a + \boldsymbol{y}_b - \boldsymbol{y}_a \cap \boldsymbol{u}_b \tag{6}$$

where the quantity vector $\boldsymbol{c} = \boldsymbol{y}_a - \boldsymbol{u}_b$ is the vector of the object quantities which are compensated or in other words disappear from the input and output vectors of the multiplication.

Multiplication is associative, but not commutative, because the product C depends in general on the order of the two factors A and B. Indeed, given A and B, two products $C_1 = BA$ and $C_2 = AB$ exist, being in general different and with different compensation vectors $\boldsymbol{c}$. In the former case A is pre-multiplied by B and it holds $\boldsymbol{c} = \boldsymbol{y}_a - \boldsymbol{u}_b$; in the latter one A is post-multiplied by B and it holds $\boldsymbol{c} = \boldsymbol{y}_b - \boldsymbol{u}_a$.

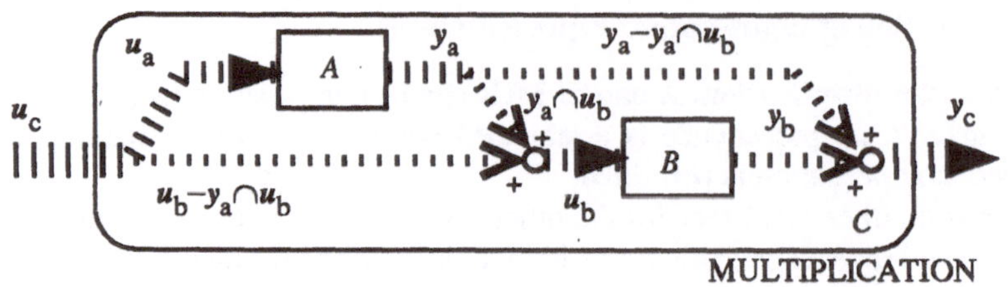

Fig. 5. Multiplication of two MOs.

Identity. An identity I is defined to be the solution of either equations: $A = AI$, $A = IA$. Therefore, one can easily verify that I has the following properties:

i) $I = (\boldsymbol{u},\boldsymbol{u})$: the identities have the same input and output vector,
ii) I is not unique and depends on the specific representation $A \in \mathcal{B}$,
iii) each representation $A = (\boldsymbol{u}_a,\boldsymbol{y}_a) \in \mathcal{B}$ has two set of identities: the left identities I_L satisfying $A = I_L A$ and the right identities I_R satisfying $A = AI_R$. The right identity $I_R = (\boldsymbol{u},\boldsymbol{u})$ is such that $\boldsymbol{u} \leq \boldsymbol{u}_a$ and the left-identity $I_L = (\boldsymbol{u},\boldsymbol{u})$ is such that $\boldsymbol{u} \leq \boldsymbol{y}_a$.

Power or series repetition. The β-th power $A^\beta = (\boldsymbol{u},\boldsymbol{y})$ of an input-output representation $B = (\boldsymbol{p},\boldsymbol{q})$, with β rational, is defined by

$$\boldsymbol{u} = \beta \boldsymbol{p} - (\beta - 1)(\boldsymbol{p} \cap \boldsymbol{q}), \quad \boldsymbol{y} = \beta \boldsymbol{q} - (\beta - 1)(\boldsymbol{p} \cap \boldsymbol{q}) \tag{7}$$

Properties of multiplication. The main properties of the multiplication are hereafter proved.

Associative property. Let $A_1 = (\boldsymbol{u}_1,\boldsymbol{y}_1)$, $A_2 = (\boldsymbol{u}_2,\boldsymbol{y}_2)$ and $A_3 = (\boldsymbol{u}_3,\boldsymbol{y}_3)$ be input-output representations in $\mathcal{B}$; the multiplication is *associative* since it holds

$$(A_1\, A_2)\, A_3 = A_1\, (A_2\, A_3) \tag{8}$$

To prove the assertion, denote by $(\boldsymbol{r},\boldsymbol{s})$ the input-output vectors of $(A_1A_2)A_3$ and by $(\boldsymbol{p},\boldsymbol{q})$ the same vectors of $A_1(A_2A_3)$. Using Equation (6), the following equalities are obtained:

$$\begin{aligned} &\boldsymbol{r} = \boldsymbol{u}_1+\boldsymbol{u}_2+\boldsymbol{u}_3-\boldsymbol{c},\ \boldsymbol{s} = \boldsymbol{y}_1+\boldsymbol{y}_2+\boldsymbol{y}_3-\boldsymbol{c},\ \boldsymbol{c} = \boldsymbol{u}_1\cap\boldsymbol{y}_2+(\boldsymbol{u}_1+\boldsymbol{u}_2-\boldsymbol{u}_1\cap\boldsymbol{y}_2)\cap\boldsymbol{y}_3 \\ &\boldsymbol{p} = \boldsymbol{u}_1+\boldsymbol{u}_2+\boldsymbol{u}_3-\boldsymbol{d},\ \boldsymbol{q} = \boldsymbol{y}_1+\boldsymbol{y}_2+\boldsymbol{y}_3-\boldsymbol{d},\ \boldsymbol{d} = \boldsymbol{u}_2\cap\boldsymbol{y}_3+(\boldsymbol{y}_2+\boldsymbol{y}_3-\boldsymbol{u}_2\cap\boldsymbol{y}_3)\cap\boldsymbol{u}_1 \end{aligned} \tag{9}$$

The associative property follows by proving the equality $\boldsymbol{c} = \boldsymbol{d}$ between the compensation vectors of the pair of multiplications in Equation (8). The equality can be easily demonstrated by applying the distributive property of vector addition (+) with respect to logic intersection (∩) (see Section 2.1). The associativity property says that the input-output quantity vectors of a series of MOs are the same whatever be the association made between consecutive MOs appearing in the series.

Commutative property. As already said, the multiplication (or series composition) is not commutative; in other words it is always possible to find out in the set $\mathcal{B}$ at least a pair $A_1 = (\boldsymbol{u}_1,\boldsymbol{y}_1)$ and $A_2 = (\boldsymbol{u}_2,\boldsymbol{y}_2)$ such that the equality $A_1A_2 = A_2A_1$ does not hold. To prove the assertion it is sufficient to demonstrate that the compensation vectors $c = y_1 - u_2$ of A_2A_1 and $d = y_2 - u_1$ of A_1A_2 are different, i.e. that it exists at least an object type k such that $min(u_1(k), y_2(k)) \neq min(u_2(k), y_1(k))$. That is clearly possible, since all the four components $u_1(k)$, $y_2(k)$, $u_2(k)$ and $y_1(k)$ can be made arbitrarily different. The failure of the commutative property says that the input-output quantity vectors change if the order of the MOs is changed in a series composition.

Distributive property. The distributive properties do not hold: neither the distributive property of the addition with respect to multiplication nor the distributive property of the multiplication with respect to addition.

Example. The implications of the above properties are better explained through an example. Consider the three interacting MOs $A = (u_a, y_a)$ $B = (u_b, y_b)$ and $C = (u_c, y_c)$ shown in Figure 6, where the quantity of each input and output object type is indicated near the corresponding arrow.

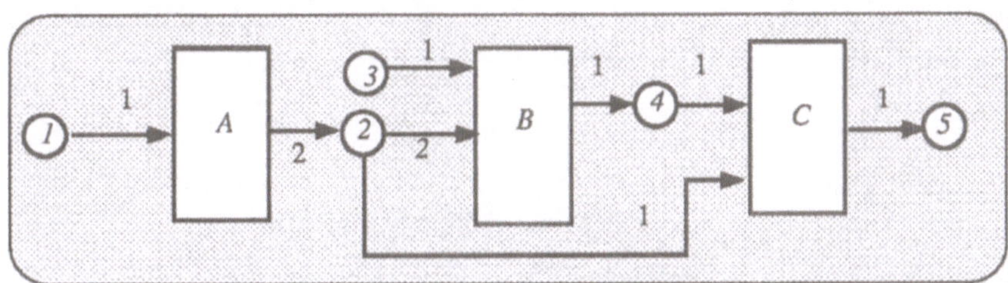

Fig. 6. Three interacting MOs

In Section 2.7 it will be shown that a set of MOs like $\{A,B,C\}$ is capable of classifying their input-output object types into three classes: raw materials (the object types 1 and 3), semifinished objects (2 and 4) and finished products (5). The semifinished objects are those objects produced by the MOs of the set $\{A,B,C\}$ and therefore they should not need to be supplied by other MOs. The peculiarity of the set $\{A,B,C\}$ is that the semifinished object 2 is produced by A in two items, i.e. $y_a(2) = 2$, but three items are requested in total by B and C, namely $u_b(2) = 2$ and $u_c(2) = 1$. Therefore the simple series composition $S = CBA = (u_s, y_s)$ will still request the semifinished object 2 as an input object, although with a quantity $u_s(2) = 1$. One can say that the above series does not make an exact compensation of the semifinished object 2 (the semifinished object 4 is instead exactly compensated) and for this reason is considered as a unbalanced MO. The problem of balancing MOs will be treated in Section 2.8 and in this case is solved by the composition $S = (CB+CB)(A+A+A)$.

In order to better explain the meaning of the associative property, let us consider again the simple series $S = CBA$ of the MOs shown in Figure 6. According to such a property the following different executions of the series, $S = (CB)A$ and $S = C(BA)$, shown in Figure 7, are equivalent from an input-output point of view, i.e. their input and output vectors are equal. The associative property does not assert any other property about the series $S = CBA$. For instance, one can see with the aid of Figure 7 that the two previous executions make a different use of the semifinished object 2. Specifically, in the latter execution, $S = C(BA)$, the object types 2 produced by A are used by B (indeed BA is balanced for what concerns the object type 2), while in the former execution, $S = (CB)A$, the object type 2 produced by A can be used either by C or by B. The different use of the object 2 is due to a different order in which the elementary MOs A, B and C are performed and consequently employ their semifinished objects. It is of capital importance for production control to have some degrees of freedom in scheduling different MOs without modifying their input-output quantities.

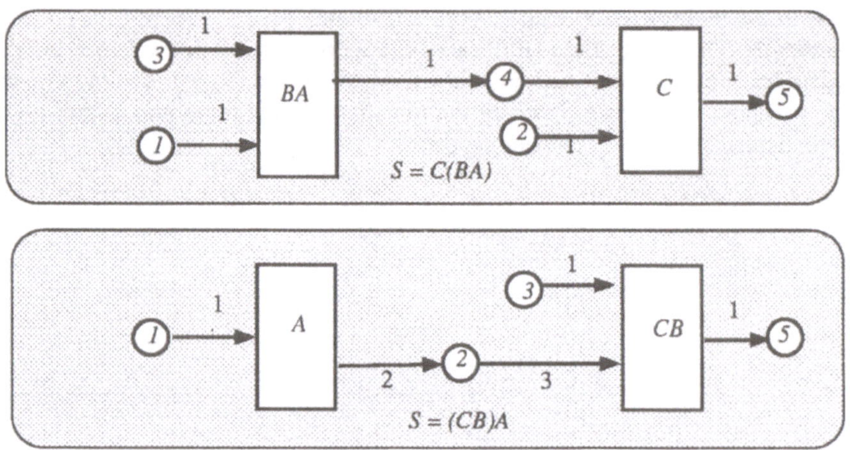

Fig. 7. Two different execution orders, input-output equivalent, of the series $S = CBA$.

2.6 Input-output representations: examples

As a first example of the different ways addition and multiplication work, consider the MOs pair A and B already shown in Figure 3. $D = (\mathbf{r},\mathbf{s})$ will denote the result of their composition (addition or multiplication) and is shown in Figures 8 and 9.

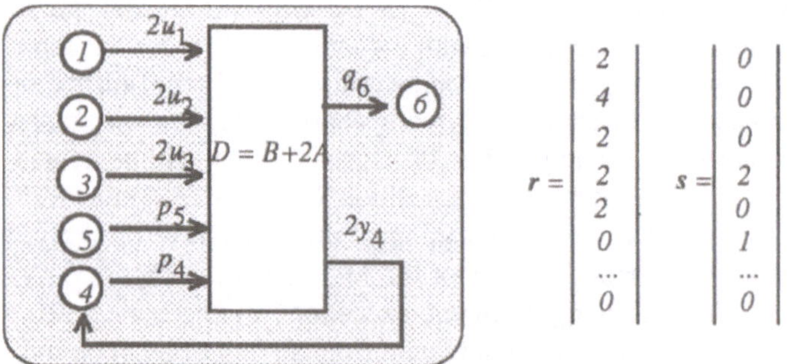

$$r = \begin{vmatrix} 2 \\ 4 \\ 2 \\ 2 \\ 2 \\ 0 \\ \cdots \\ 0 \end{vmatrix} \quad s = \begin{vmatrix} 0 \\ 0 \\ 0 \\ 2 \\ 0 \\ 1 \\ \cdots \\ 0 \end{vmatrix}$$

Fig. 8. Addition of two input-output representations.

One of the goals of the algebraic compositions previously defined is to provide a formal tool for balancing the quantity of the object types, the so-called semifinished objects, produced by some MOs and used by other MOs (see also Section 2.8). Only one semifinished object, the object type 4, exists between the MOs A and B shown in Figure 3 and its balance can be obtained by repeating A twice. Then A can be repeated either in series (by applying power) or in parallel (by applying scalar multiplication). Since A is not self-interacting, i.e. $\boldsymbol{u} \cap \boldsymbol{y} = \boldsymbol{0}$, parallel and series repetitions provide the same result, which means $C = 2A = A^2$. On the contrary by composing C and B, given they are interacting, the resulting composition D will be different when addition or multiplication are applied. For instance by applying multiplication one obtains the MO $D = BA^2 = B(2A)$ and the compensation of the semifinished object 4, which disappears from the input and output vectors of D.

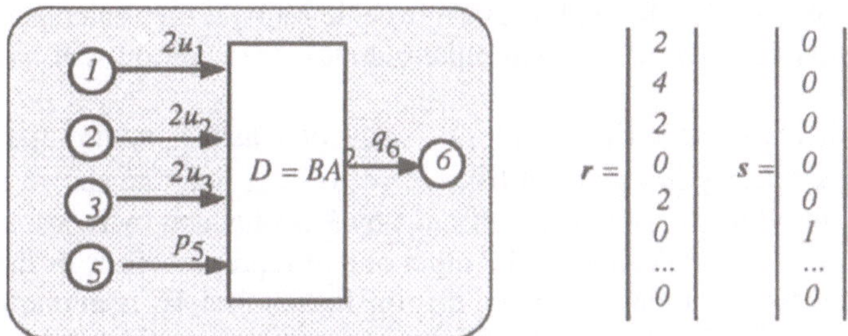

Fig. 9. Multiplication of two input-output representations.

When a pair of interacting operations, for instance $B = (\boldsymbol{p},\boldsymbol{q})$ and $C = (\boldsymbol{r},\boldsymbol{s})$ illustrated in Figure 10, are composed in the series CB, the object quantity which can be compensated is the largest quantity of the common semifinished objects defined by the intersection $\boldsymbol{q} \cap \boldsymbol{r}$, which, in this case, equals $min(q_2,r_2) = 7$.

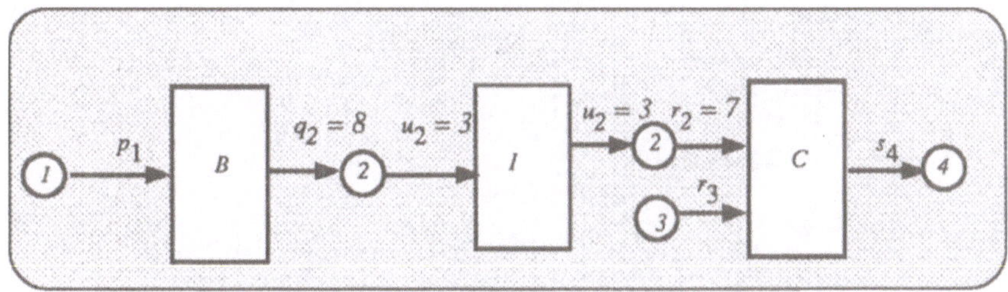

Fig. 10. Three operations concerning a different use of the object 2.

One can ask whether it be possible to compensate a quantity less than $min(q_2,r_2)$, say 3. This is easily obtained by interposing between B and C an identity $I = (\boldsymbol{u},\boldsymbol{u})$ such that $u_2 = 3$ and $u_k = 0,\ k \neq 3$. The series composition $D = CIB$ clearly satisfies the constraint. In practice, identities are used to describe transport operations. Note that the symbols in Figure 10 only show the interaction between B, C and I, and they are not meant to illustrate addition rather than multiplication.

2.7 Feasible representations and object type classification

Feasible input-output representations. As already explained for the balance vectors, not all the input-output representations can describe feasible MOs. Seemingly, also in the input-output set $\mathcal{B}$, feasibility will be obtained by restricting the input and output vectors to be integer-valued and non negative and the scalars α and β, employed in scalar multiplication and power, to be integer and non negative.

A MO $A = (\boldsymbol{u}_a, \boldsymbol{y}_a) \in \mathcal{B}$ is said to be feasible, if the input and output vectors $\boldsymbol{u}_a$ and $\boldsymbol{y}_a$ are integer-valued, that is $u_a(k) \in \mathcal{Z}$ and $y_a(k) \in \mathcal{Z}$ whatever be $k \in \mathcal{K}$. The set $\mathcal{M} \subset \mathcal{B}$ of the feasible MOs will be the set of the pairs of integer n_k-tuples. Addition and multiplication of feasible MOs still yield feasible MOs. Scalar multiplication and power of feasible MOs still provide feasible MOs, if the scalars α and β are integer and non negative. A feasible input-output representation will have a feasible balance vector.

Consider a finite set $\mathcal{F} = \{A_1, \ldots, A_h, \ldots, A_n\}$ of feasible input-output representations, having linearly independent balance vectors $\boldsymbol{b}_h$, like that shown in Figure 11. The set $\mathcal{M}_f$ of all the feasible MOs spanned by $\mathcal{F}$ is obtained by repeated application of addition and multiplication to the input-output representations of the elementary MOs A_h of the set $\mathcal{F}$. It can be shown that the feasible set $\mathcal{M}_f$ is in one-to-one correspondence with the feasible set of the balance vectors spanned by $\mathcal{F}$.

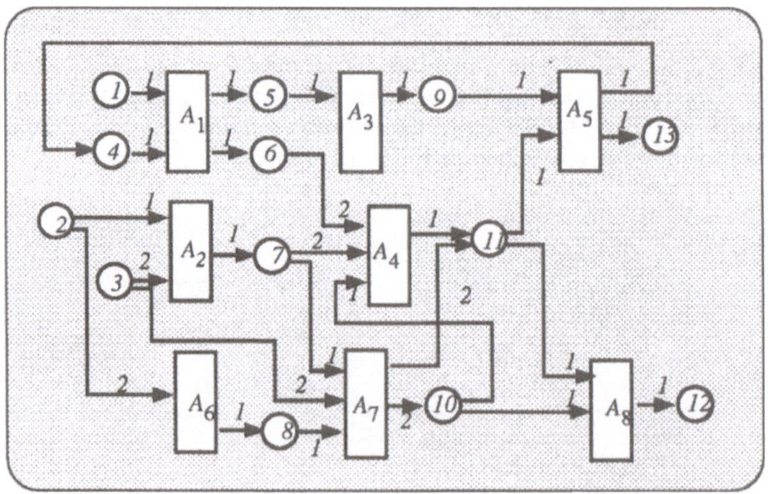

Fig. 11. A basis of feasible MOs.

Object type classification. The material parts of a real manufacturing process are usually subdivided with respect to their technological characteristics into raw materials, semifinished products, finished products, reusable tools. A similar subdivision of the object types of the set $\mathcal{K}$ is implied by any feasible basis $\mathcal{F}$ of MOs. Consider the feasible set $\mathcal{M}_f$ spanned by $\mathcal{F}$; the following subsets of $\mathcal{K}$ are defined:

i) The subset $\mathcal{K}_4$ of the *non used types* whose quantity is zero in any input and output vector of $\mathcal{M}_f$.

ii) The subset $\mathcal{K}_1$ of the *raw materials* whose quantity is zero in any output vector of $\mathcal{M}_f$.

iii) The subset $\mathcal{K}_3$ of the *finished products* whose quantity is zero in any input vector of $\mathcal{M}_f$.

iv) The remaining objects are collected in a subset $\mathcal{K}_2$ and are called *semifinished objects;* their quantity has to be non zero in at least one input and one output vector of $\mathcal{M}_f$.

If the list of the object types is ordered and partitioned according to the above four subsets, also the input and output vectors can be partitioned into four sub-vectors, one for each subset.

The feasible set of eight MOs $\mathcal{F} = \{A_1,...,A_8\}$ illustrated in Figure 11 and built over a set $\mathcal{K}$ of $n_k = 13$ object types, is linear independent, as it can be easily verified from the balance matrix B shown in Equation (10) and partitioned into the blocks of the raw materials, semifinished and finished products. No non-used object type exists. The object types 1, 2 and 3 are raw materials. The object types 12 and 13 are finished products. All other object types are semifinished.

$$B = \begin{vmatrix} B_1 \\ B_2 \\ B_3 \end{vmatrix} = \left|\begin{array}{rrrrrrrr} -1 & 0 & 0 & 0 & 0 & 0 & 0 & 0 \\ 0 & -1 & 0 & 0 & 0 & -2 & 0 & 0 \\ 0 & -2 & 0 & 0 & 0 & 0 & -2 & 0 \\ \hline -1 & 0 & 0 & 0 & 1 & 0 & 0 & 0 \\ 1 & 0 & -1 & 0 & 0 & 0 & 0 & 0 \\ 1 & 0 & 0 & -2 & 0 & 0 & 0 & 0 \\ 0 & 1 & 0 & -2 & 0 & 0 & -1 & 0 \\ 0 & 0 & 0 & 0 & 0 & 1 & -1 & 0 \\ 0 & 0 & 1 & 0 & -1 & 0 & 0 & 0 \\ 0 & 0 & 0 & -1 & 0 & 0 & 2 & -1 \\ 0 & 0 & 0 & 1 & -1 & 0 & 2 & -1 \\ \hline 0 & 0 & 0 & 0 & 0 & 0 & 0 & 1 \\ 0 & 0 & 0 & 0 & 1 & 0 & 0 & 0 \end{array}\right| \begin{array}{l} \text{Raw materials} \\ \\ \text{Semifinished} \\ \\ \text{Finished products} \end{array} \tag{10}$$

2.8 Balanced input-output representations

Definitions. Given a feasible basis $\mathcal{F} = \{A_1,...,A_n\}$ and given a product mix, i.e. a quantity vector $\boldsymbol{p}$ of finished products, it has to be found out a MO $P \in \mathcal{M}_f$ capable of producing the mix and leaving unchanged the input quantities of semifinished objects, even if they would be zero. Such a MO will be called *balanced.*

A *balanced MO* is any MO $P = (\boldsymbol{u},\boldsymbol{y}) \in \mathcal{M}_f$ whose balance vector $\boldsymbol{b} = \boldsymbol{y}-\boldsymbol{u}$ contains only raw materials counted as negative quantities and finished products counted as positive quantities. In other words, the sub-vector $\boldsymbol{b}_2$ containing the semifinished quantities of $\mathcal{M}_f$ must be identically zero. Balanced vectors of this sort will be called *normal* since in the industrial practice all feasible MOs are usually balanced. Each normal balance vector $\boldsymbol{b}$ defines in $\mathcal{M}_f$ an equivalence class $N(\boldsymbol{b})$ of input-output representations.

The following definition selects out of $\mathcal{M}_f$ balanced MOs having optimality properties. A balanced MO $P = (\boldsymbol{u},\boldsymbol{y}) \in \mathcal{M}_f$ is said to be *minimal*, when no MO $B = (\boldsymbol{r},\boldsymbol{s})$ exists in $\mathcal{M}_f$, whose balance vector $\boldsymbol{q} = \boldsymbol{s}-\boldsymbol{r}$ is an integer submultiple of the balance vector $\boldsymbol{b} = \boldsymbol{y}-\boldsymbol{u}$. In other words, the greatest common divisor (GCD) of the elements b_k of $\boldsymbol{b}$ must be equai to 1. Hence, given an equivalence class $N(\boldsymbol{b})$ of balanced MOs in $\mathcal{M}_f$, there exists a class $N_0(\boldsymbol{b}_0)$ of minimal balanced MOs, $\boldsymbol{b}_0$ being the minimum submultiple of $\boldsymbol{b}$. Vice versa, given a class $N_0(\boldsymbol{b}_0)$ of minimal balanced MOs, all the non-minimal ones are defined by $\boldsymbol{b} = \alpha \boldsymbol{b}_0$, for $\alpha > 1$ and integer.

Since in the manufacturing practice, MO*s* are usually balanced, it is of interest, given a feasible basis $\mathcal{F}$, to ask whether balanced MOs exist in the feasible spanned set $\mathcal{M}_f$ and under which conditions. Consider the balance matrix B of $\mathcal{F}$ and extract the sub-matrix B_2 listing the quantities of the semifinished objects. Then, since B is by assumption full rank and rank$(B) = n$, there is a one-to-one relation $B\boldsymbol{m} = \boldsymbol{b}$ between the balance vectors $\boldsymbol{b}$ and the bill-of-manufacturing operations $\boldsymbol{m}$. Therefore the set of all the balanced MOs is defined as the set of all the solutions $\boldsymbol{m}$, if they exist, of the following linear equation

$$B_2 \boldsymbol{m} = \boldsymbol{0} \tag{11}$$

Existence of balanced input-output representations. Assume that B_2, the second block of the balance matrix B, has dimension $n_2 \times n$, where n_2 is the cardinality of the subset $\mathcal{K}_2 \subset \mathcal{K}$ of the semifinished types and n is the cardinality of the feasible basis $\mathcal{F}$. Standard results of the linear algebra allow to state the following two corollaries.

Corollary 1. A necessary and sufficient condition for a feasible basis $\mathcal{F}$ of n MOs to admit in its spanned set $\mathcal{M}_f$ at least a single balanced MO having a non zero BOMO $\boldsymbol{m}$, is that $\text{rank}\{B_2\} < n$.

Corollary 2. All the balanced MOs of the set $\mathcal{M}_f$ possess a bill-of-manufacturing-operations $\boldsymbol{m} \neq \boldsymbol{0}$ belonging to the null space $\mathcal{N}\{B_2\}$ of B_2. The null space dimension is denoted by μ, being $\mu = n\text{-rank}\{B_2\}$. Moreover, if $M = [\boldsymbol{m}_1,\ldots,\boldsymbol{m}_\mu]$, $n \times \mu$, is the matrix of a suitable vector basis of the null space $\mathcal{N}\{B_2\}$, then the BOMO $\boldsymbol{m}$ of any balanced MOs can be computed from $\boldsymbol{m} = M\boldsymbol{n}$, $\boldsymbol{n}$ being an integer vector of dimension μ. So the bill-of-manufacturing-operations $\boldsymbol{m}$ of a balanced MO can be obtained as a linear combination of a finite number of BOMOs, $\boldsymbol{m}_1,\ldots,\boldsymbol{m}_\mu$, where μ is the rank deficiency of B_2.

Since given a BOMO $\boldsymbol{m}$ the corresponding balance vector $\boldsymbol{b}$ is unique, it is also possible to define a basis $B^* = BM = [\boldsymbol{b}_1^* \ldots \boldsymbol{b}_\mu^*]$ for the balance vectors of the balanced MOs. In this way for each balanced MO there exists an integer vector $\boldsymbol{n}$ of dimension μ allowing to write $\boldsymbol{b} = BM\boldsymbol{n}$. At the end we have found a set $\mathcal{P}^* = \{P_1^*,\ldots,P_\mu^*\}$ of μ balanced MOs, which are defined either by M or by B^* and act as a basis for all the balanced MOs in $\mathcal{M}_f$. Moreover, if each element P_j^*, $j = 1,\ldots,\mu$, of the basis is selected to be minimal, the corresponding basis $\mathcal{P}^*$ can be considered like a unit vector basis in $\mathcal{M}_f$ and it can be called the *minimal basis of the balanced MOs*. Furthermore the balance matrix B^* of such a basis defines the minimal and independent mixes of products in number of μ which can be produced by the feasible basis $\mathcal{F}$ without modification of the initial stock of semifinished objects.

As an example consider the feasible basis $\mathcal{F} = \{A_1,...,A_8\}$ illustrated in Figure 11. The balance matrix B reported in Equation (10) is full rank and the sub-matrix B_2 has rank deficiency $\mu = 2$. Therefore B_2 has a bi-dimensional null space, spanned by two minimal balanced MOs P_1^* and P_2^*, whose BOMOs define the minimal basis $M = [\boldsymbol{m}_1\ \boldsymbol{m}_2]$. The graphical symbols of the minimal balanced MOs P_1^* and P_2^* are shown in Figure 12.

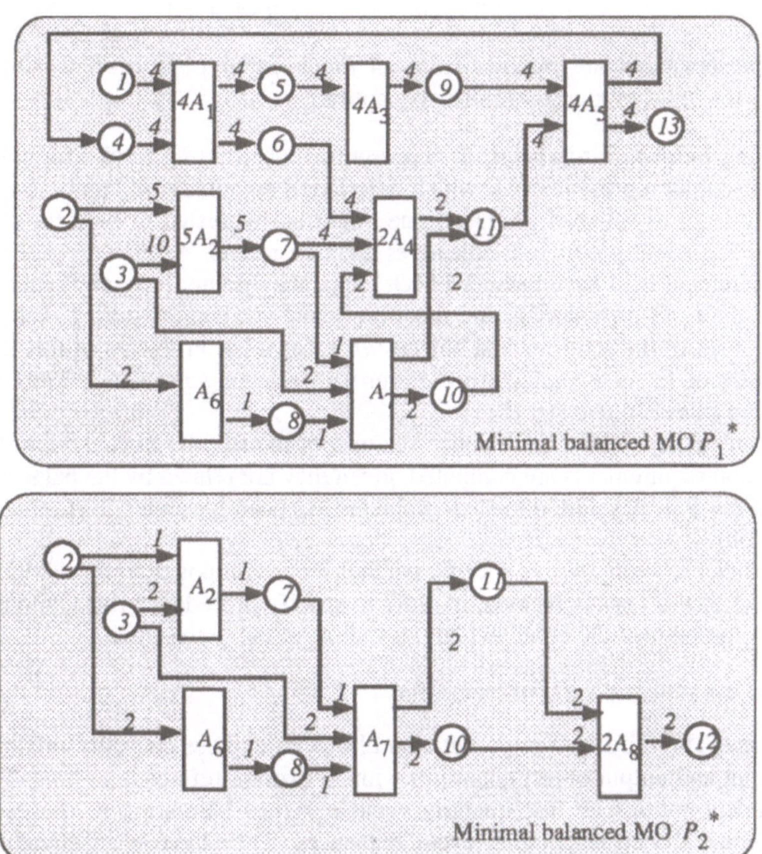

Fig. 12. The minimal balanced MOs defining a minimal basis for the balanced MOs of the feasible basis $\mathcal{F}$ of Figure 11.

The balanced vector $\boldsymbol{b}_j^*$ of each minimal balanced MO P_j^*, $j = 1,2$, defines an equivalence class $N(\boldsymbol{b}_j^*)$ of input-output representations in $\mathcal{M}_f$. All of them can be obtained by applying a suitable set of parallel and series compositions to the elementary MOs of $\mathcal{F}$. The selection of one composition instead of another, defines precedence rules among the elementary MOs. Input-output representations having optimal properties can be obtained among the elements of the equivalence classes $N(\boldsymbol{b}_j^*)$. For instance the following representation of $N(\boldsymbol{b}_2^*)$ has no semifinished object in the input and output vectors

$$R_2 = (2A_8)A_7(A_6+A_2) \tag{12}$$

A representation having the same properties as R_2 cannot be obtained for the class $N(\boldsymbol{b}_1^*)$ due to the presence of the semifinished object 4 acting as a reusable object. But the object type 4 is the only semifinished object in the input and output vectors of the following representation

$$R_1 = (4A_5)(2A_4+4A_3)(4A_1+A_7)(5A_2+A_6) \tag{13}$$

Both representations are optimal from an input-output point-of-view since they minimize the input-output semifinished quantities.

Computing balanced input-output representations. Starting from a feasible basis $\mathcal{F}$ of input-output representations, which describes a manufacturing process aimed at producing a set of finished products, one might be interested to look for complex MOs meeting some production objectives and constraints. As already said, a common requirement is to have balanced MOs, since they do not have any semifinished object in their bill-of-materials and therefore leave unchanged, during their execution, the stock of the semifinished objects whatever it be. Problems of this kind are at the basis of many manufacturing design and planning problems. They can be tackled and solved in two steps:

i) *Planning step.* Either the balance vector $\boldsymbol{b}$ or the BOMO $\boldsymbol{m}$ satisfying the production requirements are computed, given they are related by the balance equation $\boldsymbol{b} = B\boldsymbol{m}$. A solution to such problems has been presented in (Canuto et al., 1996b).
ii) *Precedence rules step.* A specific parallel/series composition of the MOs listed in the BOMO $\boldsymbol{m}$ is obtained, in order to specify precedence rules among them and to optimize the input and output vectors of the representation.

2.9 The event sequence representation

Definitions. The third MO representation allows to express causality relations between input and output object quantities with an explicit reference to time. The key mathematical entities are the quantity event $e_q = (t,\boldsymbol{q})$ belonging to the event set $\mathcal{E}_q = \mathcal{R} \times \mathcal{Q}$ and the corresponding event sequences. They have been already introduced in section 2.1.

A MO A is represented by a finite event sequence $\sigma_a = \{e_a(i)\}$, $i \in (1,n_a)$, where the i-th event is defined by $e_a(i) = (t_a(i),\boldsymbol{q}_a(i)) \in \mathcal{E}_q$ and it describes the drawings and the deliveries of the input-output objects of A occurring at time $t_a(i)$. The set $\mathcal{C}$ of the event sequence representations is the set $\Sigma(\mathcal{E}_q)$ of all the finite event sequences corresponding to the event set $\mathcal{E}_q$. Input (or drawn) quantity vectors $\boldsymbol{q}_a(i) = \boldsymbol{q}_{au}(i)$ are negative, i.e. $\boldsymbol{q}_{au}(i) \leq \boldsymbol{0}$; output or delivered quantity vectors $\boldsymbol{q}_a(i) = \boldsymbol{q}_{ay}(i)$ are positive, i.e. $\boldsymbol{q}_{au}(i) \geq \boldsymbol{0}$.

It is no more necessary, in order to formulate causality, to distinguish between input and output sequences (the former ones including only drawing events and the latter ones only delivery events), because the events are naturally ordered by time and the input-output quantities are distinguished by the sign. Of course, when useful, the input and output sequences $\sigma_{au} = (t_a(i), \mathbf{q}_{au}(i))$ and $\sigma_{ay} = (t_a(i), \mathbf{q}_{ay}(i))$ could be extracted from σ_a, with the constraint that $\sigma_a = \sigma_{au} + \sigma_{ay}$. For instance the input-output representation of A can be simply obtained by summing the quantity vectors of the input and output sequences

$$\mathbf{u}_a = \Sigma_{i=1}^{n_a} \mathbf{q}_{au}(i), \quad \mathbf{y}_a = \Sigma_{i=1}^{n_a} \mathbf{q}_{ay}(i) \tag{14}$$

The occurrence times of all sequences in $\mathcal{C}$ have a same time origin equal to zero. For such a reason, when a MO has to be actuated, all the occurrence times have to be delayed by a fixed interval corresponding to the MO starting time. To distinguish between the occurrence times of the event sequence model and the actuation times, the former ones are referred to as *relocatable times*.

Operations. The composition of different MOs is now obtained by scheduling the starting time of each elementary MO or, what is the same, by time-shifting the occurrence time of the relevant event sequence, and then by combining the sequences of different MOs into a single one, according to the rules of the event sequence addition explained in the Appendix. To this end two operations between the event sequence representations are introduced in $\mathcal{C}$.

Time shift. Given a MO A represented by the event sequence $\sigma_a = \{e_a(i) = (t_a(i), \mathbf{q}_a(i))\}$, $i \in (1, n_a)$, and given a time shift T, the time-shifted MO B is defined by the event sequence $\sigma_b = \{e_b(i) = (t_a(i)+T, \mathbf{q}_a(i))\}$, $i \in (1, n_a)$.

Addition. Given a pair of MOs A and B described respectively by the event sequences $\sigma_a = \{e_a(i)\}$, $i \in (1, n_a)$, and $\sigma_b = \{e_b(j)\}$, $j \in (1, n_b)$, the addition $C = A+B$ is defined by the event sequence $\sigma_c = \sigma_a + \sigma_b$ according to the rules given in the Appendix.

The *null sequence* O, defined by the equation $A = A+O$ whatever be A, is not unique. O can be any event sequence $\sigma_0 = \{e_0(i) = (t_0(i), \mathbf{q}_0(i))\}$, $i \in (1, n_0)$ where $\mathbf{q}_0(i) = \mathbf{0}$ for any i. The additive inverse $B = -A$ is defined by the event sequence $\sigma_b = \{e_b(i) = (t_a(i), -\mathbf{q}_a(i))\}$, $i \in (1, n_a)$. Addition is commutative and associative.

Also in the set $\mathcal{C}$ the feasible MOs are defined as those having integer quantity vectors $\mathbf{q}(i)$ in their event sequence. Addition and time-shifting of feasible MOs still yields feasible MOs. The set of the feasible MOs spanned by a finite basis $\mathcal{F}$ of feasible and linearly independent MOs is obtained by the repeated application of time-shift and addition to the elementary sequences. In other words the feasible set will be the set of all possible *schedules* of the elementary MOs.

2.10 Correspondence between MO representations

Let us consider the three MO representation sets $\mathcal{A}$, $\mathcal{B}$ and $\mathcal{C}$. The following correspondence functions are introduced.

The correspondence $\Lambda_{cb} : \mathcal{C} \rightarrow \mathcal{B}$ simplifies an event sequence representation into an input-output one. It can be formally defined with the help of the *event adder operator* Σ introduced in the Appendix 1. The event adder creates a step-like continuous-time function by adding at each occurrence time the corresponding event quantity. The value of such a time function at the last occurrence time $t(n_a)$ of the event sequence corresponds to the input and output quantity vectors. It is now useful to treat separately the input and output sequences σ_{au} and σ_{ay} of the model. Formally $u_a = -\Sigma[\sigma_{au}]$ and $y_a = \Sigma[\sigma_{ay}]$. Each event model in $\mathcal{C}$ corresponds to a single input-output model in $\mathcal{B}$, but each input-output model in $\mathcal{B}$ is the image of a subset of models belonging to $\mathcal{C}$. Therefore, the function $\Lambda_{cb}:\mathcal{C}\rightarrow\mathcal{B}$ partitions $\mathcal{C}$ into a set of equivalence classes having a one-to-one correspondence with the input-output models of the set $\mathcal{B}$.

The correspondence $\Lambda_{ba}:\mathcal{B}\rightarrow\mathcal{A}$ simplifies an input-output representation into a balance vector. The balance vector is obtained by summing the input and output quantity vectors. One and only one balance model belonging to $\mathcal{A}$ corresponds to each input-output model in $\mathcal{B}$. Each model belonging to $\mathcal{A}$ is the image of a subset of models belonging to $\mathcal{B}$. The function $\Lambda_{ba}:\mathcal{B}\rightarrow\mathcal{A}$ partitions the set $\mathcal{B}$ into a set of equivalence classes, having a one-to-one correspondence with the balance models of the set $\mathcal{A}$.

The above correspondences are graphically represented in Figure 13.

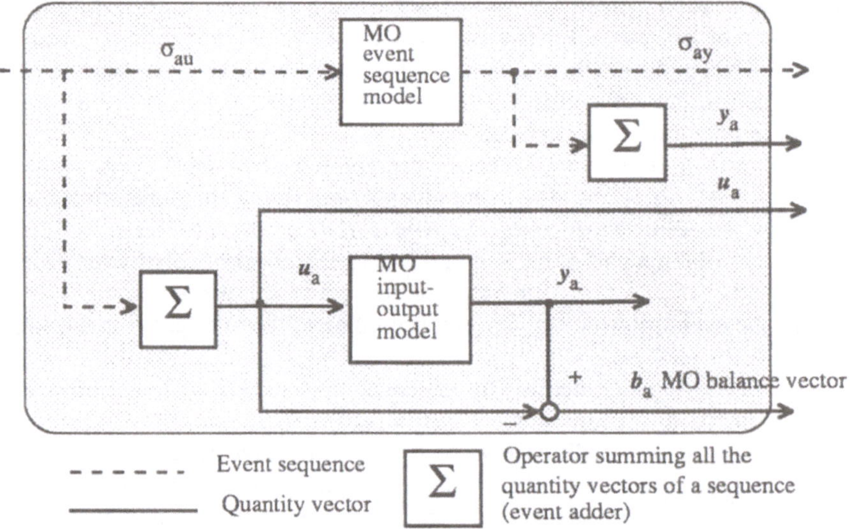

Fig. 13. The MO representations and their correspondence.

3 The factory dynamics

The *factory* is the mathematical model of the production plant where the set of MOs which have been defined as elements of some manufacturing models will be actuated. From a mathematical viewpoint, the factory is defined as a network of dynamic operators: *Storage Units* (SU), *Production Units* (PU) and *Resource Units* (RU), capable of transforming object quantities through appropriate MOs, under the management of *Production Control Units* (CU). SUs, PUs and RUs are interconnected to form a network describing the factory layout. A set of CUs manages and controls the factory production during time. The ensemble of the factory and its control units will be called *production system*. In this chapter the dynamic operators, Storage Unit, Production Unit, Resource Unit and Control Unit will be defined together with their interconnecting signals. Their dynamic evolution will be formulated in terms of discrete-event state equations.

3.1 The storage units

A *storage unit* is a mathematical operator describing the time evolution of the object quantities in any factory area where they are stored after having been delivered and before being drawn. The available and finite-dimensional space of the factory, where objects can be located, is represented by a finite set $\mathcal{S}$ of SU*s* having cardinality n_s, A function $loc{:}\mathcal{O}\rightarrow\mathcal{S}$, written as $s = loc(o)$, is defined to assign one and only one element $s\in\mathcal{S}$ to each object $o\in\mathcal{O}$. In any SU $s\in\mathcal{S}$, objects of different types $k\in\mathcal{K}$ can be located.

The SU dynamics is described by the *event adders*, a class of linear operators defined in the Appendix, which transform event sequences into time functions. The SU input sequence is a quantity sequence σ_q of drawing/delivery events $e_q(i) = (t_q(i),\boldsymbol{q}(i))$, where $\boldsymbol{q}(i)\in\mathcal{Q}$. The SU output is a step-like function $\boldsymbol{x}(t)$ denoting the object quantities stored in the factory SUs at time t., where $\boldsymbol{x}(t)\in\mathcal{Q}$. Using the event adder, the input-output relation of a SU can be written as

$$\boldsymbol{x}(t) = \Sigma[e_q(i)] \tag{15}$$

The relation is represented in Figure 14, together with the symbol of a SU. Each SU has a finite capacity and at the same time each quantity component cannot be less than zero (physical bound). Capacity and physical bounds can be described by a convenient weighted norm $||\boldsymbol{x}(t)||$ and expressed by the inequality

$$0\leq||\boldsymbol{x}(t)|| = c \tag{16}$$

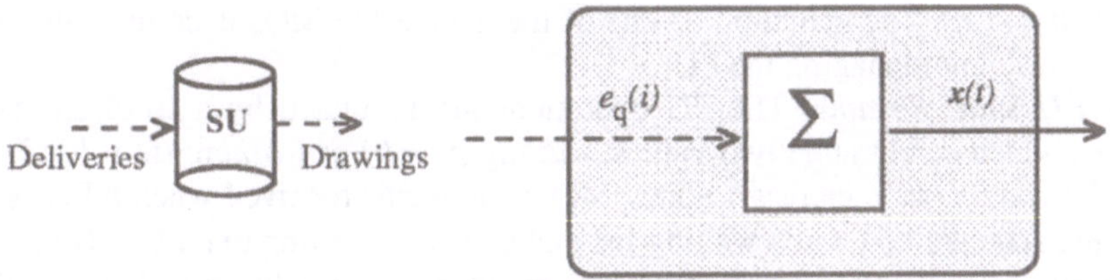

Fig. 14. Dynamic model and symbol of a storage unit.

3.2 The production units

Definition and state variables. The key elements of the factory dynamics are the *production units*, describing the actuation of all MOs which are necessary to complete the planned finished products, according to their manufacturing models presented in Chapter 2. Each PU will perform only one MO at a time according to the command received by its Control Unit. Each factory CU coordinates a set of PUs and can command them to actuate only their admissible MOs. The actuation of a MO by a PU is described by the occurrence of two different event sequences.

i) The sequence of the drawing/delivery events regarding the input and output objects of the actuated MO. An actuated MO is completed only when the last event of its sequence representation has occurred. The time interval between the MO starting time and its conclusion is called the *production (or lead) time,* being specific of the pair (MO,PU).

ii) A start and end event, during which the PU cannot actuate other MOs. The time interval between two successive start/end events will be called *working time,* being specific of the pair (MO, PU).

The PU dynamic model, according to a specific aggregation methodology developed as a part of the Manufacturing Algebra (see Canuto et al., 1998), can be used to describe single workstations and machines or more complex production plants like flexible cells and in the limit the whole factory. When going from simple to more complex production plants, the production and working times of each admissible MO can become very different, and specifically the production time can become very long with respect to the working time. Indeed, PUs describing complex plants can work out several products at a time, but still with the modelling assumption that the MOs can be commanded only one at a time. Therefore the production time measures the time interval employed by a product to be manufactured, while the working time measures the inverse of the production rate, i.e. the time interval between the completion of two successive products. Think for instance of a car factory producing 1,000 vehicles each 8-hour shift, which means a working time of about 30s. If the factory employs four 8-hour shifts to complete each vehicle from the first operation to final delivery, the production time is equal to 32 hours. Note that when the production rate tends to become (or its assumed as) constant, the working time is called *cycle time.*

In a factory there exists a finite set $\mathcal{P}$ of PUs having cardinality n_p. Each PU, indexed by p, can only actuate a finite set of MOs, called the admissible set, $\mathcal{J}_p = \{A_h\}$, $h = (1,n)$. The actuation of one of the admissible MOs is commanded by a single control unit managing the PU.

The PU state variable. The PU operations are regulated by a Boolean state variable w, which can assume two values, *waiting* $(w = 0)$ or *working* $(w = 1)$. Only a PU in the waiting state can start a new MO. Commands received when a PU is in the working state are taken in a waiting list and served according to a FIFO (first-in-first-out) rule. When a PU can actuate a commanded MO, the PU switches into the working state and remains there for a time interval *(the working time)* which is typical of the actuated MO but not necessarily equal or related to the time interval *(the production time)* required to release the last output object which completes the MO. The graphical symbol of a production unit is illustrated in Figure 15.

A PU is always connected to input and output SUs represented in Figure 15 by cylinders. It draws from them the input objects and delivers to them the produced objects. A PU is also connected to a CU from where it receives the actuation commands defined later on.

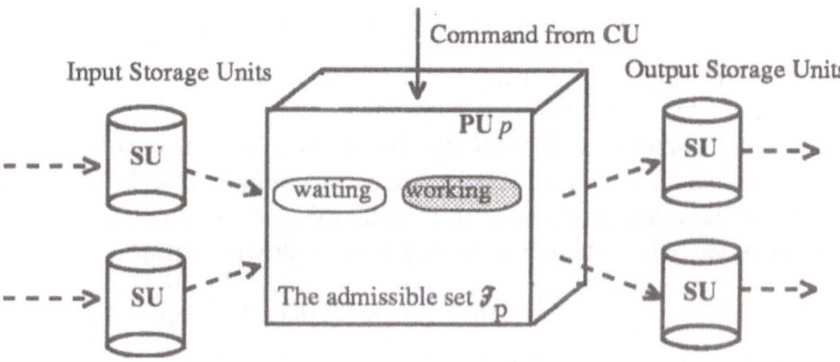

Fig. 15. Symbol of the *p-th* PU.

The PU input-output model. A PU is a dynamic element which can be described by discrete-event equations. First we shall define its input and output sequences. For simplicity we shall assume a time invariant-model, but more general models can be formulated.

The input sequence. The set $\mathcal{U}_p$ of the admissible commands u of a PU p corresponds one-to-one to the set $\mathcal{J}_p$ of the admissible MOs, meaning that u refers to a specific MO $A_h \in \mathcal{J}_p$. Then, let $e_u = (t_u, u)$ be the event describing the command $u \in \mathcal{U}_p$ occurring at time t_u and let $\sigma_u = \{(e_u(j) = (t_u(j), u(j))\}$ be the countable command sequence, j being the ordinal of the input events. The command sequence is the input signal of each PU, which generates the pair of proper output sequences of any PU.

The output sequences. The occurrence of an input event e_u generates two different event sequences describing the actions that a PU performs to start and complete the commanded MO. Let us assume for simplicity that $t_u = 0$ and that the PU initial conditions are *zero*; in other words, that the PU is waiting and all the previous commanded MOs have been completed.

i) The first output sequence $\sigma_w(0,u)$ is the start/end sequence, describing the state transitions from waiting state ($w = 0$) to working state ($w = 1$) and vice versa. Its events will be indicated as w-events and denoted by e_w.
ii) The second sequence $\sigma_q(0,u)$ describes the succession of the drawing/delivery events of the commanded MO. The events will be indicated as q-events and denoted by e_q.

The output of a generic command event $e_u = (t_u, u)$, when the initial conditions are not zero, is obtained by shifting in time the above zero-time sequences and by adding them to the PU output sequences whose events did not yet occur. The time shift will be just t_u if the PU is waiting and the list of the pending MOs is empty. On the contrary the time shift will depend on the actuation of the pending MOs.

The start/end sequence. The events of the start/end sequence are defined by the binary set of facts $\mathcal{W} = \{1,-1\}$, where the fact 1 denotes the transition to the working state $w = 1$ and the fact -1 denotes the transition to the waiting state $w = 0$. Threfore the zero-time sequence in response to an input event $e_u = (t_u = 0, u)$ can be defined as

$$\sigma_w(0,u) = \{(e_{wu}(1) = (\tau_{wu}(1),1)), (e_{wu}(2) = (\tau_{wu}(2),-1))\} \tag{17}$$

The starting time $\tau_{wu}(1)$ is typically close to zero, meaning that a PU switches to working state as soon as it receives a command. The end time $\tau_{wu}(2)$ defines the MO *working* (or *cycle*) *time* for that PU. As already said, it must be distinguished from the MO *production time* equal to the occurrence time of the last delivery event.

The drawing/delivery sequence. It is equal to the event sequence representation of the commanded MO, which is rewritten in the following form:

$$\sigma_q(0,u) = \{e_{qu}(i) = (\tau_{qu}(i),\boldsymbol{q}_u(i))\}, i \in [1,n_u] \tag{18}$$

The quantity vector $\boldsymbol{q}_u$ is now a two index vector whose generic component $q_u(i,s)$ defines the object quantity of type $i \in \mathcal{K}$ drawn from or delivered to the SU $s \in \mathcal{S}_p$, where $\mathcal{S}_p$ is the subset of the input and output storage units of the PU p. The last occurrence time $\tau_{qu}(n_u)$ is equal to the MO *production time*.

Forced response to commands. It is now possible to characterize completely the forced response of a PU to a generic input event $e_u(j) = (t_u(j),u(j))$ or more generally to a command sequence $\sigma_u = \{(e_u(j) = (t_u(j),u(j))\}$, under the assumption of a PU at zero state at time $t = 0$. Moreover we shall assume that the PU under consideration is *linear* and *time-invariant*. The forced response to a single command is obtained by shifting the zero-time output sequences by the command time $t_u(j)$, owing to the time-invariance assumption. The following output sequences result:

$$\begin{aligned} \sigma_w(e_u(j)) &= \{(e_{wu(j)}(1) = (t_u(j)+\tau_{wu(j)}(1),1)), (e_{wu(j)}(2) = (t_u(j)+\tau_{wu(j)}(2),-1))\} \\ \sigma_q(e_u(j)) &= \{e_{qu(j)}(i)=(t_u(j)+\tau_{qu(j)}(i),\boldsymbol{q}_{u(j)}(i))\}, i \in (1,n_u) \end{aligned} \tag{19}$$

The forced response to a command sequence is obtained by adding the single forced responses, under the simplifying assumption that each command event $e_u(j) = (t_u(j),u(j))$ occurs when the PU is waiting. Owing to linearity assumption the effects of all the commands overlap. The following output sequences result:

$$\begin{aligned} \sigma_w(\sigma_u) &= \Sigma_j \sigma_w(e_u(j)) = \{e_w(l) = (t_w(l),\delta(l))\} \\ \sigma_q(\sigma_u) &= \Sigma_j \sigma_q(e_u(j)) = \{e_q(k) = (t_q(k),\boldsymbol{q}(k))\} \end{aligned} \tag{20}$$

In Equation (20) the following notations have been used: $e_w(l) = (t_w(l),\delta(l))$ denotes a generic event of the start/end output sequence and $e_q(k) = (t_q(k),\boldsymbol{q}(k))$ denotes a generic event of the drawing/delivery output sequence.

Remarks. The integer variables j, k and l which are respectively the ordinals (or the counters) of the input sequence σ_u and of the output sequences σ_q and σ_w, have to be taken as different. Moreover the three sequences are completely asynchronous and do not contain equal numbers of events. The output sequence σ_q contains the drawing and delivery events of the input and output SUs interconnected with the PU p under consideration. However it is possible to decompose σ_q into a finite number of sub-sequences σ_{qs}, where $s \in \mathcal{S}_p$. The PU state variable $w(t)$ and the SU state variables $\boldsymbol{x}_s(t)$ are obtained by "integrating" through event adders respectively the output sequences σ_w and σ_{qs}, as shown in Figure 16. The dashed arrows indicate event sequences and the continuous arrows indicate step-like time functions.

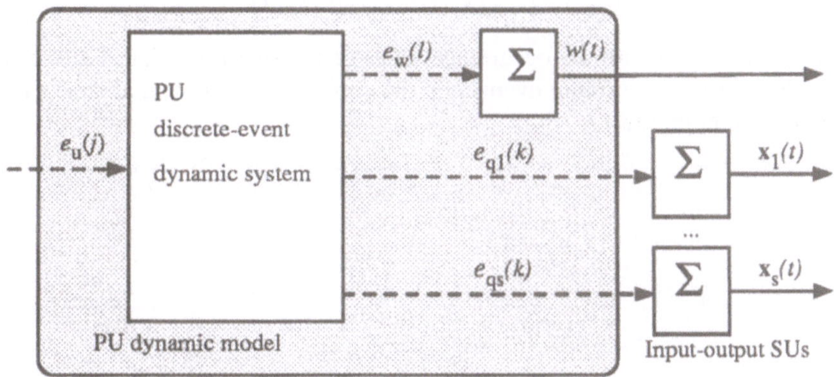

Fig. 16. The PU linear time-invariant model.

The PU state equations. The above linear time-invariant model can be given a state equation form upon definition of suitable state variables. Here only the state variables definition will be provided. The details of the state equations are presented in (Donati et al., 1997 and 1998).

The future potential event. The PU state at time t is expressed by the list $\boldsymbol{x}$ of all the future potential events which, in absence of new input events, will occur and appear in the output event sequences. Therefore, the future potential events of a PU will be of the same type of output events, namely either start/end events (or w-events) describing the PU waiting/working transitions or drawing/delivery events (or q-events) describing the consumption and production of object quantities.

A list $\boldsymbol{x}$ of future potential events will be formulated as a sequence of events belonging to a specific sequence set $\Sigma(\mathcal{E})$. A list $\boldsymbol{x}$ ends with the *null event* $(t = \infty, 0)$ corresponding to the null fact occurring at infinite time (in practice never occurring). One can therefore define the *empty list* $\boldsymbol{x}_0$ as the list just containing the null event, in such a way that also an empty list is a true event sequence belonging to $\Sigma(\mathcal{E})$. To denote the list $\boldsymbol{x}$ at the current time t, the letter $\boldsymbol{x}$ will be indexed by an integer $r(t)$ i.e. $\boldsymbol{x}(r(t))$, $r(t)$ being the integer counting each time the list is updated. The list $\boldsymbol{x}$ is updated each time the first event of the list will occur or an input event $e_u(j)$ adds to the list other potential events.

The PU state events. To describe the PU state two lists are defined:

i) the list $\boldsymbol{x}_w \in \Sigma(\mathcal{E}_w)$ of the future potential w-events,

ii) the list $\boldsymbol{x}_q \in \Sigma(\mathcal{E}_q)$ of the future potential q-events.

Each list, for instance $\boldsymbol{x}_q$, is in turn considered as the fact occurring at the time $t_{xq}(s)$ of the first event e_q contained in the list. Then the pair (time, list) defines a new event called *potential state event*, describing the PU state. Two potential state events are introduced:

$$e_{xw}(r) = (t_{xw}(r), x_w(r)),\ t_{xw}(r) = t_{wr}(1)$$
$$e_{xq}(s) = (t_{xq}s), x_q(s)),\ t_{xq}(s) = t_{qs}(1) \quad (21)$$

where $t_{wr}(1)$ and $t_{qs}(1)$ are the occurrence times of the first potential events of the two lists. The lists of the future events and the corresponding potential state events are summarized in Figure 17.

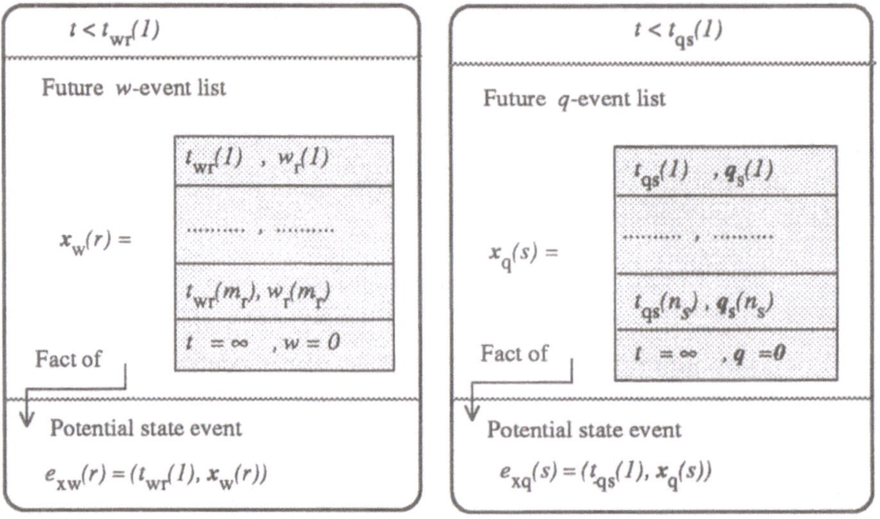

Fig. 17. The PU state variables at time t.

State event sequences. When a potential state event either $e_{xw}(r)$ or $e_{xq}(s)$ occurs, it becomes an actual state event belonging to the respective state event sequences

$$\sigma_{xw} = \{\mathbf{e}_{xw}(l) = (t_w(l), \mathbf{x}_w(l))\},\ \sigma_{xq} = \{\mathbf{e}_{xq}(k) = (t_q(k), \mathbf{x}_q(k))\} \quad (22)$$

Let us remark that not all the potential state events will occur. They will occur only if they remain unchanged until their occurrence time equal to the time of first event. When an input event forces a modification of the list, a state transition occurs from one potential state event, say $e_{xw}(r)$, to another one $e_{xw}(r+1)$, without the former one having occurred and having been registered as an event of the corresponding sequence σ_{xw}. For this reason the events $\mathbf{e}_{xw}(l)$ and $\mathbf{e}_{xq}(k)$ of the state sequences defined in equation (21) must be indexed with a pair of counters, l and k, different from the potential event counters r and s. Therefore the events of the state sequences (here denoted in boldface) are a subset of the potential state events $e_{xw}(r)$ and $e_{xq}(s)$.

Forced and free evolution. From the above discussion, it should be clear that the forced evolution of a PU corresponds to the transition of the actual potential state events into other ones. In such a case, the actual potential event does not occur, but changes into a new event, whose fact will depend both on the actual list and on the input event. Instead, the free evolution of a PU will happen when the first events of the actual list can occur owing to the absence of input events. Of course the occurrence of such an event will force a transition into a new list. The occurrence of a potential event will force the occurrence of output events.

To make the forced evolution sensitive to the actual potential events, e_{xw} and e_{xq}, they must be registered as facts in a time function *S(t)*, called state functions and defined at each time instant *t*. A PU has two state functions defined by

$$S_w(t) = e_{xw}(r),\ S_q(t) = e_{xq}(r) \tag{23}$$

The above considerations can be translated into a block diagram shown in Figure 18, describing the discrete-event state equation of a generic PU.

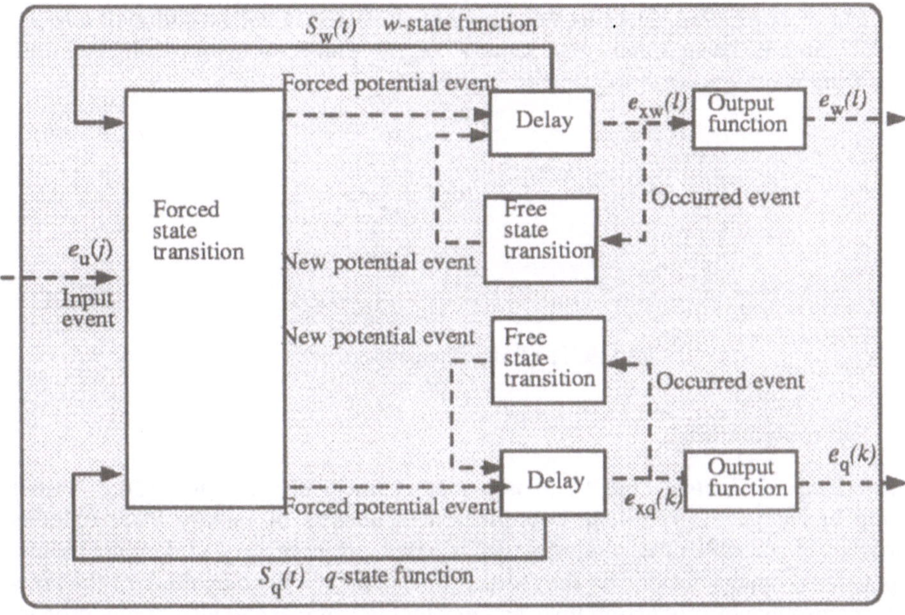

Fig. 18. Block diagram of the PU dynamic model.

In Figure 18 the output functions, the free and forced state transitions are static event functions transforming events into other events. The delay function is a dynamic event function defined in the Appendix. Upon reception of an event, it registers the fact into the value of a time function and then actuates the event at the occurrence time.

3.3 The resource units

To actuate and complete the commanded MOs, factory PUs need input objects, man-power, machines, services and, generally speaking, *resources*. In the usual cases all the necessary resources to complete a commanded MO are specific of that MO and are available to those PUs having that MO in their admissible set, or are made available, like for instance the input objects, before the relevant command is dispatched. As a matter of fact also the PUs themselves are factory resources. In some particular cases the factory could be organized so that some resources, which are elements distinct from the PUs, exist and are necessary to perform some sets of MOs. In such cases a new mathematical element, the *resource unit* (RU) is introduced. Therefore to perform MOs also the necessary RUs must be allocated, beside the input objects and the PU itself. The mathematical model and the symbol of a RU are illustrated in Figure 19.

The RU mathematical model corresponds to an *event adder* operator having as output function the step-like function $W(t)$ whose values are:

i) 0 or *waiting state*, denoting that the RU is free and not allocated to the execution of any MO,

ii) 1 or *working state*, denoting that the RU is involved in the execution of a MO.

The RU input is the start/end event sequence $e_w(i)$, whose events force the RU transition from waiting to working state or vice versa.

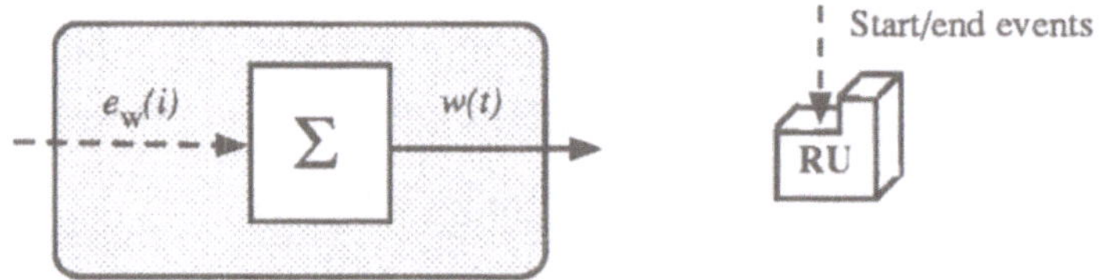

Fig. 19. The RU mathematical model.

3.4 The control units

A *(production) Control Unit* (CU) is a mathematical element describing entities having in charge the *real-time* coordination of subsets of factory PUs. In other words a CU schedules the admissible MOs which are necessary to complete a mix of planned products decided by the factory management. One may think of the automatic controllers of production lines and machining centres, and also of any production manager or shop-floor foreman. Managers and foremen will be assisted by a computer node equipped with suitable input-output man-machine interfaces.

A CU receives production commands from an upper-level CU (or from the factory management) and has the task of real-time scheduling the feasible MOs of the PUs in charge in order to punctually fulfill the received commands. In the Manufacturing Algebra real-time scheduling means that no a priori (or open-loop) schedule is available but that the decisions concerning the MO precedence, routing and dispatching are made in closed-loop and are based upon the best real-time knowledge of the status of the factory production process.

The sequence of commands dispatched by a CU is formulated as an event sequence {(starting time, MO)}, where the MO is one of the admissible MOs of the PUs to be controlled. In Figure 20 the factory is modelled as a set of PUs, RUs and a single SU. All the PUs are controlled by a single CU. Only the connections between the CU and its controlled PUs are shown in Figure 20.

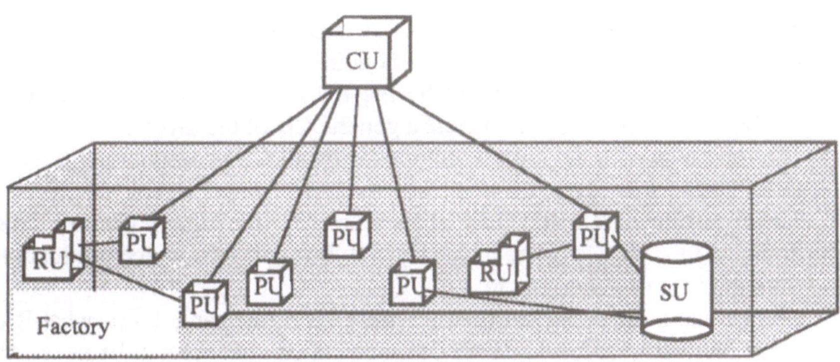

Fig. 20. A factory and its Control Unit.

The principles of hierarchical control. The Manufacturing Algebra permits to conceive and realize hierarchical and real-time control systems to meet the complexity of the manufacturing systems. To this end, a specific hierarchical control theory has been developed in (Canuto et al., 1998) starting from the concept of PU aggregation

PU aggregation principle. Any subset of PUs controlled by the same CU can be modelled as a single PU, an aggregated PU of higher level. By applying such a principle any aggregated PU will have its own set of feasible Manufacturing Operations (MOs) obtained by aggregating the admissible MOs of the lower level PUs. For instance a set of factory PUs describing the shop-floor workstations can be aggregated into a new PU describing a shop where the real-time control system corresponds very often to a foreman. The principle allows to decompose a factory into a reduced set of aggregated PUs (f.i. the shops), which can be coordinated by higher-level CUs. Such a decomposition is illustrated in Figure 21, where the resource units have been partitioned between the shops and there is still a single SU for all shops.

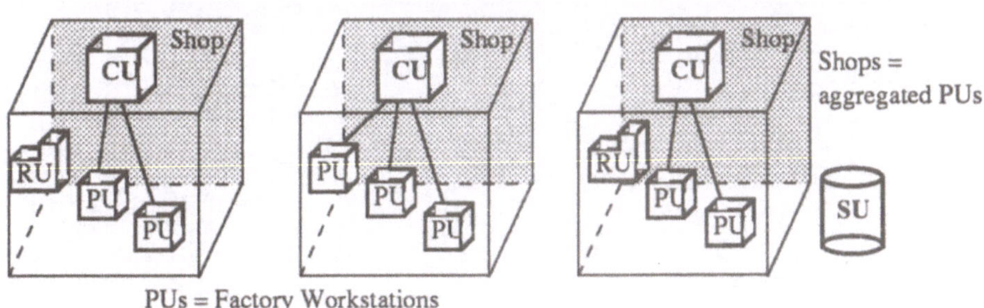

Fig. 21. The factory decomposition into aggregated PUs.

The PU aggregation principle permits to build up a hierarchy of aggregated PUs and a hierarchy of CUs, controlling subsets of aggregated PUs at different levels.

The CU hierarchy. The CU hierarchy has the following rules:

i) The real-time control system of a factory is decomposed into a finite number of layers, $k = 1,2,...,K$, each layer corresponding to the set of the CUs being at the same hierarchical level, defined by the PU hierarchy, explained below. The symbol CU_k will be used to denote a generic CU of the layer k.

ii) The lowest layer is the layer 1, composed by all CUs coordinating a set of factory PUs.

iii) The highest layer K will consist of a single CU_K (the factory Control Unit), coordinating the CU_{K-1} of the lower layer $K-1$; it will receive the production plan from the factory management.

iv) No negotiation is permitted between the CUs of the same level, what is called direct hierarchy; accordingly no communication channel will be necessary between the same level CUs.

v) Any CU_k can communicate only with the higher level CU_{k+1} in charge of its control and with the lower level CU_{k-1} to be coordinated. Each communication channel between higher and lower level CUs will be two-ways: the top-down channel transmitting the command events, the bottom-up channel transmitting the production events.

The main goal of this solution is the uniformity of the different CUs. Their functionality and the type of input-output data exchanged will be strictly the same at any level, hence favouring modular implementations. Figure 22 illustrates a three-level hierarchy.

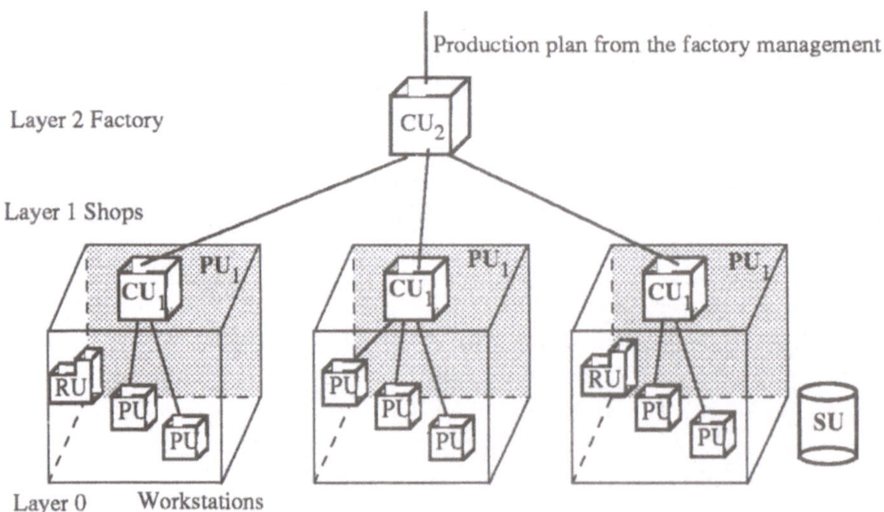

Fig. 22. A three-level control hierarchy.

The PU hierarchy. At the same time, the PU aggregation principle provides a way for constructing a bottom-up hierarchy of aggregated PUs.

i) The lowest PU layer, the layer 0, consists of the factory PUs corresponding to the shop-floor workstations, whose real-time schedule is the objective of the control system.

ii) At an intermediate layer, a PU_k is the aggregation of a set of lower level Production Units, denoted with PU_{k-1}, which are controlled by the same CU_k.
iii) At the highest layer *K*, there is only one aggregated PU_K, the factory itself, consisting of the factory CU_K and its controlled PU_{K-1}.
iv) The number of levels of the control hierarchy equals the number of the PU layers: $K+1$.

The CU functionality. In terms of classical control theory, a CU_k acts as the multivariable controller of a plant composed by a subset of PU_{k-1}s. The mathematical model of a CU has been presented in (Canuto et al., 1998). Any CU includes the following main functions:

i) *Decomposition of the aggregated MOs.* When a CU receives the command asking to actuate an aggregated MO, it has to decompose the command into the elementary MOs to be actuated by the subset of its controlled PUs. Then the list of the elementary MOs to be still actuated (the pending list) is updated.
ii) *Observer of the PU, SU and RU state.* By processing the drawing/delivery events and the start/end events transmitted by the controlled PUs, a CU can reconstruct the state of the set of the controlled PUs together with their input-output SUs and RUs.
iii) *Real-time scheduler.* After having verified the feasibility of the pending MOs, in terms of availability of the input objects and of PU and RU waiting state, a CU selects the MO to be actuated and dispatches the relevant comand to a PU, according to some real-time scheduling algorithm.

4 A case study

Within the HIMAC project a simulated case was developed to test and demonstrate the modelling and the control design capabilities of the Manufacturing Algebra in the field of real-time production controls. The case study aimed at the following objectives:

i) applying the hierarchical theory to a three-level PU hierarchy,
ii) verifying the relevant modelling and data problems,
iii) applying very simple real-time algorithms to understand limits and advantages of the hierarchical approach,
iv) assessing the hierarchical approach against unpredictable plan variations, like rush orders imposed by the high level management,
v) verifying how the hierarchical approach operates fluently in front of long queues of planned products, such to put into operation just the right number of finished products saturating the plant capacity,
vi) assessing the hierarchical approach against shop-floor micro irregularities, like those due to transportation delays, workstations irregularities, short breakdowns, worker variabilities.

The case study was adapted from the Machine Tool Division (MTD) of an German industry, starting from the analysis performed by the University of Karlsruhe and reported in (Reithofer at al., 1998). Figure 23 shows the MTD factory layout.

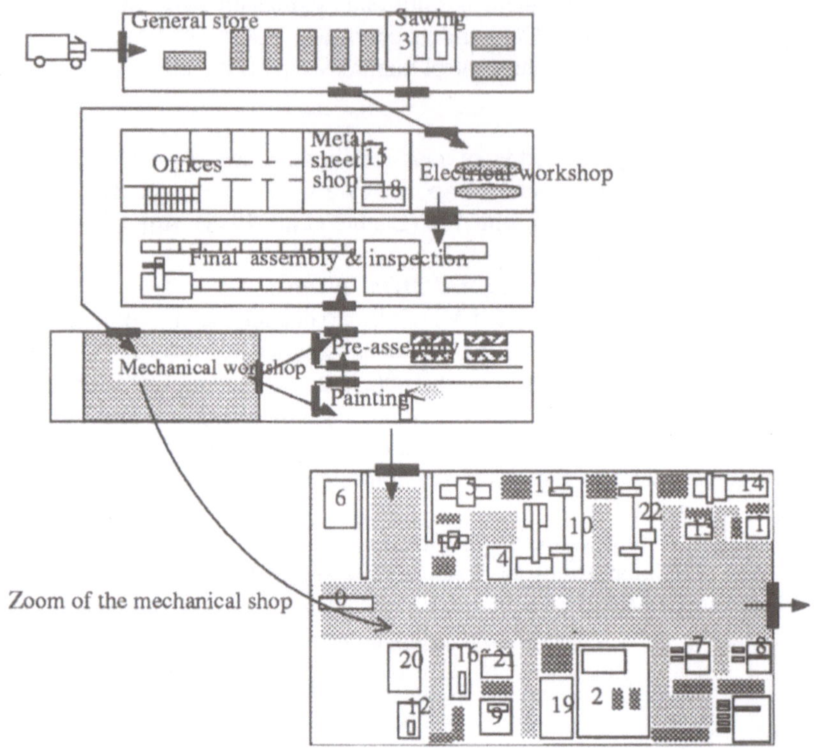

Fig. 23. The factory layout of Machine Tool Division, from where the case study was adapted.

The division employs about 120 people and manufactures universal machining centres (see Figure 24). Usually only simple milling machines are made regularly in lots of five units per year. All other centres are produced on order, therefore the production can be considered as one-of-a-kind. To simplify manufacturing, all the centres are assembled from standard modules, which can be partly customized. The module production starts in pre-production shops, then moves to the mechanical shop and finally to the pre-assembly shop. The modules are then assembled together and electrically connected in the final assembly and inspection shops. Electrical parts are prepared in the electrical shop.

The most critical and complex shop to be modelled and controlled was the mechanical shop, committed to different productions: the mechanical working of the module parts of the on-order machining centres and sub-contracted orders in lots of hundreds of equal parts (in total about six hundred orders per month). The shop includes automatic machining centres and manual machine tools (MTs).

Fig. 24. A standard milling machine produced by Machine Tool Division.

First the manufacturing models were studied. Then the factory model was developed, and the hierarchical levels defined. The most critical issue was the definition of the PUs. Three alternatives were analyzed: skilled worker, machine, machine+skilled worker. Since shop-floor decisions and operations were completely in the hands of the skilled workers, the equality PU = skilled worker was assumed to model manual MTs, assembling stations and also automatic machining centres. MTs (manual and automatic) were modelled as resource units.

4.1 The factory and manufacturing models

All models were derived having in mind a three-level hierarchy (workstations, shops and the factory itself) reflecting the actual MTD production organization. In the following the factory model will be detailed first, to better clarify the hierarchical approach. Then the manufacturing models will be presented, defining for each hierarchical level the admissible MOs of the different PUs.

The factory model. The factory hierarchy was modelled by three layers of Production Units, having more or less the same name and functionalities of the MTD departments and workstations. The top-down list of the layers is the following.

The layer 2: it includes the whole factory (code 0) modelled as a single PU_2 defined by the admissible set $\mathcal{F}_0$ of MO_2s.

The layer 1: it includes four shops modelled as four PU_1s:

i) one mechanical shop (code 1) defined by the feasible set $\mathcal{F}_1$ of MO_1s,
ii) two assembly shops (codes 2 and 3) which were assumed to be completely equivalent, hence having the same feasible set $\mathcal{F}_2 = \mathcal{F}_3$ of MO_1s,
iii) one electrical shop (code 4) defined by the feasible set $\mathcal{F}_4$ of MO_1s.

The layer 0: it includes twenty-five production units, subdivided among the four shops. As already said, each PU is the model of a single worker. The feasible sets of MOs were defined by grouping the workers of each shop into different teams. A team was defined as a set of PUs of the same shop (PU_1) having the same admissible set and the same set of Resource Units. Resource Units, corresponding to manual and automatic MTs, were assigned only to the teams of the mechanical shop. The factory PUs were assigned to the different shops in the following way:

i) the electrical shop (codes 41x and 42x) includes a pair of teams with a total of seven PUs and two admissible sets $\mathcal{F}_{41}$ and $\mathcal{F}_{42}$,

ii) the assembly shops (codes 21x, 22x, 31x and 32x) include four teams with a total of eight PUs and four admissible sets $\mathcal{F}_{21}$, $\mathcal{F}_{22}$, $\mathcal{F}_{31}$ and $\mathcal{F}_{32}$,

iii) the mechanical shop (codes 11x, 12x, 13x and 14x) includes four teams with a total of ten PUs, eighteen RUs and four admissible sets $\mathcal{F}_{11}$, $\mathcal{F}_{12}$, $\mathcal{F}_{13}$ and $\mathcal{F}_{14}$.

The complete detail of the PU hierarchy is shown in Figure 25

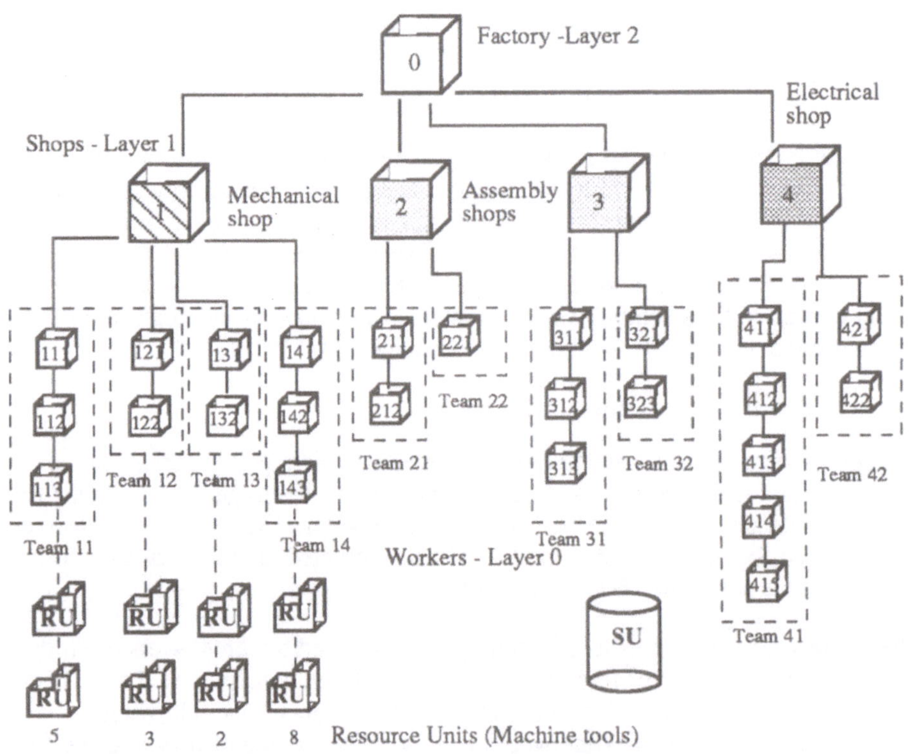

Fig. 25. The PU hierarchy with the teams and the Resource Units.

Further assumptions concerning the factory model are:

i) The factory has a single storage unit. In other words transport units and their operations were not explicitly modelled, but their events and delays were taken into account as irregularities of the delivery/drawing events of the factory MOs.

ii) Each RU of the mechanical shop can be used by a single PU at a time.

The manufacturing model. The manufacturing model describes the set of MOs which are necessary for manufacturing the MTD finished products. Three sets of models were developed: the model at the workstation level describing the admissible MOs of the PUs, the model at the shop level describing the admissible MO_1s of the aggregated PU_1s, the model at the factory level describing the admissible MO_2s of the factory PU_2. All MOs of the higher levels were obtained by aggregating the MOs of the lower level. Each MO_k, $k=0,1,2$, was then assigned to one or more PU_ks, thus defining the admissible sets of all Production Units at each hierachical level.

Owing to the specific MTD manufacturing process, each model can be graphically represented as a tree of MO_ks, which, starting from the MOs processing the raw materials, progressively builds up semifinished objects and then a single finished product. The graphical trees have been simplified with respect to the symbolism introduced in section 2. Each object type has been indicated by an arrow and not by arrows and circles, since each type is always produced in a unitary quantity $q_k = 1$ and is employed by a single MO. The different models are hereafter detailed top-down, starting from the factory layer.

The factory layer. The manufacturing model provides the list of the finished products and of the aggregated MO_2s, having as input only raw materials and as output a single finished product. All MO_2s belong to the feasible set $\mathcal{F}_0$ of the factory PU_2. For each MO_2 the input-output and event sequence models were created. Nominal cycle times, production times and the percentage of capacity utilization for three typical MO_2s, are given in Table 1.

Table 1. Summary of the nominal MO_2 parameters

Finished product	Cycle time	Production time	PU capacity utilization	
			Percentage	Comment
1001	546	2033	47	Severe bottleneck
1501	1000	4035	99	Perfectly balanced
1701	966	3061	81	Bottleneck

Times are expressed in an arbitrary unit which, for this kind of manufacturing process, amounts to few minutes. The capacity utilization is measured in steady state (by repeating the same MO_2) and under nominal conditions (without irregularities) as the ratio of the average PU working time with respect to the MO_2 cycle time (the percent ratio is given in Table 1). As far as the capacity utilization decreases, there will be one or more PUs working at full rate (the so-called bottlenecks) and all the other ones will have more or less long idle periods. Three typical finished products with different capacity utilization are considered in Table 1.

The shop layer. The manufacturing model describes the set of the aggregated MO_1s, connecting raw materials to finished products through a set of semifinished objects of level 1. Figure 26 illustrates the tree of the MO_1s and of the relevant semifinished objects for a typical finished product (1501). Each MO_1 is assigned to one or more PU_1s.

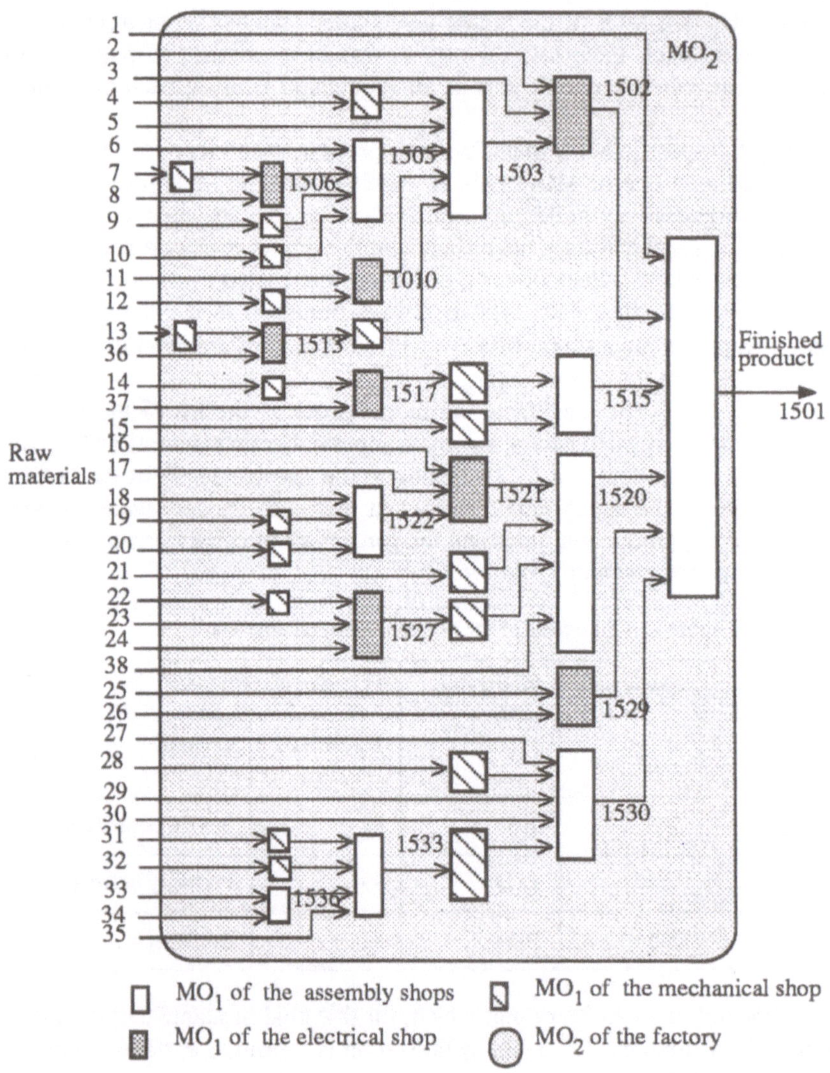

Fig. 26. The level-2 and level-1 manufacturing models of a typical finished product.

The workstation layer. Each MO_1 was further decomposed in a tree of MOs (shop-floor operations), having as input objects either raw materials or semifinished objects of level 1 and as output objects still semifinished objects of level 1 or finished products. Each tree defines a new set of semifinished objects of level 0. The ten admissible sets of level 0 were built by assigning the MOs to one or more worker teams. The assembly tree of each MO_1 of the assembly shops are shown in Figure 27. Their structure is quite similar to the trees of the eight aggregated MO_1s of the electrical shop, not shown in Figure 27. Each of the nineteen MO_1s of the mechanical shop was just described by variable-length sequences of elementary MOs, for a total of eighty-eight MOs.

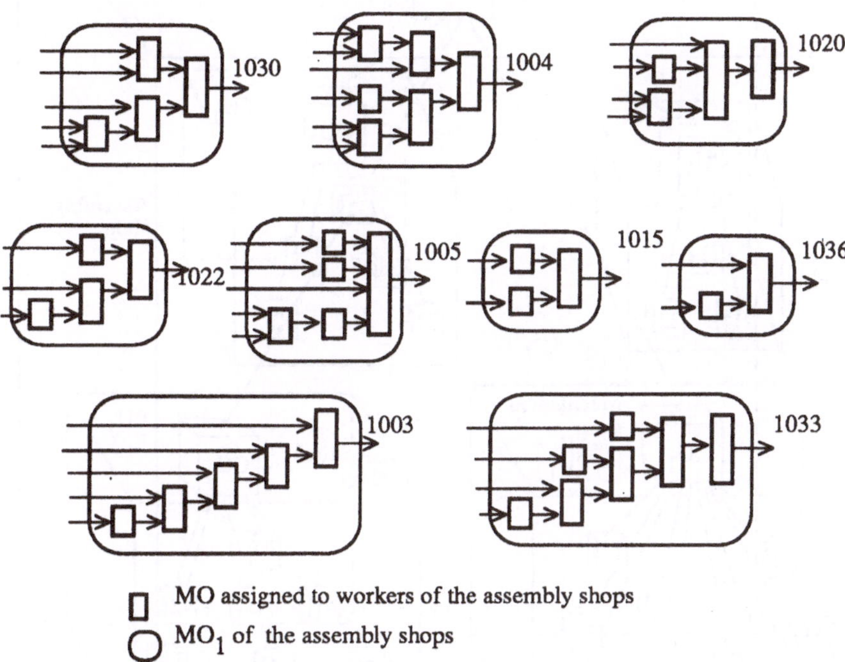

Fig. 27. The level-0 and level-1 manufacturing models of the assembly shops.

A summary of the manufacturing models of three typical products employed in the case study is provided in Table 2, together with the total number of events included in the event sequence models of the different MOs.

Table 2. Summary of the manufacturing models

Finished product	Aggregated MO_1s	Elementary MOs	Events
1001	26	100	832
1501	36	208	1621
1701	36	176	1407

4.2 The hierarchical control

The control hierarchy shown in Figure 28 strictly follows the factory hierarchy and consists of five Control Units and twenty-five PUs.

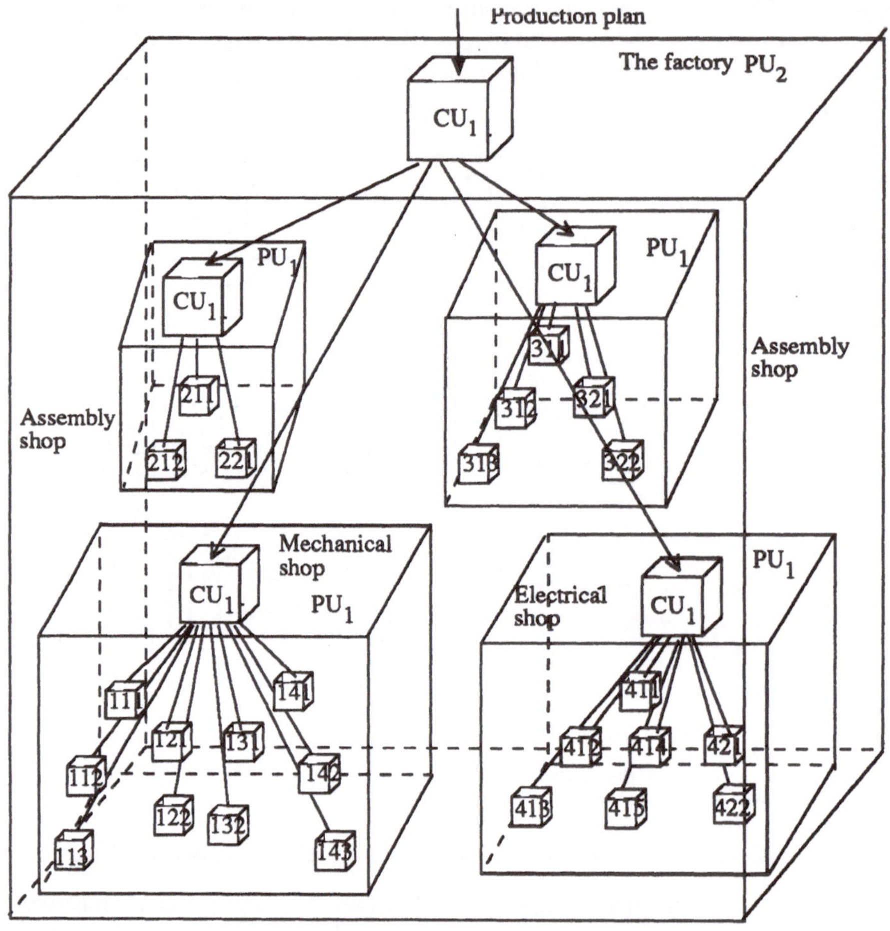

Fig. 28. The control hierarchy of the case study.

The detail of the control hierarchy is the following:

i) one factory Control Unit, denoted with CU_2, receiving the production plan and dispatching commands to four CU_1s,
ii) four shop Control Units. denoted with CU_1,
iii) twenty-five Production Units, denoted with PU, modelling the skilled workers.

The real-time control was designed and operated according to the lines described in Section 3 and in (Canuto et al., 1998). The main points are:

i) The production plan was described as an ordered list of aggregated MO_2s each one corresponding to a single finished product. As mentioned in Section 4.1, factory MO_2s with different utilization of the factory capacity were considered.

ii) Production plans were just provided in order not to stop the factory production. Any ordered sequence of finished products was allowed, the order defining the MO_2 priority. Whenever a feasible MO_2 has not yet been actuated by the factory CU_2, its priority can be modified. In other words the pending plan of the MO_2s can be completely modified at any time to allow rush orders.

i) For each aggregated operation, either factory MO_2 or shop MO_1, an event sequence model was built by providing the relocatable start and end events of all the lower-level MOs and the end event of the aggregated MO itself. The relocatable events are made absolute by time-shifting the sequence when the actual starting time of the aggregated MO has been computed by the relevant CU. Such a model, being a simple open-loop time-schedule, was used to establish a time priority between the lower-level MOs of each aggregated MO. The priority was employed by the real-time control to create an ordered sequence of pending MOs.

ii) The state observer was employed to update the SU state (object quantities), the PU state (waiting/working state and the future potential events) and the RU state (waiting/working) on the basis of the actual event occurrence transmitted by the set of the controlled PUs.

iii) The real-time schedulers of the CU_ks, $k = 1,2$, operated in the following way: they first selected the feasible MO_{k-1} from the list of the pending ones, then dispatched the highest priority MO_{k-1} to the waiting PU having the dispatched MO in its admissible set.

The hierarchical real-time control was successfully tested on a very detailed factory model, implemented on a discrete-event simulator. The simulator was developed by EICAS Automazione within the HIMAC project in order to simulate factory production and control on the basis of the mathematical models of the Manufacturing Algebra.

5 Conclusions

The Manufacturing Algebra was developed during several years as a new mathematics specifically conceived and tailored for modelling and designing control systems of complex manufacturing systems. Major features are:

i) A simple and easily understandable mathematics not only oriented to manufacturing systems but more generally to complex discrete-event systems.

ii) Few and simple building blocks mathematically formulated such to describe, at various levels of detail, any complex manufacturing system from finished products to production processes.

iii) A hierarchical control theory for designing, realizing and validating real-time production control systems and coordinating any complex production process at shop-floor level. The control systems are conceived with a hierarchical structure, to meet the complexity and heterogeneity of the production processes and actuate "true" closed-loop control strategies. Their decisions, when and where to actuate a MO, are taken in real-time based on the real occurrence of the production events (material flow, plant capacity, breakdowns, ...) and of the production plans dispatched by the management.

iv) The control hierarchy is mathematically formulated on the basis of the aggregation methodology originally developed within the Manufacturing Algebra and specifically on the PU aggregation principle, which allows to simplify factory, manufacturing models and the relevant control strategies.

v) A fairly complex case was simulated, out of a real case, to demonstrate the simplicity, the efficiency and the performance robustness of the Manufacturing Algebra and of the relevant control methodologies.

Appendix. Event sequences and their operations

Event sequences. Let Ξ be a set of elements $\xi\in\Xi$ called *facts* and let the real variable t denote the *time*. The generic finite or infinite time interval $\mathcal{T}=[t_1,t_2]$ is called *time set*. Given a time set $\mathcal{T}$ and a fact set Ξ, the elements $e=(t,\xi)\in\mathcal{E}$ of the Cartesian product $\mathcal{E}=\mathcal{T}\times\Xi$ are called *events*. An event is therefore described by the pair (t = the occurrence time of the event, ξ = the fact associated with the event).

An *event sequence* σ is a countable (finite or infinite) set of events $\{e(i)=(t(i),\xi(i))\}$, $i\in[1,m]$, belonging to the event set $\mathcal{E}$. The set $\mathcal{E}$ must be completely ordered by the time variable t; it means that the constraint $t(i)>t(j)$ for $i>j$ must hold. Given a segment of natural numbers $[1,m]$, $m\in\mathcal{N}$, an event sequence can be expressed as a function $\sigma:[1,m]\rightarrow\mathcal{E}$, establishing a one-to-one correspondence between the integer numbers i of the segment $[1,m]$, called the sequence support, and the events $e(i)=(t(i),\xi(i))$ of the set $\mathcal{E}$. An event sequence will be denoted by σ, when the entire sequence is considered, that is

$$\sigma=\{e(i)=(t(i),\xi(i))\}\ i\in[1,m] \tag{24}$$

The *i-th* element of the sequence is denoted by $e(i)$ or $(t(i),\xi(i))$. When i varies over the sequence support $[1,m]$, the single event $e(i)$ may be used to denote the entire sequence to which it belongs.

By assumption, two different events cannot occur at the same time: therefore simultaneous events are excluded. But it is accepted that more than one fact be associated to the same event. When a plurality of facts can be associated to the same event, the plurality itself is made equivalent in some way to a single fact belonging to the set Ξ. By summarizing: (i) an event sequence cannot have simultaneous events; (ii) all facts associated to a single event of an event sequence can always be described by a single element of the fact set Ξ.

Two different kinds of event sets, having different properties and operations, are considered.

The fact set Ξ is a *linear space* over $\mathcal{R}$ (the set of the real numbers) and two operations, *addition* and *scalar multiplication*, are defined over the set Ξ. The addition of two or more facts defines a new fact Ξ. It means that whenever different facts have to be associated to the same event, the equivalent fact is obtained through their addition. For instance the Manufacturing Algebra treats a class of events whose facts are object quantities drawn from or delivered to Storage Units. Then the fact set is the linear quantity vector space $\mathcal{Q}$ with the vector addition.

The fact set Ξ is a *finite set* which does not have other specific properties. In this case different facts are not mutually compatible since they cannot be made equivalent to a single fact and hence associated to the same event. For instance the Manufacturing Algebra treats some events whose facts are commands dispatched to a PU, ordering the actuation of a specific MO. The fact set is the finite set of the admissible MOs of the commanded PU and only a single MO can be commanded at a time. The set of all countable (finite and infinite) sequences which can be constructed over the event set $\mathcal{E} = \mathcal{T}\times\Xi$, is indicated by $\Sigma(\mathcal{E})$.

Operations between event sequences. The following operations are defined over the set $\Sigma(\mathcal{E})$ of the event sequences.

Restriction of an event sequence. The restriction σ' of an event sequence σ to a time interval $t(i) > t$ is a sub-sequence $\sigma' \in \Sigma(\mathcal{E})$ of σ which includes only the events $e(i)$ occurring at times $t(i) > t$. The restriction operation will be indicated by $\sigma' = \sigma(t(i) > t)$.

Time-shift of an event sequence. Given an event sequence $\sigma = \{e(i) = (t(i),\xi(i))\}$, $i \in [1,m]$, a time-shift T can be applied producing a new event sequence $\sigma' = \{e(i) = (t(i)+T,\xi(i))\}$, $i \in [1,m]$, where all the occurrence times are increased by T and the corresponding facts are unchanged. The shift operation is indicated by $\sigma' = \sigma(t+T)$.

Addition of event sequences. Two cases are considered:

i) The fact set Ξ is a linear space. Given two sequences $\sigma_1,\sigma_2 \in \Sigma(\mathcal{E})$, let us consider the union $\sigma_1 \cup \sigma_2$ of all the events of the two sequences. Simultaneous events are assumed to be the same event and their facts are added. The sum $\sigma_3 = \sigma_1 + \sigma_2$ is obtained by ordering in an increasing way the occurrence times of the union $\sigma_1 \cup \sigma_2$.

ii) The fact set Ξ is a generic finite set. Two sequences $\sigma_1,\sigma_2 \in \Sigma(\mathcal{E})$ are said *summable* if and only if the union $\sigma_1 \cup \sigma_2$ does not have simultaneous events. When two event sequences are summable, addition is obtained as above.

The addition of event sequences defined in $\Sigma(\mathcal{T}\times\Xi)$ is commutative and associative. If the fact set Ξ is a linear space, the corresponding sequence set $\Sigma(\mathcal{T}\times\Xi)$ is again a linear space. Addition is defined as above and the properties of scalar multiplication and addition are those induced by the linear space Ξ.

Event sequences and time functions. In this section four operators are defined which transform event sequences into time functions and vice versa. Two operators are defined to transform event sequences into time functions: the *event register R* and the *event adder* Σ, the latter being of capital importance in Factory Dynamics.

The event register R. Given a time interval $\mathcal{T}$ and a suitable fact set Ξ, let $z(t)$ be a time function $z:\mathcal{T}\rightarrow\Xi$ having values $z(t)$ in Ξ and let σ be the event sequence $\{e(i) = (t(i),\xi(i))\}$ belonging to the set $\Sigma(\mathcal{T}\times\Xi)$. The *event register R* operating the transformation $z(t) = R[e(i)]$ is defined by the following relation and the symbol is ilustrated in Figure 29.

$$z(t) = \xi(i) \text{ for } t(i)<t\leq t(i+1) \tag{25}$$

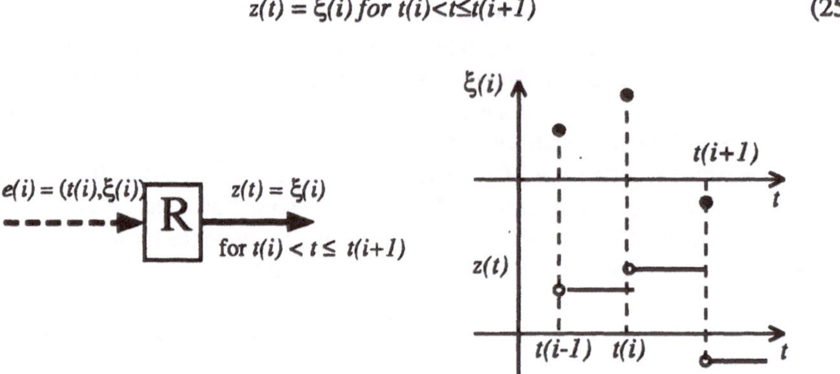

Fig. 29. The event register transforming event sequences into time functions.

The input event sequence σ is denoted in Figure 29 by a dashed arrow and the values of its facts $\xi(i)$ by full dots located over the occurrence times $t(i)$. The output time function $z(t)$ is denoted by a continuous arrow and its values as step-like drawings. Note that the value of $z(t)$ at each event occurrence (jump) is well defined and is equal to the left limit. The fact that the right limits are not values of $z(t)$ is indicated by empty dots.

The event adder Σ. Given the time interval $\mathcal{T}$ and a suitable fact set Ξ, let us assume that Ξ is a linear space. The *event adder* Σ operates the transformation $z(t) = \Sigma[e(i)]$ of an event sequence σ into a continuous-time step-like function $z(t)$, having values equal to the successive sums of the input facts $\xi(i)$, as follows (see also Figure 30):

$$z(t) = z(t(i))+\xi(i) \text{ for } t(i)<t\leq t(i+1) \tag{26}$$

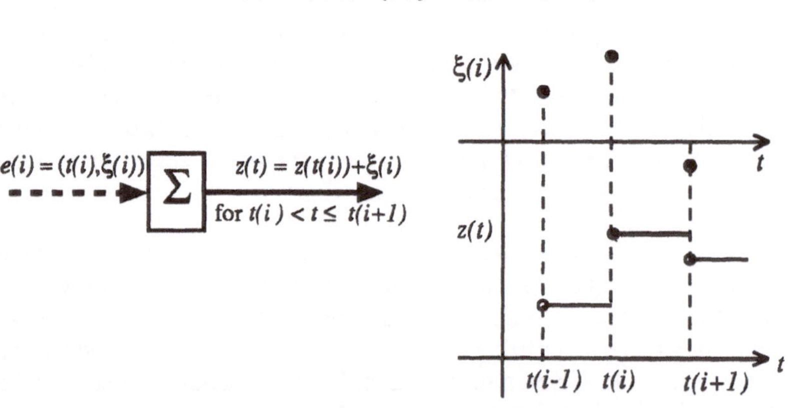

Fig. 30. The event adder transforming event sequences into time functions.

The event actuator. The operator transforming time functions into event sequences is the event actuator. Given an event set $\mathcal{E} = \mathcal{T}\times\Xi$, with a generic element $e = (\tau,\xi)$, let $\mathcal{F}$ be the class of the time functions $f: \mathcal{T}\to(\mathcal{T}\times\Xi)$ with values $f(t) = (\tau(t),\xi(t))$. The event actuator is the operator transforming, as time t increases in $\mathcal{T}$, a time function $f(t) \in \mathcal{F}$ into the event sequence $\sigma = \{e(i)\} = \{A(f(t))\}$ according to the following rule:

$$\textit{if } t = \tau(t) \textit{ then } i = i+1,\ e(i) = f(t) \tag{27}$$

The symbol is shown in Figure 32, with an illustration of the transformation from the time function $f(t)$ into the event sequence $e(i)$.

Note that the value $f(t) = (\tau(t),\xi(t))$ of the time function is equal at any time to the future potential event $e(i)$ not just to the fact $\xi(t)$. As such, the time function $f(t)$ could be sometimes equal to not occurring events (*dreams*), as far as their occurrence time $\tau(t)$ is updated before its occurrence. This fact is shown in Figure 32 by a time function $\tau(t)$ which does not remain constant between successive occurrences, nor is monotonic. Our experience is full of dreams which although expected do not occur at all, or in the favourable cases occur either anticipated or delayed.

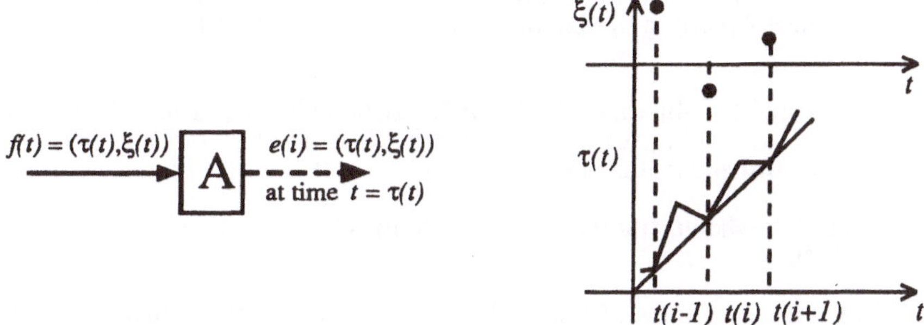

Fig. 31. The event actuator from time functions to event sequences.

The delay operator. The cascade of an event register R and of an event actuator A (see Figure 33) allows to formulate the concept of time delay for the event sequences. That is done by introducing the *event delay D* operator receiving as input a future potential event $e(i+1)$ to be actuated when its time $t(i+1)$ occurs. To this end the event has to be saved by an event register R and then actuated by an event actuator A.

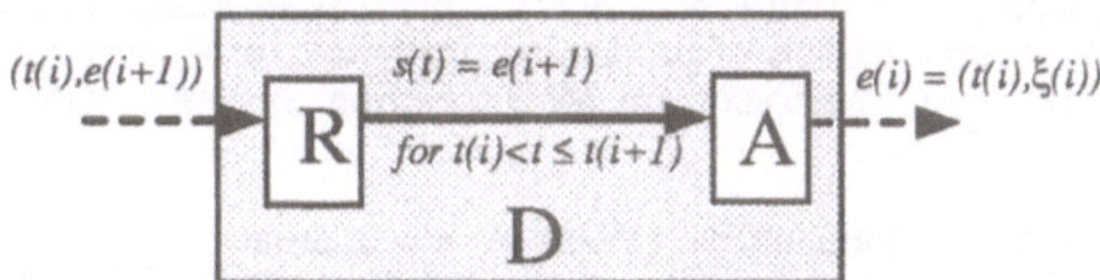

Fig. 32. The symbol of the delay operator.

Note that the input to the register is an event $(t(i),e(i+1))$ whose fact is the future potential event. The register saves in the function $s(t)$ the future potential event

$e(i+1) = (t(i+1), \xi(I+1))$, according to its definition. Then the saved event is actuated if and only if $t = t(i+1)$ and will be lost if for any reason is changed before $t(i+1)$.

References

Canuto, E., Donati, F., Vallauri, M. (1993b): Factory modelling and production control. International Journal of Modelling and Simulation, 11: 162-166.

Canuto, E., Donati, F., Vallauri, M. (1995a): An algebra for modelling manufacturing processes. In: Proceedings of the 3rd IEEE Mediterranean Symposium on New Directions in Control and Automation, Vol.II, Limassol, Cyprus. IEEE, pp.438-445.

Canuto, E., Donati, F., Vallauri, M. (1995b): An approach to factory dynamics based on Manufacturing Algebra. In: Proceedings of the 3rd IEEE Mediterranean Symposium on New Directions in Control and Automation, Vol.II, Limassol, Cyprus. IEEE, p.446-453.

Canuto, E., Donati, F., Vallauri, M. (1996a): Factory models based on Manufacturing Algebra. In: Vallauri, M. (ed.): Proceedings of the 1st HIMAC Workshop: A New Mathematical Approach to Manufacturing Engineering. CELID, Torino, pp.33-40.

Canuto, E., Donati, F., Vallauri, M. (1996b): Production planning using Manufacturing Algebra. In: Proceedings of the 4th IEEE Mediterranean Symposium on New Directions in Control and Automation, Chania, Greece. IEEE, pp.177-182.

Canuto, E. (1997): Manufacturing Algebra. Intelligent Automation and Soft Computing, 2: 389-406.

Canuto, E., Christodoulou, M., Chu, C., Donati, F., Gaganis, V., Janusz, B., Proth, J.-M., Reithofer, W., Vallauri, M. (1997): The ESPRIT Basic Research HIMAC: Hierarchical Management and Control in Manufacturing Systems", in: Kopacek, P. (ed.): Preprints of the 1st IFAC Workshop on Manufacturing Systems, MIM'97, Vienna, Austria, p. 337-342.

Canuto, E., Donati, F., Vallauri, M., Richard, F. (1998): Theory of real-time hierarchical control of manufacturing systems. In: Vallauri, M. (ed.): Proceedings of the 2nd HIMAC Workshop, New Design Methodologies of the Production Management and Control. CELID, Torino, pp. 59-102.

Donati, F., Canuto, E., Vallauri, M. (1997): Advances in Manufacturing Algebra: discrete-event dynamic models of production processes. In: Kopacek, P. (ed.): Preprints of the 1st IFAC Workshop on Manufacturing Systems, MIM'97, Vienna, Austria, pp. 461-467.

Donati, F., Canuto, E., Vallauri, M. (1998): A new approach to Discrete-Event Dynamic System Theory. In: Proceedings of the 2nd Workshop on Trends in Theoretical Informatics, Budapest, Hungary (to appear).

Reithofer W., Janusz B., Raczkowsky J. (1998): HIMAC Test Cases. In: Vallauri, M. (ed.): Proceedings of the 2nd HIMAC Workshop, New Design Methodologies of the Production Management and Control. CELID, Torino, pp. 103-118.

List of contributors

Rudolf F. Albrecht
Informatik, Universität Innsbruck
Technikerstraße 25, A-6020 Innsbruck, Austria
Rudolf.Albrecht@uibk.ac.at

Fernando J. Barros
Universidade de Coimbra, Departamento de Engenharia Informática
Pólo II, P-3030 Coimbra, Portugal
barros@dei.uc.pt

Enrico Canuto
Dipartimento di Automatica e Informatica, Politecnico di Torino
Corso Duca degli Abruzzi 24, I-10129 Torino, Italy

Xiaohong Chen
School of Business Administration, Central South University of Technology
Changsha, Hunan, China
cxh@csut.edu.cn

Mario Dal Cin
Institut für Mathematische Maschinen und Datenverarbeitung (Informatik)
Universität Erlangen-Nürnberg
Martensstraße 3, D-91058 Erlangen, Germany
dalcin@informatik.uni-erlangen.de

Francesco Donati
Dipartimento di Automatica e Informatica, Politecnico di Torino
Corso Duca degli Abruzzi 24, I-10129 Torino, Italy

Gérard Dray
Nonlinear and Uncertain Systems Group, LGI2P, EMA-EERIE,
Parc Scientifique Georges Besse, 30000 Nimes, France
dray@eerie.fr

Gillian Hill
Imperial College, Department of Computing
180 Queen's Gate, London SW7 2BZ, United Kingdom
gah@doc.ic.ac.uk

Wolfgang Kreutzer
University of Canterbury, Department of Computer Science
Christchurch 1 / Newzealand
wolfgang@cosc.canterbury.ac.nz

Christian Märtin
Fachhochschule Augsburg
Baumgartnerstr. 16, D-86161 Augsburg, Germany
maertin@informatik.fh-augsburg.de

David W. Pearson
Nonlinear and Uncertain Systems Group, LGI2P, EMA-EERIE,
Parc Scientifique Georges Besse, 30000 Nimes, France
pearson@eerie.fr

Franz Pichler
Institut für Systemwissenschaften, Universität Linz
A-4040 Linz, Austria
pichler@cast.uni-linz.ac.at

Charles Rattray
Department of Computing Science and Mathematics, University of Stirling
Stirling FK9 4LA
cr@cs.stir.ac.uk

Yasuhiko Takahara
Department of Industrial and Systems Engineering
Chiba Institute of Technology
Tsudanuma, Narashino, Chiba, Japan
takahara@cc.it-chiba.ac.jp

Krzysztof Tchoń
Institute of Engineering Cybernetics, Wroclaw University of Technology
ul. Janiszewskiego 11/17, 50-372 Wroclaw, Poland
tchon@ict.pwr.wroc.pl

Maurizio Vallauri
Dipartimento di Automatica e Informatica, Politecnico di Torino
Corso Duca degli Abruzzi 24, I-10129 Torino, Italy

Horst D. Wettstein
Institut für Betriebs- and Dialogsysteme, Technische Universität Karlsruhe
P.B. 6980, D-76128 Karlsruhe, Germany
wettstein@ira.uka.de

Bernard P. Zeigler
University of Arizona, AI and Simulation Research Group
Department of Electrical and Computer Engineering
Tucson, AZ 85721, USA
zeigler@ece.arizona.edu

SpringerWienNewYork

Springer-Verlag
and the Environment